Essential Molecular Biology, Volume I

The Practical Approach Series

SERIES EDITORS

D. RICKWOOD

Department of Biology, University of Essex
Wivenhoe Park, Colchester, Essex CO4 3SQ, UK

B. D. HAMES

Department of Biochemistry and Molecular Biology,
University of Leeds, Leeds LS2 9JT, UK

Affinity Chromatography
Anaerobic Microbiology
Animal Cell Culture (2nd edition)
Animal Virus Pathogenesis
Antibodies I and II
Biochemical Toxicology
Biological Data Analysis
Biological Membranes
Biomechanics—Materials
Biomechanics—Structures and
  Systems
Biosensors
Carbohydrate Analysis
Cell–Cell Interactions
Cell Growth and Division
Cellular Calcium
Cellular Neurobiology
Centrifugation (2nd Edition)
Clinical Immunology
Computers in Microbiology
Crytallization of Nucleic Acids and
  Proteins
Cytokines
The Cytoskeleton
Diagnostic Molecular Pathology I and
  II
Directed Mutagenesis
DNA Cloning I, II, and III
Drosophila

Electron Microscopy in Biology
Electron Microscopy in Molecular
  Biology
Electrophysiology
Enzyme Assays
Essential Molecular Biology I and II
Experimental Neuroanatomy
Fermentation
Flow Cytometry
Gel Electrophoresis of Nucleic Acids
  (2nd Edition)
Gel Electrophoresis of Proteins
  (2nd Edition)
Gene Transcription
Genome Analysis
Glycobiology
Growth Factors
Haemopoiesis
Histocompatibility Testing
HPLC of Macromolecules
HPLC of Small Molecules
Human Cytogenetics I and II
  (2nd edition)
Human Genetic Diseases
Immobilised Cells and Enzymes
Immunocytochemistry
In situ Hybridization
Iodinated Density Gradient Media
Light Microscopy in Biology

Essential
Molecular Biology
Volume I
A Practical Approach

Edited by
T. A. BROWN

Department of Biochemistry
and Applied Molecular Biology
UMIST, Manchester M60 1QD, UK

OXFORD UNIVERSITY PRESS
Oxford New York Tokyo

Oxford University Press, Walton Street, Oxford OX2 6DP

Oxford New York Toronto
Delhi Bombay Calcutta Madras Karachi
Kuala Lumpur Singapore Hong Kong Tokyo
Nairobi Dar es Salaam Cape Town
Melbourne Auckland Madrid

and associated companies in
Berlin Ibadan

Oxford is a trade mark of Oxford University Press

A Practical Approach 🛇 is a registered trade mark
of the Chancellor, Masters, and Scholars of the University of Oxford
trading as Oxford University Press

Published in the United States by
Oxford University Press Inc., New York

© Oxford University Press, 1991

First published 1991
Reprinted 1992, 1993, 1994

British Library Cataloguing in Publication Data
Essential molecular biology.
Vol. I
1. Molecular biology
I. Brown, T. A. (Terence Austen) 1953–   II. Series
574.88

Library of Congress Cataloging in Publication Data
Essential molecular biology: a practical approach/edited by T. A. Brown.
(Practical approach)
Includes bibliographical references and index.
1. Molecular Cloning.   2. Recombinant DNA.
I. Brown, T. A. (Terence A.)   II. Series: Practical approach series.
QH442.2.E77   1991   574.87'3282—dc20   90–49865

ISBN 0 19 963110 7 (Hbk: v. 1)
ISBN 0 19 963111 5 (Pbk: v. 1)

Printed in Great Britain by Information Press Ltd., Eynsham, Oxon.

Preface

There are now a number of molecular biology manuals on the market and the editor of an entirely new one has a duty to explain why his contribution should be needed. My answer is that although there are some excellent handbooks for researchers who already know the basic principles of gene cloning there are very few that cater for the absolute beginner. Unfortunately, everyone is a beginner at some stage in their careers and even in an established molecular biology lab the new research student can spend a substantial amount of time not really understanding what is going on. For the experienced biologist expert in a discipline other than molecular biology, and perhaps without direct access to a tame gene cloner, guidance on how to introduce recombinant DNA techniques into his or her own research programme can be very difficult to obtain. For several years I have run a basic gene cloning course at UMIST and I have continually been impressed by the number of biochemists, botanists, geneticists, cell biologists, medics, and others who want to learn how to clone and study genes.

The contributors to *Essential Molecular Biology: A Practical Approach* were asked to write accounts that combine solid practical information with sufficient background material to ensure that the novice can understand how a technique works, what it achieves, and how to make modifications to suit personal requirements. Where appropriate the reader is also given advice on more advanced or specialized techniques. In all cases the authors have responded to the challenge and produced chapters that make concessions to the beginner without jeopardizing scientific content or practical value. I hope that the result is a handbook that will guide newcomers into molecular biology research.

The book is split into two parts. Volume I deals with the fundamental techniques needed to carry out DNA cloning experiments. The emphasis is on coming to grips with the necessary practical skills and understanding the background in sufficient detail to be able to adjust to circumstances as the project progresses. In Volume II procedures for preparing gene libraries and identifying genes are described, together with methods for studying the structure of a cloned gene and the way it is expressed in the cell. It is assumed that the basics from Volume I are now in place, but the procedures are still described in the same down to earth fashion with protocols complemented by background information and troubleshooting hints.

I must thank a number of people for their help with this book. First, I am grateful to the authors who provided the manuscripts more or less on time and were prepared in many cases to make revisions according to my requests. I would especially like to thank Paul Towner for stepping in at the last minute after I had been let down with one chapter. The participants and assistants on the recent

UMIST gene cloning courses helped me formulate the contents of the book and my colleague Paul Sims provided valuable advice. The Series Editors and publishers made encouraging noises when the going got tough, and sorted out a number of problems for me. I am also grateful to my research students for giving their opinion on the practical content of parts of the book. Finally, I would like to thank my wife Keri for proving that even an archaeologist can learn how to clone genes.

1990   T.A.B.

Contents

Contents

6 Recovery of DNA from electrophoresis gels 127

Paul Towner

7 Construction of recombinant DNA molecules 143

Frank Gannon and Richard Powell

Contents

Contributors

A. T. ANDREWS
AFRC Institute of Food Research, Reading Laboratory, Shinfield, Reading RG2 9AT, UK

B. W. BAINBRIDGE
Division of Biosphere Sciences, Kings College, University of London, Campden Hill Road, London W8 7AH, UK

T. A. BROWN
Department of Biochemistry and Applied Molecular Biology, UMIST, Manchester M60 1QD, UK

F. GANNON
National Diagnostics Centre and Department of Microbiology, University College, Galway, Ireland

C. KELLER
Cold Spring Harbor Laboratories, PO Box 100, Cold Spring Harbor, NY 11724, USA

P. MYERS
Cold Spring Harbor Laboratories, PO Box 100, Cold Spring Harbor, NY 11724, USA

P. H. POUWELS
TNO Medical Biological Laboratory, PO Box 45, Rijswijk, The Netherlands

R. POWELL
Department of Microbiology, University College, Galway, Ireland

R. J. ROBERTS
Cold Spring Harbor Laboratories, PO Box 100, Cold Spring Harbor, NY 11724, USA

P. TOWNER
Department of Biochemistry, University of Bath, Claverton Down, Bath BA2 7AY, UK

M. WILKINSON
Microbiology and Immunology Department and the Vollum Institute, Oregon Health Sciences Laboratory, Portland, Oregon 97201, USA

Abbreviations

A_{260}	absorption at 260 nm
BAP	bacterial alkaline phosphatase
CHEF	contour clamped homogeneous electric fields
CIP	calf intestinal phosphatase
CTAB	cetyl trimethyl ammonium bromide
DEP	diethyl pyrocarbonate
DMAPN	3-dimethylaminopropionitrile
DMSO	dimethylsulphoxide
DNase	deoxyribonuclease
EDTA	ethylenediaminetetraacetate
FIGE	field inversion gel electrophoresis
IPTG	isopropyl-β-D-thiogalactopyranoside
mRNA	messenger RNA
OD	optical density
OFAGE	orthogonal field alternation gel electrophoresis
PACE	programmable autonomously-controlled electrode
PCR	polymerase chain reaction
PEG	polyethylene glycol
PFGE	pulsed field gradient gel electrophoresis
PVP	polyvinylpyrrolidone
RBS	ribosome-binding site
RF	replicative form
RFE	rotating field electrophoresis
RFLPs	restriction fragment length polymorphisms
RGE	rotating gel electrophoresis
RNase	ribonuclease
rRNA	ribosomal RNA
SDS	sodium dodecyl-sulphate
TAE	Tris–acetate–EDTA buffer
TAFE	transverse alternating field electrophoresis
TBE	Tris–borate–EDTA buffer
TE	Tris–EDTA buffer
TEA	triethanolamine
TEMED	N,N,N',N'-tetramethylethylenediamine
TLC	thin-layer chromatography
TNM	tetranitromethane
Tris	tris(hydoxymethyl)amino methane
tRNA	transfer RNA
UV	ultraviolet
X-gal	5-bromo-4-chloro-3-indolyl-β-D-galactopyranoside

1

The essential techniques in molecular biology

T.A. BROWN

In several respects molecular biology is not a user-friendly discipline. The newcomer is faced with a bewildering array of techniques, some unhelpful jargon, and the impression that only complicated experiments are worth doing. In fact, most of the apparent problems are illusory and there is no reason why any competent research worker should not be able to clone a gene and obtain information concerning its structure and mode of expression. This is true not only for people in established molecular biology laboratories but also for researchers from other areas wishing to apply gene analysis techniques in their own projects.

The chapters which follow describe the procedures for purifying DNA and RNA, constructing recombinant DNA molecules, and obtaining clones (Volume I), preparing gene libraries, identifying genes, and studying gene structure and expression (Volume II). However, before tackling the detailed practical aspects of molecular biology the reader should take some time to think in more general terms about molecular biology techniques and what they can achieve. The aim of this introductory chapter is to provide such an overview for the techniques described in Volume I, and to emphasize several general points relevant to successful gene cloning and analysis.

1. Gene cloning in outline

The basic steps in a gene cloning experiment are shown in *Figure 1*. They are as follows (1):

(a) DNA is prepared from the organism being studied and fragments inserted into vector molecules to produce chimaeras called recombinant DNA molecules.

(b) The recombinant DNA molecules are introduced into host cells.

(c) Within the host cells the recombinant DNA molecules replicate and are passed on to progeny cells during cell division.

(d) Continued replication of the host cells results in clones, each of which

1

1 Construction of a recombinant DNA molecule

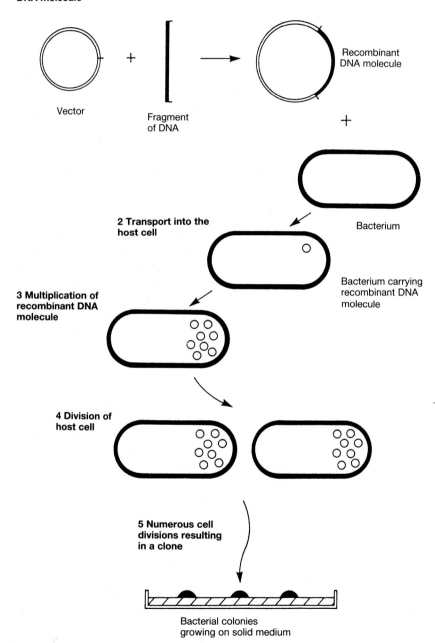

Vector

Fragment of DNA

Recombinant DNA molecule

+

Bacterium

2 Transport into the host cell

Bacterium carrying recombinant DNA molecule

3 Multiplication of recombinant DNA molecule

4 Division of host cell

5 Numerous cell divisions resulting in a clone

Bacterial colonies growing on solid medium

Figure 1. The basic steps in a gene cloning experiment. (Reproduced from ref. 1 with permission.)

consists of identical cells all containing copies of a single recombinant DNA molecule.

Although straightforward in outline the procedures involve a number of different techniques. These are introduced in the following sections.

1.1 Handling bacteria (Chapter 2)

The first practical skill required by a molecular biologist is the ability to grow pure cultures of bacteria, including cultures infected with bacteriophages. This is because all gene cloning experiments involve the use of a bacterium, almost invariably *Escherichia coli*, as the host organism. Even if the objective is to express a cloned gene in a higher organism, the initial manipulations will be carried out in *E. coli* because of the ease with which this bacterium can be handled and because of the numerous cloning vectors specific for it.

The key aspect of practical microbiology is sterile technique, which is the means by which cultures are kept pure and contamination of work surfaces and personnel is avoided. A rigorous sterile technique is the first thing that the molecular biologist must learn, because without it things will not progress very far. Even when carrying out manipulations that do not involve bacteria it is advisable to use sterile materials (e.g. pipette tips, glassware, microfuge tubes) and a clean technique (wear gloves, observe standard biological lab practice) as bacterial contamination and secretions from the skin can degrade DNA and RNA, and inactivate restriction and other enzymes.

1.2 Preparation of DNA (Chapter 3)

For many the first real molecular biology to be attempted is purification of DNA from the organism being studied. This will provide the material to be fragmented and inserted into the cloning vector. The main consideration is to obtain DNA that has not been broken up too extensively during the purification procedure and which is free from contaminants that will interfere with the reactions involved in constructing the recombinant molecules. In general terms, the procedure used is the same regardless of the organism involved, though appropriate steps have to be taken to obtain an adequate cell extract and to remove specific contaminants such as lipid and carbohydrate.

DNA preparations will also be carried out to purify vector DNA and recombinant molecules. Vector DNA can be purchased from commercial suppliers (see *Appendix 7*) but most molecular biologists maintain stocks of bacteria containing the vectors they are working with and prepare DNA as and when needed. This has the advantage of cheapness (1 litre of *E. coli* can yield about £2000 of M13 vector) and assured quality control. Even if you do not make your own vector at the outset of the project you will at a later stage need to purify recombinant DNA in order to study the gene you have cloned.

1.3 Preparation of RNA (Chapter 4)

The preparation of pure, intact RNA is relatively difficult because of the ubiquity of contaminating enzymes that degrade RNA molecules. RNases present in the cells from which RNA is being extracted can be inactivated during the purification procedure by a suitable RNase inhibitor, but considerable problems can be presented by glassware and solutions contaminated with RNases derived from skin secretions. These enzymes are very resistant to physical treatments, some even withstanding autoclaving, and their control can be a real problem. However, for some types of project RNA preparation is of equal or greater importance than DNA preparation. Often the most suitable approach to obtaining a clone of your gene is to start not with DNA but with RNA. This is because the mRNA molecules present in a particular tissue or population of cells represent only those genes that are actually being expressed, which may of course be a very small subset of the total gene content of the organism. Purification of the mRNA fraction followed by conversion into cDNA (Volume II, Chapter 3) will provide fragments suitable for insertion into a cloning vector, and may result in a clone library from which identification of the desired gene is much easier than from a library representing the whole genome. If this is the route that you need to take in your project then it is advisable to purchase some yeast RNA (or similar) and establish that your ribonuclease-free technique is sufficiently sound before starting to work with your own, more precious RNA samples.

1.4 Separating DNA and RNA by gel electrophoresis (Chapter 5)

Once upon a time, separating nucleic acid molecules of different sizes was a laborious business. The development of gel electrophoresis systems for DNA and RNA was an essential pre-requisite for the gene cloning revolution of the 1970s. The techniques are now standard in all molecular biology labs but are understood in very few. The theoretical basis to gel electrophoresis is complicated but has to be tackled in order to gain a real appreciation of the behaviour of DNA and RNA in electrophoretic systems and the influences of factors such as gel type, buffer, size of molecule, and electrical parameters. Several important applications (e.g. separation of chromosomes, Chapter 5 of Volume I; DNA sequencing, Chapter 6 of Volume II) are pushing gel electrophoresis to its limits, and satisfactory results demand a full understanding of the technology.

Agarose gels are routinely run to estimate the sizes of DNA and RNA molecules, which enables one to assess the success of a preparative technique, to plan a strategy for constructing recombinant molecules, to determine if *in vitro* manipulations have had their desired affect, and to analyse the DNA molecules contained in recombinant clones. These applications will be continually referred to throughout the two volumes of this book. Polyacrylamide gels, which are used to resolve small DNA molecules in procedures such as DNA sequencing

(Volume II, Chapter 6), are more difficult to set up than agarose gels, but because of their use in protein biochemistry are often more familiar to the biochemist starting to clone genes.

1.5 Purifying DNA molecules from electrophoresis gels (Chapter 6)

In some areas of molecular biological research there are a number of alternative procedures for achieving the same objective. The purification of DNA molecules from gels is a case in point. Frequently an agarose or polyacrylamide gel will provide a band that can be confidently identified as containing a DNA fragment of interest. The next step might be to excise the band and purify the DNA molecules prior to construction of recombinant DNA molecules and further study. In practice this is not so easy, as even the 'purest' agarose contains innumerable contaminants that will co-purify with the DNA fragment and quite possibly prevent, or at least retard, subsequent *in vitro* manipulations. Various methods have been devised to get around the problems and their individual strengths and limitations should be understood before choosing one for a particular operation.

1.6 Construction of recombinant DNA molecules (Chapter 7)

So far we have dealt only with procedures involved in preparation and handling of DNA and RNA molecules. The first real real step in a gene cloning experiment (see *Figure 1*) is construction of recombinant molecules by insertion of DNA fragments, including the gene to be cloned, into vector molecules. At this stage one of the characteristic features of molecular biology research, the use of purified enzymes to carry out controlled manipulations of DNA molecules, is first encountered. The two main types of enzymes used in recombinant DNA construction are:

• restriction endonucleases

• ligases

Restriction endonucleases (*Appendix 5*) make double-strand breaks at specific recognition sequences within DNA molecules. They enable DNA molecules to be cut ('restricted') in a reproducible fashion to provide a discrete family of fragment sizes. A circular DNA molecule that possesses just one restriction site for a particular endonuclease will be converted into a linear molecule by treatment with that enzyme. As can be seen in *Figure 1* this is the essential manipulation required to prepare the cloning vector for insertion of the DNA fragments to be cloned.

DNA ligases (*Appendix 6*) can join together DNA fragments prepared by restriction. To be ligated a pair of ends must be 'compatible', which means they must both be flush-('blunt'-)ended, or both must have single-strand overhangs ('sticky ends') that are able to base-pair. The DNA fragments do not have to be prepared with the same restriction enzyme in order for their ends to be compatible; two enzymes with different recognition sequences may still produce compatible single-strand overhangs. In addition, one type of sticky end can be converted to another type of sticky end or into a blunt end, and blunt ends given sticky overhangs, by use of the more specialized manipulative enzymes described in Chapter 7 and *Appendix 6*.

These considerations form the basis to the restriction and ligation manipulations that are used to make pre-determined constructions *in vitro*. Your understanding of the range of manipulative enzymes and your ingenuity in applying them in your research will have a strong bearing on how successful your work is.

1.7 Introduction of recombinant molecules into host cells and recombinant selection (Chapter 8)

Once recombinant DNA molecules have been constructed they must be introduced into the host *E. coli* cells. This is quite straightforward and standard techniques exist for different types of vector. More complex is the means used to identify colonies containing recombinant molecules. Only a proportion of the cells in the culture will take up any DNA molecule at all, and of these a fraction will usually take up a self-ligated vector that carries no insert DNA. Identification of recombinant colonies (i.e. those containing a recombinant DNA molecule) makes use of the genes carried by the cloning vector. One of these will have been inactivated by insertion of the cloned DNA, leading to a phenotypic change that can be directly selected (e.g. antibiotic resistance/sensitivity) or easily scored (e.g. ability to use a chromogenic substrate).

Under some very favourable circumstances it may be possible at this stage to identify recombinant clones that contain the gene of interest. This would be the case if the recombinant molecules were constructed in such a way that each one must contain the correct gene, or if the cloned gene itself confers an identifiable phenotype on the host cell. However, this is not usually the case, and more frequently the cloning experiment will simply provide a collection of recombinants from which the desired one must be identified by a second series of experiments, using the strategies described in Volume II, Chapter 1.

1.8 Cloning vectors (Chapter 9)

By now it will be clear that the choice of cloning vector is an important one. It will determine the nature of the procedure used to introduce the recombinant molecules into the host cells and the way in which recombinants are selected. It is also important to choose a vector that is suitable for the size of DNA fragments that you wish to clone, as some vectors are designed for relatively small fragments (5 kb and less) whereas others require fragments above a certain size.

There are many different cloning vectors for use with *E. coli* but all are derived from natural bacterial plasmids or from the genomes of bacteriophages λ or M13. Plasmids and phage genomes possess replication origins and either code for replicative enzymes or can use host enzymes to direct their own replication. They therefore possess the means for existing and multiplying within the host cells, providing the basic features needed to obtain a recombinant clone.

Choice of cloning vector is difficult and advice from different sources can be contradictory. Over a hundred different types are available commercially, and a whole host of applications are catered for. Often it is tempting to use a sophisticated vector even though it carries a number of functions that are unnecessary for the experiment being conducted. As a rule of thumb choose the simplest vector suitable for your requirements, but be prepared to change if you encounter problems with handling or recombinant selection. Once you have success with a vector, try to use it for future experiments whenever possible: better the vector you know than the vector you don't.

2. Practical requirements

2.1 Experimental skills

The newcomer to molecular biology should have no concerns about the manipulative skills needed to clone and analyse genes. Certainly any researcher who has purified an active enzyme, micro-manipulated yeast asci or carried out a patch-clamping experiment will have few problems with the procedures described in this book. Many manipulations in molecular biology require nothing more demanding than pipetting microlitre amounts of solutions from one tube to another.

It would be a mistake though to believe that sloppy work will lead to good results. Accuracy is obviously important but so is scrupulous cleanliness to avoid contamination of bacterial cultures, degradation of nucleic acid samples, and inactivation of enzymes. The following are the key aspects of laboratory practice for a successful molecular biologist:

- learn how to pipette down to 1 μl accurately and reproducibly;
- develop a steady hand for loading samples on to gels and other manipulations: give up caffeine if it makes you shake;
- do not allow stocks of enzymes to warm up to room temperature: keep in the freezer and take out only for as long as it takes to remove an aliquot;
- autoclave plasticware (pipette tips, microfuge tubes, etc.) and keep sterile before use; do not handle pipette tips;
- always use double-distilled water; make sure the glassware cleaning programme includes a final rinse in double-distilled water;
- get in the habit of wearing disposable gloves at all times.

2.2 A word on kits

A popular recent development amongst commercial suppliers has been the development of molecular biology kits that provide all the reagents and other materials needed for a particular experiment, together with protocols and sample DNA. So, for instance, you can buy DNA labelling kits, cDNA synthesis kits, plus many others. These are undoubtedly a great help in many applications but certain factors should be borne in mind.

First, on the positive side, kits usually provide reasonably good results and in most cases these results are easier to obtain than by the traditional method of purchasing the materials separately, making up one's own buffers, and suchlike. On the negative side is the fact that the traditional route, though slower, will usually provide better results in the long run, as experience gained along the way can be used to modify steps and buffer compositions to suit the particular DNA molecules being studied. A twofold improvement in a procedure can have a major effect on the outcome of the project. However, the main argument against kits, as raised in the literature (2), is the way they can provide a substitute for laboratory skills. You will not understand much about a procedure by slavishly following a protocol using pre-mixed reagents. Without understanding a procedure you will not be able to use it to its full advantage, know what to do when things go wrong, or appreciate when a result is artefactual or aberrant. The best advice is to learn and understand a technique first, then when you have it working see if a commercial kit is just as efficient and if so use the kit for repetitive experiments. If you are a research student then do not get the idea that using kits is the same as being a molecular biologist.

2.3 Equipment and materials

Appendix 1 provides a list of the equipment and other materials needed to carry out molecular biology experiments. Much of the equipment will already be present in the 'well-found' biology laboratory, but if you are moving into molecular biology from some other area then you will almost certainly need to obtain some specialist items.

2.4 Safety

The major hazards encountered in molecular biology research are:
- microorganisms
- radiochemicals
- organic solvents
- mutagens/carcinogens
- toxic chemicals
- ultraviolet radiation
- high voltage electricity

Guidance on how to deal with these hazards is given in *Appendix 2*. Bear in mind that these are only broad guide-lines and that it is the individual research worker's responsibility to ensure that working practices are safe and do not endanger other personnel. Consult your department safety representatives for advice and follow local and national regulations.

3. Conceptual awareness

To gain success in molecular biology it is clearly important to understand the techniques and attain the necessary practical skills. It is just as important though to understand what can be achieved by a particular research programme. Gene cloning is not a panacea; in fact, it can be very difficult to interpret in a meaningful way the results of a molecular biology project if there is no complementary biochemical or genetical information on the system being studied.

To illustrate the potential pitfalls of molecular biology projects I will take a fairly typical scenario. A research group has studied an enzyme for a number of years and now wish to clone the gene, sequence it, and study its expression. Before starting this new project several questions should be asked. Similar questions are appropriate before projects with other starting points or objectives. The points will be expanded on when strategies are discussed in Volume II, Chapter 1, but it is important that the questions are asked right at the beginning of the project.

3.1 Will it be possible to identify the correct clone?

Only under special circumstances will it be possible to devise the cloning experiment in such a way that the desired gene can be identified straightaway. Normally the primary objective will be to prepare a clone library, either from cDNA or from genomic DNA, and to identify the clone containing the relevant gene from this library. To do this a probe specific for the desired gene will be needed. Is such a probe available?

In general terms there are four ways of obtaining a suitable probe:

- The amino acid sequence of the enzyme, if known, can be used to dictate the sequence of a specific oligonucleotide hybridization probe.
- If the gene for the enzyme has already been cloned from a related organism, then it can be used as a heterologous hybridization probe.
- If the enzyme is known to be abundant in a particular tissue then it may be possible to identify the relevant clone by virtue of its abundance in a tissue-specific cDNA library.
- If the enzyme has been purified then an antibody raised against it can be used to screen recombinant cells for the presence of the enzyme.

These are generalizations, but they cover most of the practical ways of identifying a gene in a clone library. Certainly, the more sophisticated techniques, such as chromosome walking, are not recommended to the newcomer. If one of the above possibilities cannot be satisfied then it would be pointless to embark on a gene cloning project.

3.2 Will the DNA sequence provide any new information?

A DNA sequence on its own is not much use. If you do not know what to expect, it may not even be possible to assign unambiguously the coding region of a gene. Intron boundaries can be guessed from their similarities to concensus sequences but often two or more alternatives cannot be discounted purely from sequence information. So it may not even be possible to determine the coding sequence of the gene.

If the amino acid sequence of the enzyme is already known, then of course the codon sequence will be identifiable, and any introns can be located with precision. But if the amino acid sequence is already known then what new information will you get from the gene sequence?

The most difficult problems arise if the nature of the gene product is unknown. It may be possible to identify the function of a gene sequence by comparing it with other sequences held in the various databases, but nine times out of ten a sufficiently convincing match will not be found. How to go from DNA sequence to identification of a gene product is one of the great problems of recombinant DNA research and can provide an effective block to the progress of a project. This should be borne in mind if cloning and sequencing are to be used to study, for instance, tissue-specific genes for which translation products have not been identified, or genetic loci whose phenotypic effects are not clearly associated with a known gene product.

So far we have not got very far with our project. What about expression of the gene?

3.3 Will studies of gene expression be informative?

Before you even start to purify DNA you can assume that your gene will have a promoter and, if it is differentially-expressed, other upstream control sequences to mediate gene regulation. Will your expression studies take you any further than this?

The answer is yes, probably, though as with DNA sequencing a biologically-meaningful interpretation of the results depends on being able to relate the data provided by molecular biology with information obtained from biochemistry or genetics. Without being too adventurous it is possible to:

- determine which tissues a gene is expressed in;
- determine the approximate sizes of transcripts and whether the transcripts are the same size in all tissues;
- determine the precise 5'- and 3'-termini of the transcripts and of any introns within the transcripts.

More ambitious experiments can:

- locate upstream control sequences that mediate gene regulation;
- isolate regulatory proteins that bind to these control sequences.

However, if results along these lines are achieved in a vacuum then their importance will relate primarily to molecular biology and only vaguely to the biology of the cell as a whole. Can gene regulation be related to control of activity of your enzyme, or are transcriptional events masked by allosteric controls and protein activation events? Are you aware of external and internal biochemical stimuli that influence gene expression? Are there clearly-defined defects in gene product activity that may point to impairments in gene regulation? If you can ask specific questions along these lines then you will be able to make the difficult transition from the genes to the rest of the cell: so get cloning!

In Section 3, I have deliberately taken the stance of devil's advocate, not to dissuade people from embarking on molecular biology projects, but to try to ensure that at the outset the foundations are laid for a successful research programme. Molecular biology can provide a wealth of information and the chapters in this book will help you obtain that information. It is up to you to make sure the information is useful.

References

1. Brown, T.A. (1990). *Gene Cloning: An Introduction*, 2nd edn. Chapman & Hall, London.
2. Davies, J. and Pugsley, A. (1990). *Trends in Biochemical Science*, **15**, 137.

2

Microbiological techniques for molecular biology: bacteria and phages

BRIAN W. BAINBRIDGE

1. Introduction: techniques for handling microbes

1.1 Basic microbiological techniques

Recombinant DNA technology depends on the manipulation of particular strains of bacteria and bacteriophages (phages). Expertise in isolating, checking, growing, and analysing these strains is crucial for the success of genetic engineering. Many experiments are now permitted under conditions of 'good microbiological technique' and anybody embarking on research in this area should check exactly what is meant by this phrase. Essentially, all of the basic techniques depend on the culturing of a particular microbe in the absence of other organisms (sterile or aseptic technique), the isolation of a genetically-pure culture or clone derived from a single cell (single colony or single plaque isolation) and the characterization of the known genetic features of the strain. From a safety point of view, experiments are done in such a way that recombinant microbes do not escape into the environment or infect the experimenter or other people. The basis of sterile technique is the efficient sterilization of all equipment and growth media followed by the protection of the media from contamination during manipulations. If contamination of a culture is suspected then colony morphology, smell of the culture or staining of the cells followed by microscopic examination can be useful. Growth of the strains in liquid or solid media allows larger numbers of cells to be cultured and it is important to understand the various phases of the growth curve as different types of cells are required for different experiments (1). Plating techniques such as dilution and streak plates can give rise to single colonies or plaques derived from single-cell or phage particles. Selective techniques are often used to isolate particular strains or to detect rare events, for example in transformation (see Chapter 8), and such techniques are very sensitive to low levels of contamination by unrelated strains which happen to have the same characteristics as the desired strain. This may be resistance to antibiotics or nutritional independence.

1.2 Basic techniques of microbial genetics

Genetic purity of strains, plasmids or bacteriophages is critical and it should be clearly understood that no strains can ever be completely stable (2). Growth can lead to rearrangements of DNA, mutation of essential genes, or infection by foreign DNA. Strains should therefore be carefully checked to make sure that they are of the correct genotype before starting crucial experiments. Chemicals and enzymes are not self-replicating and they are therefore usually purchased from suppliers. In contrast strains of bacteria and phages are frequently subcultured many times and passed from one research group to another. Quality control and frequent checking is necessary or significant alterations in the genome of the organism may occur. It is particularly important to make a careful note of any deliberate changes or improvements in particular strains. This means recording the pedigree of the strains and using a unique numbering system and detailing the full genotype. The dangers here are that essential gene mutations such as *recA* (deficient for the major recombination system) may have reverted to wild-type and transposons may have caused rearrangements. An empirical approach can be adopted such that if the technique works the strain must be correct, but this can delay the problem until a later stage. Unless you are absolutely sure of the origin of the strain it is much safer to obtain the correct culture from a reputable supplier. When the correct strain has been successfully characterized it is essential that it is preserved efficiently. Any growth can lead to mutation and any dessication or contamination can lead to death of the culture. The basis of preservation is to stop growth by lyophilization or freezing in the presence of cryoprotectants such as glycerol. Duplicate cultures should be kept in different locations to avoid dangers of apparatus failure.

1.3 Safety in the molecular biology laboratory

Although *Escherichia coli* K12 is not considered to be pathogenic, there are other strains of *E. coli* which are dangerous and can cause septicaemia or kidney infections. The development of disease symptoms depends on the type and number of organisms which gain access to the body as well as on the efficiency of the immune system of the laboratory worker. Most healthy people can cope with small numbers of microbes but depression of the immune system by illness or infection or the use of immunosuppressive drugs can markedly increase the risks of a serious infection. It is safer to assume that all microbes are potential pathogens and to treat them with respect. The basic safety precautions, modified from GMAG note 15 (3), are shown below, and should be followed along with local and national regulations:

(a) Laboratory overalls, which should ideally be side- or back-fastening, must be worn.

(b) Avoid all hand-to-mouth operations such as licking pencils or labels.

(c) Make a habit of washing your hands before leaving the laboratory.

(d) Before starting to work, protect all cuts with adequate waterproof dressings.

(e) Cultures spilt on the bench, floor, apparatus or on yourself or others, should be treated with disinfectant. Material used to wipe up should be discarded for incineration or sterilization.

(f) Do not eat, drink, smoke, or apply cosmetics in the laboratory.

(g) All contaminated apparatus should be sterilized before washing or disposal.

(h) All contaminated glassware such as pipettes and tubes should be discarded into disinfectant prior to sterilization and washing.

(i) All contaminated disposable plasticware should be discarded into strong autoclavable bags for sterilization or incineration.

(j) All apparatus for autoclaving or incineration should be carried in leak-proof containers.

(k) All apparatus used for the culture of microbes should be clearly labelled before inoculation. Apparatus to be left in communal areas should show your name, the organism, and the date.

(l) All contaminated syringe needles should be discarded into special 'sharps' containers for special sterilization and disposal.

(m) Every effort should be made to avoid the production of aerosols. These are produced, for example, by blenders, centrifugation, ultrasonication, and movement of liquids against surfaces. If the microbes present a hazard then the equipment used should be placed in a suitable safety cabinet.

(n) Records should be kept of the storage and transport of all microbes.

A more detailed treatment of this subject will be found in reference works by Collins (4, 5). A recent development in the UK has been the regulations for the Control of Substances Hazardous to Health (6) which include 'microorganisms which create a hazard to the health of any person'. The regulations include a clause which states that an assessment should be made of 'the risks created by that work to the health of those employees and of the steps that need to be taken to meet the requirements of these regulations'. Essentially, exposure to microbes by inhalation, ingestion, absorption through the skin or mucous membranes or contact with the skin must be either prevented or controlled. Thus safety should be treated seriously and precautions taken to limit the access of microbes to the environment, laboratory workers, and the public. Safety is not the blind following of regulations but an awareness of the hazards and the methods which can be used to minimize them.

This chapter is intended to provide a basic introduction to the techniques involved in the handling of *E. coli* and its phages lambda and M13. It does not include techniques in microbial genetics such as mutant induction, gene mapping and replica plating which will be found in a variety of other manuals (7, 8).

1.4 Sterilization and disinfection

It is important to distinguish between these two processes. The aim of sterilization is to eliminate all microbes from laboratory equipment or materials. Disinfection aims to eliminate organisms which may cause infection. Space does not permit a theoretical treatment of this subject (4) but the important practical principles will be given.

Sterilization can be achieved by heat, chemicals, radiation, or by filtration. Nichrome loops are sterilized by flaming in a bunsen burner, and disposable plastics can be sterilized by incineration. Glassware can be sterilized either by autoclaving or by dry heat. Autoclaving which uses wet heat is much more efficient than dry heat as hydrated microbes are killed more easily. Autoclaves vary from domestic pressure cookers to large, industrial-size motorclaves. It is very important that you follow the operating instructions and in particular do not overload the autoclave, as the central region may not reach the necessary temperature. It is important to remove all of the air from the autoclave because the presence of air depresses the final temperature reached. Autoclave tape which changes colour after the correct time and temperature is a useful check. Loosen the caps of all bottles and do not autoclave completely-sealed bags in small autoclaves. Always make sure that there is sufficient water in pressure cookers and check that the correct procedure for autoclaving and recovery of materials is followed. Autoclavable plastic tubes such as pipette tips and microfuge tubes should be wrapped in autoclavable nylon bags. Dry-heat sterilization is normally used for flasks and glass pipettes, which should be left on a 6–12 h cycle at 160°C. Sterilization with chemicals and radiation are not practical methods in the average laboratory. Sterile plasticware has normally been produced by gamma irradiation.

Disinfection procedures will vary from laboratory to laboratory but a general-purpose disinfectant is 2% (v/v) Hycolin which turns from green to blue when no longer effective. Chlorine-based disinfectants such as Chloros are more effective against spores and viruses but have the disadvantage that they need to be checked more frequently. Any contaminated glassware or unwanted cultures should be immersed in disinfectant before autoclaving and washing. Contaminated disposable plasticware should be placed in autoclavable bags or bowls, which must be leak-proof. This material should be disposed of by incineration. Detailed procedures are the responsibility of the local safety officer.

1.5 Basic principles of aseptic technique

There are two basic principles of aseptic technique:

- protection of yourself and of others.
- Protection of cultures and apparatus from contamination by unwanted microbes.

In normal laboratory areas, microbes are everywhere—in the air, in dust, on your fingers. It is in fact very difficult to produce an environment completely free from microbes, and special equipment such as lamina-flow cabinets and sterile areas is required. In a clean laboratory, with reasonable precautions, it is not necessary to use inoculating cabinets for the preparation of media and the manipulation of cultures. However, if you have trouble with contamination or your cultures are particularly slow-growing then a cabinet can be useful. Safety cabinets should not be needed for experiments classified as needing only 'good microbiological technique'. Experiments at higher levels of containment need different pre-cautions in different countries (e.g. ref. 9) and you should consult your local safety officer for guidance. See also *Appendix 2.*

Unless laboratory air has been efficiently filtered, it will contain many suspended bacterial cells and spores, fungal spores, and, in some laboratories, air-stable phage particles. This population of particles is added to by air movements which resuspend dust particles from bench surfaces. These airborne particles will settle on to any exposed surface and this is a major source of contamination, therefore anything which is to be kept sterile must be exposed to the air for a minimum period of time. Dust and aerosol particles tend to settle rather than drift sideways unless there is a draught of air. Consequently, containers and Petri dishes should not be left open, surface upwards, although tubes opened at an angle are less at risk. All apparatus which cannot be flamed in a bunsen burner immediately prior to use should be left in the wrappers or containers in which they have been sterilized, until actually needed. No sterile equipment should be allowed to come into contact with unsterile surfaces. Plugs and caps from sterile tubes and bottles should not be placed on the bench, although they can be placed on a tile swabbed with a disinfectant.

The commonest source of contamination in the laboratory is the access of unsterile air to the apparatus. This is increased by draughts and general movement of air and it follows that every effort should be made to work in still-air conditions. Windows and doors should be closed and all rapid movements in the laboratory should be eliminated. It is obvious that the laboratory should be free from dust which could be resuspended by air movement but it should be remembered that cleaning techniques such as sweeping and dusting can be a serious source of aerial contamination. The major advantage of inoculating cabinets is that they give protection from these air movements and allow a small volume of air to be sterilized by UV radiation or by a 70% (v/v) alcohol spray. The principles of spraying the air in a cabinet or over a bench is that it settles dust particles on to the bench, from where they can be removed with a paper towel. Fungal spores are adapted to aerial transmission and particular care should be taken when handling fungi or disposing of apparatus contaminated by fungi. Fungal spores released into the environment can take up to 7 h to settle and therefore can be a source of contamination for a considerable period.

Skin, hair, breath, and clothing are all sources of microbes and it is particularly important that you do not touch sterile surfaces such as the tips of pipettes and the inside of containers. Do not bend over your equipment such that skin scales or dust from your hair might fall into your cultures. Problems have arisen from contamination with yeasts traced to home baking of bread. Where strict asepsis is required (as in operating theatres) sterile caps, gloves, and gowns are worn.

2. Culturing of *Escherichia coli*

You will normally receive a culture of an *E. coli* strain as a broth culture, on a Petri dish, or as a freeze-dried culture from a supplier. The first step is to make a careful record of the strain number and genotype of the strain. From this you will be able to identify an appropriate medium on which it will grow well and any additions such as antibiotics which are necessary to ensure the stability and maintenance of plasmids. Prepare some well-dried agar plates as described in *Protocol 1*. It is necessary to dry the plates as *E. coli* is motile and will swim across the plate in the thin film of water. In addition, contaminants will also spread more easily across the plate and the desired single colonies will not be isolated. Cooling the agar at 50°C reduces the condensation of water on the inside of the lids. Drying the plates overnight at 37°C has the advantage that contaminated plates can be detected and discarded, but care should be taken to avoid micro-colonies on the plate. This is particularly critical with spread plates where the contaminant colonies will be spread over the plate. To avoid this problem it is better to dry the plates in an oven.

Protocol 1. Preparation of agar plates

1. Select the ingredients for the required agar medium.
2. Loosen the top of the agar bottle and melt the agar in a steamer or in a microwave oven. Do not use metal caps in a microwave oven.
3. Swirl the liquid gently to check that the agar is fully melted. Take care that superheated agar does not boil over.
4. Allow the agar to cool for about 10 min at room temperature and then place in a water bath at 50°C for at least 20 min.
5. Place the concentrated nutrient medium in the same water bath to equilibrate to the same temperature.
6. Flame the tops of the glass bottles containing agar and concentrated nutrients and pour the nutrient liquid into the bottle containing the agar. Screw the cap tight and shake to ensure complete mixing.
7. Return the bottle to the water bath and allow time for any air bubbles to disappear.
8. Arrange the sterile Petri dishes on a level surface and label the base of each plate to indicate the medium prepared.
9. Remove the bottle from the water bath and wipe the outside carefully with a paper towel (water baths can be contaminated).

Protocol 1. *Continued*

10. Flame the neck of the bottle and pour the required amount of medium into the plate. This will vary between 10 ml for a thin plate for short-term bacterial culture to 40 ml for phage culture where large plaques are required.

11. Allow the plates to set. Dry the surface of the plates by overnight incubation at 37°C (check for contamination) or by opening the plates and placing them, medium surface down, in an oven at 45–55°C. The lids should also face downwards separately from the base of the plate. Leave plates at 45–55°C for 15 min.

12. Most nutrient plates can be stored for at least a few weeks at 4°C following wrapping in parafilm or sealing in a plastic bag.

Escherichia coli can be cultured on slopes, on plates, in broth, or in stab cultures. The first step is to isolate single colonies usually by a streak plate method (see *Protocol 2*). A single colony is used to produce a series of identical broth cultures, and a check is made of the phenotype of the strain. The culture can then be used for experimental purposes. As soon as it is found that the culture is correct, every effort should be made to preserve it. This can be done on plates, in broths, or in stab cultures where protection from contamination and dessication is particularly important (Section 4). Freezing is necessary for long-term storage.

2.1 Single colony isolation

The principle of this technique is to streak a suspension of bacteria until single cells are separated on the plate. Each individual cell will then grow in isolation to produce a clone of identical cells known as a colony. The vast majority of these cells will be genetically identical although mutation can occur during the growth of even a single colony to give low levels of mutant cells. This technique assumes that there are no clumps of cells in the culture as it is not unknown for contaminant organisms to stick to bacterial cells via an extracellular polysaccharide. Examination of a culture under the microscope (see Section 3.8) should give evidence for these clumps. If they are present every effort should be made to disrupt them by suspension in phosphate-buffered saline (PbS) followed by agitation or gentle ultra-sonication treatment. Repeated single colony isolations should result in a pure culture.

Protocol 2. Streak-plate method for single colony isolation (refer to *Figure 1*)

1. Flame a Nichrome loop which is about 3 mm across and has a stem of about 6 cm. Allow the loop to cool, or cool by immersion in a sterile area of agar.

2. Flame the neck of an overnight broth culture and remove a loopful of cells. Alternatively, make a suspension of cells directly from an agar plate in

Protocol 2. *Continued*

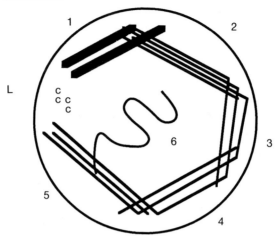

Figure 1. Procedure for the production of single colonies by the streak plate technique as described in *Protocol 2*.

    phosphate-buffered saline (*Appendix 4*) or dilute Ringer's solution. Vortex and remove a loopful of cell suspension.

3. Streak the cells at one side of a well-dried agar plate at position 1 as shown in *Figure 1*. Streak several times close together.

4. Flame the loop and cool carefully at one side of the plate (position L).

5. Streak again at position 2 on *Figure 1*.

6. Flame the loop and cool as before. Repeat steps 4 and 5 as indicated at positions 3, 4, 5, and 6 on *Figure 1*.

7. Incubate the plate at 37°C with the agar facing downwards to minimize contamination and to reduce the chance of droplets of condensation falling on the agar surface.

    The plates are then examined for colony morphology and the presence of possible contaminants. If all of the colonies are of a uniform size and appearance you can assume that you have a pure culture. Subculture of almost any colony should give you the required strain and the plate is worth keeping as a future source of a purified culture. When plasmid-containing strains have been plated on a medium supplemented with an antibiotic it is often observed that colonies of a variety of sizes are obtained. It is not always immediately obvious which is the correct strain and it may be necessary to subculture a representative range of colonies on to fresh medium for further checking. A careful note should be made of the characteristics of the correct strain for future reference. Re-streak the correct colony to give a plate containing a uniform colony size and use this as a source of purified culture.

In a number of experiments such as transformation and the preparation of plasmid or cosmid genomic libraries it is necessary to have as many single colonies on the same plate as possible. This involves a dilution series and spread plates:

Protocol 3. Dilution series and the spread-plate method for single colony isolation

1. Prepare a series of six tubes containing 9 ml of PbS (*Appendix 4*) or dilute Ringer's solution. Label the tubes −1, −2, −3, −4, −5, and −6. Prepare three LB agar plates (*Appendix 4*) labelled −4, −5, and −6.

2. Take 1 ml of the test culture (assumed to be about 10^8 cells per ml) and add to tube −1. This is a 10^{-1} dilution. Vortex the tube to mix the cells.

3. With a fresh pipette or disposable tip, take 1 ml from the −1 tube and add to the −2 tube. This is a 10^{-2} dilution.

4. Repeat this operation from −2 to −3 and so on down the series until −6 is reached.

5. Take 100 μl of the −6 dilution and add to an LB agar plate.

6. Dip a glass spreader into absolute alcohol in a glass Petri dish and pass rapidly through a bunsen burner to burn off the alcohol.

7. Cool the spreader on the agar surface and spread the suspension evenly over the plate. Repeat for tubes −5 and −4.

8. Incubate the plate overnight at 37°C. There should be about 1000, 100, and 10 colonies on the −4, −5, and −6 plates respectively. Plate lower dilutions for more dilute suspensions of cells.

9. A variation of this method which is more economical on plates is to add drops of 20 μl on marked places on the same plate without spreading. This is the Miles Misra technique. The plate should be incubated at 25°C for 2 days to give smaller colonies which are easier to count and to subculture.

2.2 Small-scale broth culture

Suspend an appropriate single colony in 0.5 ml of PbS (see *Appendix 4*) or dilute Ringer's solution and vortex the suspension. Use this to inoculate a series of three or four tubes containing 5 ml of a suitable liquid nutrient medium. Incubate the tubes at 37°C overnight and then store the tubes at 4°C until required. If higher cell densities are required the tubes can be shaken at 250 r.p.m. during overnight growth. These cultures can be used over a period of weeks to provide a source of purified cultures. Alternatively, before each experiment take a single colony from the original plate and use it to inoculate a single 5 ml broth culture, shake overnight and then use to seed a larger-scale culture.

2.3 Large-scale broth culture

These cultures are prepared when more cells are required for plasmid extraction or as a source of competent cells for transformation. The scale of the culture depends on the number of cells required. A typical yield would be 5×10^8 cells per millilitre so that a 20 ml culture would give a total of 10^{10} cells (dry weight 4 mg) whereas a 500 ml culture would produce 2.5×10^{11} cells (dry weight 100 mg). A 3–5 ml culture is suitable for a mini-preparation of plasmid DNA whereas a 500 ml culture is required for a large-scale plasmid preparation (see Chapter 3, Section 4).

Protocol 4. Large-scale broth culture of *E. coli*

1. Prepare a 5 ml broth culture from a single colony as described above (Section 2.2) and use this as a seed culture for the large-scale culture.

2. Prepare conical flasks containing the appropriate volume of sterile medium in flasks roughly ten times the volume of the medium (i.e. 25 ml in a 250-ml flask or 200 ml in a 2-litre flask). This ratio has been shown to give maximum aeration so that oxygen is not the growth-limiting factor. However, it is possible to obtain good yields of cells using 500 ml of broth in a 2 litre conical flask.

3. Dilute the seed culture 1 in 20 into the large-scale culture and shake the flask overnight at 37°C on a rotary shaker at 250–300 r.p.m.

4. Harvest the cells by centrifugation at 2000–4000 *g* depending on the tightness of the pellet required and the characteristics of the strain.

5. Resuspend the pellet of cells in a suitable buffer which will vary according to the purpose of the experiment.

6. Recentrifuge and resuspend the pellet for further use.

7. To follow a growth curve (see Section 2.3.1), wait until the culture is just visibly turbid (usually about 2–3 h) and remove a small sample (1–2 ml depending on culture size).

8. Assay optical density (OD) at 550 nm in disposable cuvettes using the uninoculated growth medium as a blank.

9. Continue sampling approximately every 30 min until the desired OD is reached.

2.3.1 The bacterial growth curve

A knowledge of bacterial growth kinetics is essential for a number of techniques in recombinant DNA technology (1). *Figure 2* shows a typical growth curve for

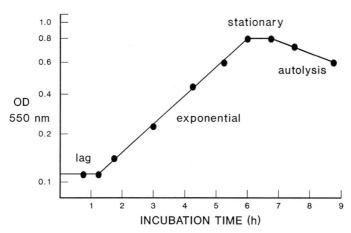

Figure 2. Typical growth curve of *E. coli* in a shake flask at 37°C. Samples are taken about every 30 min and the optical density (OD) at 550 nm measured. The length of the lag, stationary, and autolysis phases will vary.

E. coli, the basic features of which are a lag phase of about 1.5 h followed by a period of exponential growth, a deceleration phase, a stationary phase and finally a decline or autolytic phase. When an overnight culture is diluted 1 in 20 into fresh medium, there is typically a lag phase when no growth can be detected. This can vary from 1.5 h to 3 h depending on the strain, its growth rate, and the number of cells inoculated. Technically, the easiest way to monitor growth is by use of spectrophotometry. Essentially this is a measure of the turbidity (optical density, OD) of the culture estimated on instruments which record absorbance. When a beam of light is passed through a suspension of bacterial cells, light is scattered so less light passes through the suspension in comparison with a control sample. The intensity of the light decreases exponentially as the bacterial concentration increases linearly over limited range. Ideally, this should be checked for a particular strain with a known spectrophotometer. In a complete analysis there should also be a check of OD against cell numbers or dry weight of cells (*Figure 3*). In practice, it is normally sufficient to note values of OD for particular strains which give good competent cells or adequate yields of cells for plasmid extraction or growth of phage lambda. Until you are used to a particular strain it is recommended that you follow OD at intervals and plot this on semi-log paper. This will show the lag and exponential period of growth and it will allow you to predict when the cells will be ready for harvesting.

3. Characterization of bacterial strains

3.1 Genotypes and strain nomenclature

The first step in characterizing a strain is to check its phenotype and indirectly its genotype. Some expertise is required in understanding symbols, which in general

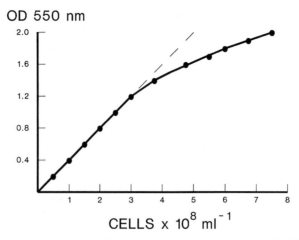

Figure 3. Typical standard curve for the correlation of cell numbers and OD. Cell numbers can be estimated by the dilution and spread-plate technique described in *Protocol 3*. The standard curve will vary with strain, cell size, and spectrophotometer used. It is useful to obtain dry weight estimates for a similar range of samples.

are based on a system proposed by Demerec *et al.* (10). A summary of the major points of this proposal follows:

(a) Each locus of a wild-type strain is designated by a three-letter, lower-case italicized symbol (e.g. *arg* is the gene determining and regulating arginine biosynthesis).

(b) Different loci, any one of which may mutate to produce the same gross phenotypic change, are distinguished from each other by adding an italicized capital letter immediately following the three-letter lower-case symbol (e.g. *argA*, *argB*).

(c) A mutation site should be designated by placing a serial isolation number after the locus symbol. If it is not known in which of several loci governing related functions the mutation has occurred, the capital letter is replaced by a hyphen (e.g. *argA1*, *argB2*, *arg-3*).

(d) Plasmids should be designated by symbols which are clearly distinguishable from symbols used for genetic loci.

(e) Mutant loci and mutational sites on plasmids should be designated by symbols of the same kind as those used for loci on the chromosome.

(f) Phenotypic traits should be described in words, or by the use of abbreviations which are defined the first time they appear in a given paper. The abbreviations should be clearly distinguished from the genotype symbols (e.g. the Arg⁻ phenotype is an arginine requirement, associated with the *argA* locus).

(g) Strains should be designated by simple serial numbers. To avoid duplications, different laboratories should use different letter prefixes. Strain designations should not be italicized (e.g. HB101).

(h) When a strain is first mentioned in a publication its genotype should be described and relevant phenotypic information should be given. The genotype includes a list of all mutant loci and/or mutant sites, a list of plasmids, and information concerning the state and location of any episomes (plasmids or prophage).

A careful record should be kept of strain designations, the origin of the strains and details of the genotype. It is not enough simply to label the strain with the plasmid which it contains as you will need to grow the strain under conditions which allow good growth without the loss of the plasmid. Maintenance of the plasmid often depends on a selective technique; for example, by incorporating an antibiotic such as ampillicin in the medium (Section 3.3). In some cases a wild-type gene on the plasmid will complement a deletion or mutant gene on the chromosome. In these cases, growth on a medium deficient in the growth factor will be necessary to ensure survival of the plasmid (Section 3.2). Similarly, the knowledge that the host strain possesses a deletion of the lactose operon may be essential in analysing plasmids which carry the normal β-galactosidase gene. These various points are illustrated by considering the genotype and phenotype of *E. coli* strain HB101 (see *Appendix 3*).

- F^- lacks the sex factor F.
- *hsdS20*   ($r_B^- m_B^-$) mutant in the site recognition gene for the strain B restriction endonuclease system. This makes the strain deficient for both restriction and modification of the DNA.
- *recA13*   deficient for major recombination protein. The strain is UV sensitive and lacks the major *E. coli* recombination system. This reduces the chance of rearrangements and transfer of recombinant DNA.
- *ara-14*   unable to utilize arabinose as the sole source of carbon and energy.
- *proA2*   requirement for proline in the medium.
- *lacY1*   mutation in the permease for the uptake of lactose.
- *galK2*   unable to utilize galactose.
- *rpsL20*   (Sm^r) resistance to streptomycin due to a mutation in the ribosome.
- *xyl-5*   unable to utilize xylose.
- *mtl-1*   unable to utilize mannitol.
- *supE44*   carries a suppressor for the amber chain-terminating triplet.
- $(\lambda)^-$ non-lysogenic for bacteriophage lambda. Does not carry the lambda prophage.

This genotype tells us that the strain carries no plasmids or phage lambda and so can be used for plasmid transformation or assay of phage. It can be transformed efficiently by plasmid DNA and then used for the production of large-scale preparations of plasmid DNA. It can grow on glucose but is unable to grow on a range of sugars as sole sources of carbon and energy. A chemically-defined medium would need to have proline added, but this would normally be supplied by peptone in complex media. Although this strain is Lac⁻ it does carry a wild-type gene for β-galactosidase (*lacZ*⁺) so this strain is unsuitable for use as a host for plasmids which depend on lactose fermentation to give a blue coloration with X-gal (Section 3.5). The *supE* locus means that it will suppress amber mutations, for example in phage lambda, thus allowing it to produce plaques.

Space does not permit a complete glossary of a wide range of symbols and the reader is referred to the *Molecular Biology Labfax* (11) which contains genotypes of all relevant strains as well as a list of gene symbols.

3.2 Characterization of nutritional mutants

It is normally only essential to check for genetic characters which you intend to use in a particular experiment. Thus if absence of β-galactosidase activity is essential then you should check that the strain is Lac⁻. Similarly if you needed to select for the Pro⁺ phenotype then you would need to check that the strain is proline requiring. A method illustrating the general points is given in *Protocol 5*. However, if you suspect that your strain has mutated, reverted or is a contaminant, it may by necessary to do a more thorough check of other markers using the same general approach.

Protocol 5. Analysis of nutritional mutants

1. Prepare a stock solution of a chemically-defined minimal agar medium lacking a carbon and energy supply or growth factors.

2. Examine the genotype of the strain(s) for analysis and identify the growth factors to be analysed. The system will be illustrated for HB101 and the *pro*, *lac*, and *mtl* markers only.

3. Label four tubes (20 ml capacity) with the letters A, B, C, and D and make additions as indicated:

	A	B	C	D
proline	+	+	+	−
glucose	+	−	−	+
lactose	−	+	−	−
mannitol	−	−	+	−

Use a 1% (w/v) proline stock solution and dilute 1 in 100; use 20% (w/v) sugar stocks diluted to 0.1% final volume.

4. Add the contents of each tube to 20 ml agar medium and pour four plates (see *Protocol 1*).

26

Protocol 5. *Continued*

5. Suspend single colonies of HB101 and a wild-type control in 0.5 ml PbS (see *Appendix 4*) or dilute Ringer's solution and vortex.

6. Streak a loopful of HB101 and the wild-type control on to all four media at previously marked positions.

7. Incubate for 16–24 h and record growth. You should always compare the relative growth of the mutant strain on the media with the growth of the wild-type strain.

8. Analysis of mutants defective in the utilization of different nitrogen, sulphur or phosphorus sources requires a minimal medium lacking these supplements.

3.3 Characterization of antibiotic resistance

The resistance of bacterial strains to antibiotics is a very useful selective technique. Genes controlling this resistance can be carried on the bacterial chromosome or on a plasmid and the latter are very useful in ensuring that a particular plasmid is present (see Chapter 8, Section 2.1). However it should be remembered that plating large numbers of sensitive cells on media containing an antibiotic can select rare spontaneously-resistant mutants due to chromosomal mutation rather than to the receipt of a resistant plasmid. Likewise contaminants may also be naturally resistant to the same antibiotic and even a low level of contamination will be readily revealed on the selective plates. As antibiotics are often thermolabile they cannot be sterilized by autoclaving, and solutions should be membrane filtered. Most antibiotics are supplied sterile and it is possible to make up solutions using aseptic technique in sterile distilled water. The actual concentration of antibiotic required depends on the conditions used. In general, higher concentrations are required for high density of cells on agar plates while low concentrations are needed for low densities of cells growing in liquid medium. Detailed techniques for checking antibiotic resistance are shown in *Protocol 6* and *Appendix 4* lists some of the points to note about particular antibiotics.

Protocol 6. Analysis of antibiotic resistance

1. Make up stock solutions of antibiotics as described in *Appendix 4* using aseptic technique and sterile distilled water. Filter through a sterile membrane, pore size 450 μm, if desired.

2. Distribute in aliquots of 200 μl into labelled 1.5-ml microfuge tubes and store at $-20°$C until required.

3. Make up a suitable nutrient medium such as LB agar (*Appendix 4*) and cool to 50°C. Pour and dry a few plates (see *Protocol 1*) without the antibiotic, to act as controls.

Protocol 6. *Continued*

4. Add the antibiotic to the remainder of the nutrient medium to give the desired concentration and mix well.

5. Pour and dry the antibiotic plates. Most antibiotic plates will keep for at least two or three weeks.

6. Mark each set of agar plates with a suitable grid and label with the strains to be tested. It is best to include control strains which are sensitive to the antibiotic to test the efficiency of the antibiotic activity.

7. Suspend a single colony of a strain to be tested in 200 μl of PbS (see *Appendix 4*) or dilute Ringer's. Vortex to suspend the cells.

8. Flame and cool a Nichrome loop about 3–4 mm in diameter and remove a loopful of cell suspension.

9. Streak this suspension over a distance of 1–2 cm over each antibiotic plate. Repeat on the medium without antibiotic.

10. Repeat steps 7–9 with each strain to be tested, including the control strain.

11. Incubate at 37°C for 16–36 h depending on the strain and the antibiotic. Compare the growth of the strains on the antibiotic and control media. Note that if antibiotic activity is low or absent (due possibly to inactivation because the agar was not cooled sufficiently in step 3) then the control strain will grow on both media.

12. For large-scale checking of transformants or other recombinants, it is more convenient to pick off individual colonies with sterile toothpicks and to streak plates with these directly.

3.4 Characterization of *recA* and UV-sensitive mutants

Recombination-deficient mutants, such as *recA*, are used to minimize recombination between cloned genes and homologous regions on the chromosome. In addition the use of *recBC* mutants can minimize the loss of repetitive DNA in gene libraries. It is difficult to check for recombination deficiency, but luckily there is an associated phenotype of UV-sensitivity which is easy to analyse. As with any genetic character, it is important to have both *recA*$^-$ and *recA*$^+$ strains so that a comparison can be made. The technique is as follows:

Protocol 7. Checking strains for the Rec$^-$ phenotype (see *Figure 4*)

1. Prepare overnight broth cultures from single colonies of Rec$^+$ and Rec$^-$ cultures.

2. Using a flamed Nichrome loop, streak each strain across a well-dried LB plate (*Appendix 4*) such that the streak is about 6 cm. Allow the liquid to dry.

Protocol 7. *Continued*

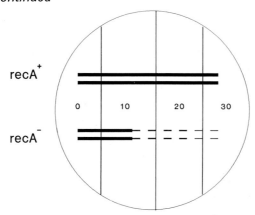

Figure 4. Procedure for characterizing *recA* strains. The marked areas show the expected growth after irradiation and incubation.

3. Mark the base of the agar plate so that each streak is divided into four segments labelled, left to right, 0, 10, 20, and 30 sec.

4. Cover the plate with a piece of card such that only the 30-sec segment is uncovered. Remove the lid of the agar plate and expose to a bacteriocidal UV lamp (254 nm) at a height of about 30 cm for 10 sec. **(Wear UV safety goggles and gloves during these operations)**.

5. Move the card across so that the 20-sec segment is exposed for 10 sec.

6. Move the card across so that the 10-sec segment is exposed for 10 sec.

7. Switch off the lamp. Wrap the plate rapidly in foil to avoid light-induced repair processes and incubate at 37°C overnight.

8. The segments have received 0, 10, 20, or 30 sec exposure to UV and there should be a clear distinction between the two phenotypes.

3.5 Characterization of the utilization of lactose: X-gal

There are several ways of detecting the utilization of sugars like lactose. One method makes use of a chemically-defined medium in which lactose is the sole source of carbon and energy. It is used to replace glucose and in this medium only Lac$^+$ strains will be able to grow. Alternatively, a nutrient medium with lactose can be used which contains the indicator dyes eosin and methylene blue. When lactose is fermented, acid is produced and the indicators change colour. Thus Lac$^+$ colonies with β-galactosidase activity give dark purple colonies with a green fluorescent sheen whereas Lac$^-$ colonies are pink. A more expensive way of analysing for this characteristic is to use the substituted β-galactoside sugar called X-gal (5-bromo-4-chloro-3-indolyl-β-D-galactopyranoside). Media are normally prepared containing this compound together with isopropyl-thio-galactopyranoside (IPTG) which fully induces the *lacZ* gene but which is not a substrate for the enzyme. Details are given in Chapter 8, Section 2.

3.6 Detection of lysogeny

Bacteria carrying the prophage of temperate phages such as λ and P2 have been exploited in a variety of ways. Lambda lysogens have been used to produce packaging extracts for isolation of genomic DNA libraries and P2 lysogens are important in the Spi$^+$ selective system for recombinant λEMBL4 phages. When it is necessary to check for lysogeny it is essential to have a strain which is sensitive to the phage. Thus, for a (P2)$^+$ lysogen it is necessary to have a second strain which is sensitive to phage P2 as well as, ideally, phage P2 itself. The simplest check is to make a single streak of each strain, such as *E. coli* Q359 (P2)$^+$ and Q358 (P2)$^-$, on an LB agar plate (*Appendix 4*). The liquid is allowed to dry and a small volume of a P2 stock is spotted on the streak. Q359 will be immune to the phage due the presence of a repressor, while Q358 will be lysed. If you do not have any P2 phage then prepare pour plates of each strain and spot overnight broth cultures of Q358 and Q359 on to each lawn. Q359 produces low numbers of free phages, due to spontaneous lysis, and these appear as a thin halo of lysis round the spot of Q359 on the lawn of Q358 (*Figure 5*). Similar techniques will also work for λ phage and lysogens.

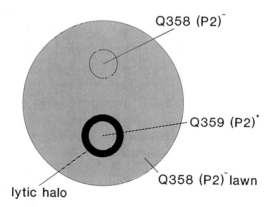

Figure 5. Detection of lysogeny by spot testing of strains on a bacterial lawn.

3.7 Screening for plasmids

It is essential to have a good accurate map of the plasmid with details of its genotype (12, 13). The most popular plasmid vectors are described in Chapter 9, and detailed compilations have been published (11, 14). The easiest characters to detect are antibiotic resistance. For example, pBR322 and pAT153 are both resistant to tetracycline and ampicillin, so a simple check can therefore be made by streaking on media containing ampicillin and tetracycline respectively. In fact, these plasmids may be lost if subcultures are made in the absence of ampicillin selection, as the *par* locus, which controls regular partition, has been deleted in these plasmids. Further preliminary screening depends on other markers carried by the plasmid, and final checking will depend on extraction, restriction and electrophoresis of the plasmid DNA.

3.8 Microscopic examination of cultures

A visual check of a culture may suggest contamination or it may be that a rare transformant simply does not look right. Rather than extracting and analysing plasmid DNA it may be worth having a look at the organism to check that it is actually Gram-negative and rod-shaped. It is even quicker to do a simple smear followed by a methylene blue stain as this will quickly eliminate budding yeasts and cocci. *Protocol 8* describes the basic methods.

Protocol 8. Techniques for the microscopic examination of strains

1. You will need the following stains and reagents: Loeffler's Methylene Blue, Crystal Violet, Lugol's Iodine, ethanol, and dilute Fuchsin.

2. Clean a microscope slide by immersing it in alcohol and flaming off in a bunsen burner.

3. Either place a loopful of a cell suspension on to the slide or suspend cells from an agar plate in a small drop of water on the slide.

4. Spread the suspension thinly and evenly over the slide. This is best done with the edge of a second microscope slide. The smear should be only just visible and the aim is to produce a monolayer of cells.

5. Leave the slide to air-dry or hold high over a bunsen burner to dry the smear slowly.

6. Fix the cells by passing the dried slide slowly three times through the bunsen flame. The bacteria adhere to the slide.

7. Place the smear on a staining rack (two glass tubes held together by short pieces of rubber tubing). Flood the slide with Methylene Blue and leave for 5 min.

8. Wash the slide under the tap and blot dry with blotting or filter paper.

9. Add a small drop of immersion oil and view down the microscope using the oil immersion lens. Look for areas where the cells are spread out well.

10. Alternatively, use the Gram stain:

    (a) Stain in Crystal Violet for 1 min, then wash well under running water and blot dry.

    (b) Wash with Lugol's Iodine and then stain for 1 min with fresh Lugol's Iodine.

Protocol 8. *Continued*

    (c) Wash with tap water and blot dry. Decolorize with ethanol until no more dye is washed out.

    (d) Wash with tap water and counterstain with dilute Fuschin for 30 sec.

    (e) Wash and blot dry. Gram-negative cells will appear red whereas Gram-positives will be dark purple. It is best to have examples of both types when you are practising the technique.

4. Preservation of stock cultures

A culture which has been checked and used successfully is a valuable laboratory asset and should be preserved as efficiently as possible. It is sensible to preserve the culture by a number of methods as you will require the strain for short-term and long-term use. The broth and plate cultures already described can be used for several weeks or even months, but there are problems inherent in using cultures stored at 4°C. Cultures may become contaminated or lose viability. They can also mutate or be infected by phages or other plasmids. Pontecorvo has been quoted as saying that the only stable strain is a dead strain. This is true, but we can get close to complete lack of growth and slow chemical change in frozen cultures. Alternatively, freeze-dried cultures are convenient to keep and are not susceptible to damage following electrical breakdown. There are several good reviews of preservation methods (15, 16) and detailed techniques on freeze-drying will not be given here.

4.1 Preservation of short-term cultures

Broth cultures of *E. coli* will retain viability for several weeks if stored in sealed containers at 4°C. This can be recommended for cultures which lack plasmids and show little variation, such as *E. coli* HB101. The use of broth cultures older than a few days for strains carrying unstable plasmids is not recommended as the antibiotics used to select the plasmid may have been inactivated and growth of cells without the plasmid may have occurred. It is much safer to use an agar plate with single colonies which have previously been checked. The major problems with agar plates are dessication and contamination. These can be reduced by sealing the edge of the plates with parafilm or adhesive tape. Plates like this can be used successfully for two or three months, but for particularly critical experiments you should re-check your strains by re-streaking on a selective or diagnostic medium.

4.2 Stab cultures

A very simple and effective method for keeping bacterial strains is to stab actively-growing cells into a nutrient medium containing 0.6 per cent (w/v) agar.

A conveniently-sized tube is about 4 cm by 1 cm with a water-tight cap. Wrap the tubes separately from their caps and autoclave for 15 min at 121°C (check that the caps are autoclavable). Fill the tubes two-thirds full with sterile nutrient agar at 50°C and cap the tubes lightly. When set, incubate the tubes at 45°C for 1–2 h to remove moisture. Label the tubes and inoculate with a loopful of an overnight broth culture of a checked strain. Stab the loop straight down the middle of the agar. Incubate the tube overnight at 37°C with the cap loose and check for contamination. This can be seen as disc- or lens-shaped colonies suspended in the agar. Discard contaminated stab cultures. Check that the culture has grown in the stab region. Tighten the cap, and seal with candle wax or melted paraffin wax if a more effective air-tight seal is required. Store the tubes mounted in polystyrene sheets in cupboards or drawers away from direct sunlight. Viable cultures have been isolated from such tubes after periods of up to ten years.

4.3 Preservation of cultures with glycerol or DMSO

Cultures can be preserved very effectively if frozen in the presence of a cryoprotectant which reduces damage from ice crystals (16). Temperatures above −20°C are not very effective due to the formation of eutectic mixtures exposing cells to high salt concentrations. The simplest way to preserve a culture is to add 15% (v/v) glycerol to the culture and then to store it at −20°C or −80°C in a freezer or at −196°C in liquid nitrogen. In practice, the survival of cells also depends on the freezing rate and efforts should be made to control this at the optimum for the strains used. *Protocol 9* shows a method for preserving broth cultures with glycerol. This method can be extended to nutrient plates containing 15% (v/v) glycerol overlaid with a nitrocellulose filter containing bacterial colonies. The whole system is wrapped in a nylon bag and stored in a −80°C freezer. This is particularly useful for gene libraries in bacterial cells, as a template or a dense plate of colonies can be stored conveniently. Another useful approach is to coat glass beads with the glycerol mix. On partial thawing, each glass bead is a convenient inoculum to grow up fresh cultures. Dimethylsulphoxide (DMSO) can also be used as a cryoprotectant at 5–7% (v/v).

Protocol 9. Preservation of cultures by freezing with glycerol

1. Grow a 5 ml overnight broth culture of the strain to be preserved.

2. Autoclave a 60% (v/v) solution of glycerol and add 1.6 ml to the overnight broth culture. Mix well.

3. (a) Dispense 500 μl into a number of sterile, screw-top, freezer-proof vials and place a series of vials at −20°C. Freeze the rest of vials in an ethanol/dry-ice mix and place the vials in a −80°C freezer or into liquid nitrogen.

   (b) Alternatively, wash and sterilize aliquots of 20–30 glass beads (2 mm diameter). Add the glycerol/broth mix and aspirate to remove air bubbles

Protocol 9. *Continued*

and to make sure that the beads are well wetted. Decant the excess liquid, dispense and store as before.

4. Recovery of cultures is by partial thawing and removal of a few ice crystals or a glass bead into fresh LB broth (*Appendix 4*) suitably supplemented. Freezing and thawing of the stocks will reduce viability and it is therefore best to store replicate cultures which are not touched.

4.4 Freeze-dried cultures

This is a relatively specialized technique and needs some expertise. It will not be dealt with here because the technique varies for different types of apparatus and there are specialized manuals available (15). It is worth considering if strains are not to be used for a number of years or if they need to be sent to other laboratories. Commercial suppliers such as the NCIB (Torry Research Station, Scotland) will carry out custom freeze-drying on a batch basis.

5. Culturing of bacteriophages lambda and M13

5.1 Theoretical background

Phages are viruses and this means that they are dependent on the host for their synthetic activities (17, 18). This is at the root of many techniques for handling phages. Bacterial cells have to be actively growing if phage replication is to occur so the basic processes affecting growth described in Section 2 also apply here. A major difference between phages and their host is that phage do not have a cell wall and are therefore more susceptible to ionic strength. Osmotic shock will simply burst the phage head and release the phage chromosome with an instant drop in viability. Phages are also sensitive to detergents and other chemical contaminants on glassware.

It should be remembered that lambda is a temperate phage and can exist in two cycles: the lytic cycle which involves infection, replication and lysis with the release of progeny phage and the lysogenic cycle in which the phage chromosome recombines with the host chromosome to become a prophage. A strain carrying a temperate phage is said to be lysogenic and is given the symbol $(\lambda)^+$ whereas a non-lysogenic strain is given the symbol $(\lambda)^-$. In the prophage state there is a complex system of regulation which results in the inactivation of most genes on the phage chromosome except for the repressor gene itself (19).

Lambda vectors are very different from the original wild-type phage. The region controlling the regulation system is dispensable and is not required for plaque formation, so in many vectors it has either been deleted or is replaced by recombinant DNA during the production of gene libraries (see Chapter 9, Section 2.2). Choice of hosts for particular strains of lambda is important as was mentioned in Section 3.6.

Bacteriophage M13 is a virulent, male-specific filamentous phage which contains a circular single-stranded DNA molecule. The attraction of the phage is that it produces single-stranded DNA which is useful as a template for DNA sequencing reactions. Consequently, the replication control region of M13 has been incorporated into hybrid vectors known as phagemids (see Chapter 9, Section 3.2). One unusual feature of M13 is that the host does not lyse to release phage. Instead the phage particles are released through the wall without killing the host. Plaques are still produced because there is inhibition of the growth of the host resulting in a reduction in the number of bacteria in the area of the plaque.

5.2 Factors affecting the growth and survival of phage lambda

In a suitable suspending medium (SM, see *Appendix 4*) lambda is very stable and will survive for months at 4°C assuming no contamination. The phage needs Mg^{++} and a low level of gelatin for stability. In general $MgSO_4$ or $MgCl_2$ must be present in all growth media used for lambda or greatly reduced yields will be obtained. The growth of lambda involves three main techniques:

• the assay of a phage stock to produce separate plaques

• growth in liquid culture to produce high titre stocks

• the induction of lysogens to produce a lytic cycle.

There are some variations on this, for example in plating at high density which can be used to produce phage that can be screened for specific sequences with gene probes or monoclonal antibodies.

The main factor underlying a number of these growth techniques is the ratio of phage to bacteria. This is known as the multiplicity of infection (m):

$$m = \frac{\text{no. of phage particles}}{\text{no. of bacteria}}.$$

A value of m of 0.01 means that there will be 10^8 phage for every 10^{10} bacteria. Very high values of m (> 1000) result in lysis from without which means that the cells are killed without the release of progeny phage. There is a further complication with lambda and that is that phage can infect and lysogenize rather than lyse. Lysogenic bacteria are immune to further infection due to the presence of a repressor. Consequently, these bacteria will continue to grow even in the presence of free phage. The aim of experiments to produce phage stocks is to maximize the lytic cycle by infecting actively-growing cells with an m of 1 or less. Cells growing slowly are more likely to be lysogenized and active growth is therefore essential. Some lambda phage vectors lack an active repressor gene and produce clear plaques and consequently with these lysogeny is not a major problem.

In assaying lambda phage to produce single plaques, it is essential to have a very low level of m with about 100 phage to about 10^7 bacteria. In this situation each phage particle will infect a single bacterium and act as a centre of infection.

After several rounds of replication and spread of the released phage particles a circle of lysed bacteria appears on the plate as a plaque. The size of this plaque is determined by the rate of phage replication and release, and the period for which the host is growing. As soon as the bacterium stops growing, the size of the plaque is fixed. This means that plaques can often be seen after about 6 h of incubation and well before full overnight growth. The size of the plaque can be controlled by manipulating the density of bacterial cells and the thickness of the nutrient agar plate. Thin plates are used for small plaques during screening with DNA probes and thick plates should be used during sub-culture for phage stock production.

Another factor which affects phage infection and growth is adsorption of the phage to the cell wall. The tail fibres of lambda phage bind to the *lamB* receptor (a maltose binding protein). This is induced by maltose but repressed by glucose and therefore for maximum adsorption 0.2% (w/v) maltose should be included in the medium. Mg^{++} has already been mentioned as being required for the integrity of the phage, but it is also needed for adsorption. Both maltose and Mg^{++} should therefore be present when assaying phages by the plaque method (see Section 5.3). However, when preparing phage stocks by the broth method (Section 5.5), reduction in titre can occur by adsorption of the phage on to the receptors in the wall debris formed by lysis of the cells. This effect can be minimized by omitting maltose from the medium. Sufficient receptors are produced for numerous rounds of the lytic cycle without over-producing receptors which would reduce the final titre. 5 mM $CaCl_2$ can also be added as this has been observed to increase yields (20).

5.3 General techniques for the assay of phages by the plaque method

The basic technique for the assay of phage is applicable to a variety of phages. If the initial titre is unknown then a dilution series down to about 10^{-10} is prepared as described in *Protocol 3*. This will cover most of the phage titres encountered. A bacterial lawn is prepared by pouring a cell suspension in 0.6% (w/v) agar over a dried nutrient agar plate. The phage dilutions are then spotted on the surface of the plate in the same way as for the Miles Misra technique described in *Protocol 3*, step 9. This gives an approximate titre which is used to make a full assay by suspending a known phage concentration with the bacterial suspension in the overlay. When the titre is known in advance a limited range of dilutions can be prepared. The details of these techniques are given in *Protocol 10*.

Protocol 10. Assay of phages by the plaque method

1. Add 1 ml of SM (*Appendix 4*) to a series of sterile plastic vials. Label the tubes -2, -4, -6, -8, and -10.

2. Add 10 μl of phage stock to tube -2 and mix well. This dilution is 10^{-2}. Repeat this operation to produce the dilution series.

Protocol 10. *Continued*

3. Prepare a thick agar plate of a rich medium such as DYT (*Appendix 4*) on a level area of bench and dry the plate well. Warm the plate by incubating at 37°C.

4. Prepare 3 ml of LTS top agar (*Appendix 4*), melt, and maintain at 50°C.

5. Add 200 μl of an overnight broth culture[a] of a suitable host strain to the molten agar at 50°C.

6. Mix the bacteria and the agar. Wipe the outside of the tube and pour rapidly over the DYT plate (again on a even surface). Rock and swirl the plate to spread the agar evenly over the surface. This must be done immediately.

7. Allow the agar to set, and dry the plate for about 15 min at 45°C.

8. At marked places, spot 20 μl of the six dilutions and the neat suspension on the overlay plate. Allow the liquid to dry before inverting the plate. Incubate at 37°C overnight.

9. Count the plaques at about 5–20 per drop. The titre of the phage per millilitre is 50 times the count multiplied by the dilution factor.

10. For a more accurate assay or when a complete plate at the same dilution is required:

    (a) Prepare as many nutrient plates as necessary. If large plaques are required for assay or production of stocks use thick plates (40 ml) while for screening use thin plates (20 ml).

    (b) Calculate the required volume of diluted phage for between 30 and 300 phage particles (or 1000 + for screening) and add this to 3 ml of molten agar medium at 50°C as described in step 4 above.

    (c) Add 200 μl of indicator bacteria and mix bacteria, phage and agar.

    (d) Pour quickly over the plate as before, allow to set and incubate at 37°C overnight. Count the plaques as before.

---

[a] Bacterial cultures for lambda should be grown on a medium containing 0.2% (w/v) maltose (*Appendix 4*) to induce receptors. The LTS agar should also contain 0.2% maltose. The broth culture can be stored at 4°C for up to a month. For M13 male bacterial strains (Hfr or F$^+$) must be used and better results will be obtained if the lawns are prepared from actively-growing cultures.

5.4 Purification of phage by single plaque isolation

All phage stocks should be purified periodically by single plaque isolation and checked for appropriate genotypes or DNA restriction patterns. This is equivalent to single colony isolation for bacteria (see Section 2.1). The same technique is used to purify a phage containing cloned DNA. Where the purity of a stock is critical it is best to choose well-separated plaques on fresh plates, as phages may diffuse sideways in the agar. A second purification step will avoid these problems. Stab an appropriate plaque, prepared by the technique in *Protocol 10*, with the tip of a sterile pasteur pipette and rock from side to side so that a plug of agar containing the plaque is removed in the pipette tip. With a rubber teat or pipette dispenser, blow the plug into 1 ml of SM (*Appendix 4*) in a sterile plastic tube. Leave for 1 h so that the phage diffuses into the SM. An average plaque will contain about 10^6 phages. Plate a suitable dilution of the phage suspension to obtain single plaques and then prepare a fresh phage stock by one of the methods detailed below.

5.5 Preparation of small-scale phage stocks

The plate method depends on producing almost confluent lysis on a plaque assay plate. It has the advantage that it is reliable and predictable but the disadvantage that agar has to be removed from the preparation. Agar contains inhibitors which can interfere with restriction, ligation, and other enzymes involved in DNA manipulations. If this is found to be a major problem it is better to replace the agar with agarose.

Protocol 11. The plate method for preparing a lambda phage stock

1. Prepare a series of up to five agar plates containing almost confluent lysis by the method described in *Protocol 10*, step 10. The number of phage particles required will vary depending on the size of the plaques but normally will be between 5000 and 100000. Plaques should just be visible in isolated areas. Plates with complete lysis are less reliable as lysis without phage replication may have occurred.

2. If mutant plaques are likely to be a problem (e.g. clear plaque mutants in a wild-type stock) then use the minimum number of phages required for almost confluent lysis and process the plates at either 6 h or about 15 h.

3. Add 5 ml of SM (*Appendix 4*) to each plate and scrape the 0.6% agar off the plates into a sterile 250 ml conical flask with a sterile glass spreader.

4. Add 0.5 ml of chloroform, vortex the flask for 1 min and leave to stand for 1 h so that the phage can diffuse from the agar.

5. Decant the mixture into a centrifuge tube (resistant to chloroform) and centrifuge at 5000 *g* for 10 min to bring down bacterial debris and agar fragments.

6. Remove the supernatant carefully and membrane filter to sterilize or add to a sterile plastic tube over a few drops of chloroform.

7. Assay the stock by one of the methods described in *Protocol 10*.

Notes:
(a) The method can be modified by using agarose instead of agar or by allowing the phage to diffuse into the SM by gentle shaking of the intact plate for 3 h. Both of these methods will reduce the contamination of the stock by inhibitors.

Protocol 11. *Continued*

(b) The method is useful when rapid lysis mutants or clear plaque mutants are a potential problem as diffusion is restricted by the agar and mutants will not overgrow the required phage as can happen in broth culture methods.

The broth method is cleaner as it is only necessary to remove bacterial debris. It is less predictable as it is essential to have the correct value for *m* to give maximum lysis (see Section 5.2). If a stock culture is required quickly then you will need to set up a series of cultures with different ratios of phage and bacteria.

Protocol 12. The broth method for preparing lambda phage stocks

1. This method can be used on any scale from 5 ml to 500 ml cultures. For lambda use LB broth (*Appendix 4*) containing 5 mM $CaCl_2$ but no $MgCl_2$/$MgSO_4$ or maltose. The method can also work with more complex media.

2. Grow an overnight culture of a suitable host and add 100 μl to 5 ml, or 10 ml to 500 ml, of LB in a 20-ml tube or 2-litre conical flask.

3. Incubate at 37°C at 250 r.p.m. until the OD at 550 nm is about 0.2. This should be approximately 10^8 exponentially-growing cells per millilitre. Incubation should take about 2–3 h.

4. Add phage to give an *m* of 0.01. For the 5-ml culture this means adding the phage from two or three plaques (removed as agar plugs; see Section 5.4), or for the 500 ml culture the phage from one or two plates showing almost confluent lysis (prepared by the method in *Protocol 11*). The latter can be done simply by scraping the agar overlay into the flask. Alternatively, prepare a phage stock from the plates and add the agar-free stock directly. It is important to remove all traces of chloroform by aeration before adding.

5. Continue shaking until the culture lyses, which will normally take between 2 and 4 h. This can be detected by a rapid drop in OD accompanied by the production of 'ropes' of lysed bacteria. If this does not occur after 5 h, either continue incubation overnight or add chloroform (200 μl/5 ml or 20 ml/500 ml). Continue shaking until lysis occurs.

6. The production of efficient lysis is essential for a high phage yield. The presence of clearing and 'rope' formation is a reliable indication of a high titre stock. If this does not occur it is better to repeat the procedure until good lysis is detected.

7. If not already added, add chloroform to lysed cultures as in step 5 above. Shake for 10 min and then harvest the cultures into centrifuge tubes which will resist chloroform. Centrifuge for 15 min at 11 000 *g* to remove cell debris and agar.

8. Remove the supernatant and store with a few drops of chloroform at 4°C.

5.6 Preparation of large-scale phage stocks

As described in *Protocol 12*, the broth method for preparation of small-scale phage stocks can be scaled up to 500 ml cultures in 2-litre flasks.

5.7 Purification of lambda phage particles

For many purposes phage preparations need to be free from bacterial nucleic acids and other contaminants such as bacterial carbohydrates or agar. Bacterial DNA can be incorporated into gene libraries and other contaminants can interfere with ligase or restriction enzyme activities. Intact phage particles are resistant to nucleases so a treatment with RNase and DNase can remove these contaminants. Phage particles can be purified by differential centrifugation but damage to the particles occurs. Polyethylene glycol (PEG) has been used effectively to precipitate the phage particles making centrifugation easier. The most efficient way of purifying the phage particles is on a CsCl gradient as this separates the particles from carbohydrates. A method for purifying lambda phage is given in *Protocol 13*. Extraction of DNA from these particles is straightforward (see Chapter 3, Section 3.1).

Protocol 13. Purification of phage lambda

1. Prepare a high titre stock (at least 10^{10} per millilitre) by the method described in *Protocol 12*. Measure the volume and add 1 µg/ml of both DNase and RNase to remove bacterial nucleic acids. Incubate at room temperature for 30 min.

2. Add 40 g solid NaCl per litre and dissolve by gentle agitation.

3. Add 140 g PEG 6000 per litre. Add this slowly with constant gentle mixing on a magnetic stirrer at room temperature.

4. Leave at 4°C overnight to precipitate the phage.

5. Swirl the flask gently to resuspend any sediment.

6. Centrifuge at 11 000 *g* for 10 min to collect the precipitated phage.

7. Decant the supernatant and discard. Invert the centrifuge tube and drain well. Wipe the inside of the tube with a paper tissue.

8. Resuspend the pellet in 16 ml of SM (per litre of original stock). For SM recipe see *Appendix 4*. Add an equal volume of chloroform and stir on a magnetic shaker at 37°C for 1–2 h until the pellet is fully resuspended. Losses can occur at this stage so care should be taken.

9. Centrifuge at 2000 *g* to separate the phases and carefully remove the upper aqueous layer containing the phage which can be stored as a stock.

10. Losses can occur at a number of stages and it is useful to do a Miles Misra

Protocol 13. *Continued*

assay (see *Protocol 3*, step 9) for phage at various steps, such as the original stock, the supernatant after PEG precipitation, and the final stock. Supernatants should be retained until the results are known so that further recovery can be made and phage stocks pooled.

11. Add 0.75 g CsCl per millilitre of stock and dissolve by inversion. This should give a density of 1.5 g/ml. Centrifuge at 100 000 g for 24 h and phage should produce a bluish band half-way down the tube. Remove the band with a 21-gauge syringe needle.

12. Dialyse against 2×1 litre of SM overnight and store at 4°C with a few drops of chloroform.

5.8 Induction of lambda lysogens

Traditionally, lysogenic cultures were induced by UV irradiation. Damage to the DNA occurred followed by derepression of the *recA* gene product which then destroyed the lambda *c*I repressor. The use of UV is not recommended as it can induce mutations in phage particles, and most lysogens are now induced by a temperature-sensitive system. With many lambda strains the repressor protein is thermolabile at 45°C and thus lysogenic cultures can be induced to undergo the lytic cycle by a brief treatment at this temperature. The S^+ gene product is required for natural lysis and some phage strains have amber mutations or chain-terminating triplets in this gene. If the host strain has the equivalent suppressor then natural lysis will occur, otherwise it has to be induced with chloroform.

Protocol 14. Induction of lambda lysogens such as (λgt11)$^+$

1. Prepare duplicate streak plates of the lysogenic culture on LB plates (*Appendix 4*) and incubate at 30°C and 42°C. No growth should occur at 42°C due to induction of the lytic cycle.

2. You will need two shaking water baths at 37°C and 45°C respectively. Grow an overnight 20 ml LB broth culture at 30°C and inoculate into 400 ml of LB in a 2-litre flask. Shake at 250 r.p.m. until the OD at 550 nm reaches 0.6 (about 3×10^8 cells per millilitre).

3. Rapidly increase the temperature in the broth culture to 45°C. This is best done in a large volume of water at 70°C. Monitor the temperature, and when this reaches 45°C, transfer the flask to the water bath at 45°C. Incubate with vigorous shaking for 15 min.

4. Transfer the flask to the 37°C water bath and continue shaking.

5. Monitor for lysis (if suppressor mutations are present) as in *Protocol 12*, step 5. Samples can be withdrawn and tested for lysis with a small volume

Protocol 14. *Continued*

(0.05 vol.) of chloroform. Vigorous shaking should be followed by lysis seen as translucence or rope-like debris.

6. Strains which do not suppress amber mutations will not lyse spontaneously. After about 2 h at 37°C, collect the cells by centrifugation at 3000 *g* and suspend in 10 ml of SM (*Appendix 4*). Add 500 μl of chloroform and vortex vigorously.

7. Stand the suspension for up to an hour at room temperature.

8. Centrifuge to remove cell debris and treat as for standard lambda phage stock as described in *Protocols 12* and *13*.

5.9 Techniques involving phage M13

Many of the techniques which apply to phage lambda also apply to M13. However, a few points should be noted. The bacterial host must be a male strain containing the F factor as the phage infects via the sex pilus, an appendage coded by the sex factor. Plaques can be produced either by transfecting competent *E. coli* cells with the double-stranded replicative form of the DNA (Chapter 8, Section 1.1) or by infecting cells with intact phage particles containing a single-stranded circular molecule. As phage particles do not lyse the host, it is possible to prepare stocks of M13 simply by inoculating a broth culture of an F^+ strain with phage particles and growing to stationary phase. The bacteria are removed by centrifugation at 10 000 *g* for 15 min. The supernatant contains free phage particles released through the wall of the host. The phage can be purified by a PEG precipitation method similar to *Protocol 13*, as described in Chapter 3, Section 3.2.

5.10 Methods for preserving phage stocks

The simplest method for keeping a phage stock is to seal a Petri dish containing separate plaques of a purified phage type. Parafilm or tape is used to seal the edges of the plate which is kept at 4°C. Viable phage can be isolated from single plaques for several months, but this method is not recommended if there is a variety of different phage types in the same area, since cross-diffusion may occur. Once a high-titre stock has been prepared in SM this can be kept over a small volume of chloroform at 4°C for many months. The chloroform may evaporate with time and should be replenished or the stock may become contaminated with fungi or bacteria. Stocks can also be preserved by the addition of 7% (v/v) DMSO. Add DMSO and mix gently. Freeze rapidly in liquid nitrogen or a dry-ice/alcohol mix and store at − 70°C. Scrape the surface with a sterile Nichrome loop and spot on to a overlay lawn of a sensitive host to test for viability.

6. Trouble-shooting

6.1 General principles

Techniques are always easy when you know how or when you can watch an expert using the method. However, it is a common occurrence to find that the technique is simply not working. The speed of solving these technical problems makes a big impact on the time available for research and your rate of progress. One basic principle is to include controls whenever you can, since these can show clearly what has gone wrong. Solving more difficult problems is very much like detective work: following up clues, eliminating possible suspects, and testing theories. It makes sense to ask others who have managed to get the technique working and to show them your plates or photographs of puzzling results. In microbiology, you should be fully aware of where microorganisms can grow: they can grow wherever conditions are not inhibitory. Algae and bacteria have been found growing in distilled water, and phosphate buffer with a low glucose concentration is an ideal growth medium. Solutions which were sterile last week may have been opened and left on the bench, and be heavily contaminated. If they are, you can be sure that the contaminant produces a nuclease which will destroy your precious DNA.

6.2 Contamination

6.2.1 Contaminated cultures

The source of contaminated cultures may be the air, the apparatus or the growth medium. The culture may have been already contaminated when it was sent to you. A visual inspection of a streak plate should show the most obvious contaminants and these can also be checked by their phenotypes and microscopic properties. More subtle changes are difficult to check and the failure of a technique to work may be the only clue.

If you are not sure which colony on a plate is the correct strain you may have to prepare a range of cultures and carry out some simple tests to see which are, for example, sensitive to lambda phage or which can conjugate with a known Hfr strain. Genetic variation is the most difficult to detect and may not become apparent for some time. Chromosomal rearrangements such as transposition may only become apparent on genetic mapping or on analysis of antibiotic resistance. Gene mutation or reversion will only be detected if you check for these particular genetic markers. If all else fails and the technique is still not working, ask somebody else for an alternative culture or order a substitute from a commercial supplier.

6.2.2 Contaminated media, solutions and equipment

Your strains may be correct but you may be contaminating them every time you attempt to do an experiment. The inclusion of uninoculated controls should indicate the source of the problem. Media with disc-shaped colonies probably means that one of the media components is contaminated. If you suspect that solutions and buffers are contaminated then spread them on nutrient media and look for colonies after incubation. Wherever possible solutions should be sterilized, stored at 4°C, and opened, using sterile technique. If you use a solution frequently it is much better to sterilize it in suitable aliquots and discard material once it has been opened a few times. Contaminated equipment implies that your quality control on sterilization or storage is inadequate. Autoclave tape may indicate that your use of the autoclave is at fault or it could be that the environment is contaminating the equipment after sterilization. Glass pipettes and glassware generally can be flamed before use but this is not possible for plasticware previously sterilized by gamma radiation. If you suspect a particular piece of apparatus such as a centrifuge tube, then add some sterile nutrient medium, incubate and look for turbidity.

6.3 Poor growth of bacteria or phage

Techniques often quote a time-course for a particular experiment and if your cultures are growing much slower than this then you may suspect that there are factors limiting growth. The first component to suspect is that the growth medium is inadequate. A check of the genotype of the bacterial strain should reveal if there are growth factor requirements which may be absent from the medium; for example, peptone may be deficient in arginine and vitamins may be absent from yeast extract. Alternatively, the concentration of an antibiotic may be too high and inhibition may have occurred. Components of the medium can be altered and the growth rate of the strain re-checked. A low level of contamination can also affect growth particularly if the contaminant is a streptomycete which is producing an antibiotic which inhibits the strain under study. Another factor which affects growth is the aeration of the culture as large volumes of media wth inadequate shaking can rapidly become oxygen-limited. Ideally, a flask of 250 ml capacity should have no more that 25 ml of medium to avoid oxygen-limitation. In practice the earlier phases of growth are less likely to be oxygen-limited and larger volumes can be used without affecting doubling times. A common source of problems is inhibition by chemicals contaminating the glassware. Glassware is often used many times and may have been washed in a dish-washer which has left residues of salts and detergents. These can rapidly destroy phage lambda and consequently plastic tubes are best for the handling of phages. You are strongly advised to wash your own glassware and to make sure that rinse cycles are sufficient to remove these chemical contaminants. Poor growth of phages can be due to a variety of factors such as resistance of the host to phage infection and absence of co-factors essential for adsorption and infection. Sudden lysis of cultures may be due to the presence of air-stable phages in the environment. A rich medium with a good carbon source will obviously give better bacterial growth than a chemically-defined medium where the organism must synthesize almost everything for itself. Good bacterial growth will normally give good phage growth but note should be taken of the optimum conditions for handling lambda as discussed in Section 5.

References

1. Schlegel, H.G. (1986). *General Microbiology* (trans. M. Kogut), 6th edn. Cambridge University Press, Cambridge.
2. Bainbridge, B.W. (1987). *Genetics of Microbes*. Blackie, Glasgow.
3. Genetic Manipulation Advisory Group (1981). *GMAG note* 15. GMAG, London.
4. Collins, C.H. (ed.) (1988). *Safety in Clinical and Biomedical Laboratories*. Chapman & Hall, London.
5. Collins, C.H. (1988). *Laboratory-Acquired Infections*. Butterworths, London.
6. Health and Safety Executive (1988). *Control of Substances Hazardous to Health*. HMSO, London.
7. Clowes, R.C. and Hayes, W. (1968). *Experiments in Microbial Genetics*. Blackwell, Oxford.
8. Miller, J.H. (1972). *Experiments in Molecular Genetics*. Cold Spring Harbor Laboratory, Cold Spring Harbor, New York.
9. Advisory Committee on Genetic Manipulation (1988). *Note 7*. HMSO, London.
10. Demerec, M., Adelberg, E.A., Clark, A.J., and Hartman, P.E. (1966). *Genetics*, **54**, 61.
11. Brown, T.A. (1991). *Molecular Biology Labfax*. BIOS, Oxford.
12. Grinstead, J. and Bennett, P.M. (ed.) (1988). *Plasmid Technology*, 2nd edn. *Methods in Microbiology*, Vol. 21. Academic Press, London.
13. Novick, R.P., Clowes, R.C., Cohen, S.N., Cutiss, R., Datta, N., and Falkow, S. (1976). *Bacteriological Reviews*, **40**, 168.
14. Pouwels, P.H., Enger-Valk, B.E., and Brammar, W.J. (1985). *Cloning Vectors*. Elsevier, Amsterdam.
15. Hatt, H. (ed.) (1980). *Laboratory Manual on Freezing and Freeze-Drying*. American Type Culture Collection, Rockville, Maryland.
16. Kirsop, B.E. and Snell, J.J.S. (1984). *Maintenance of Microorganisms*. Academic Press, London.
17. Eisenstark, A. (1967). In *Methods in Virology*, Vol. 1 (ed. K. Maramorsch and H. Koprowski), pp. 449–524. Academic Press, New York.
18. Lewin, B. (1977). *Gene Expression*, Vol. 3. Wiley, New York.
19. Ptashne, M. (1986). *A Genetic Switch: Gene Control and Phage Lambda*. Blackwell, Oxford.
20. Miller, H. (1987). *Methods in Enzymology*, Vol. 152 (ed. S.L. Berger and A.R. Kimmel), pp. 145–70. Academic Press, New York.

3

Purification of DNA

PAUL TOWNER

1. Introduction

This chapter explains how to purify genomic DNA and the DNA of different types of vectors. Genomic DNA can be obtained from any microorganism, plant or animal at any time during development and provides DNA which contains a copy of every gene from the organism. This material can then be used for the construction of libraries and used in Southern hybridization analysis. Vector DNA refers to plasmid, lambda, and M13 phage DNA.

The protocols presented are based on well-established methods which are virtually foolproof. In all of the manipulations described good microbiological practice should be followed and microbiological waste disposed of effectively (Chapter 2, Section 1), not only to protect individuals but also to avoid cross-contamination between experiments. The text is self-contained and describes all of the essential techniques. There are however many different ways of preparing genomic and vector DNA and it is worthwhile to compare and contrast the methodologies which have arisen (1–4).

1.1 Basis of methods for isolation of DNA

Genomic and plasmid DNA can be isolated from cells by disrupting the tissue and then exploiting the difference in properties between DNA, protein and other constituents. Phage particles are isolated from the supernatant of cell cultures, purified, stripped of their protein coats and the DNA retrieved. The techniques require the use of several chemicals to partition and precipitate the DNA sample. All reagents should be the best available and made up in high-quality water. Phenol, which is the most hazardous chemical used in DNA extractions (see *Appendix 2*), is employed to denature and dissolve proteins leaving nucleic acid in aqueous solution. The purchase of poor quality phenol, which needs redistillation, is an unnecessary hazard which should be avoided. High-quality phenol can be purchased as a liquid equilibrated with water and has a pH of 3–4. Prior to use it should be prepared and stored as described in *Protocol 1A*.

Protocol 1. Preparation of phenol and chloroform solutions

A. *Phenol*

1. Remove and discard the upper aqueous layer from a 1-litre bottle of water-equilibrated phenol. Replace the aqueous layer with 1 M Tris–HCl, pH 8.0 and mix the liquid phases by inverting the bottle several times.

2. Allow the layers to separate, then remove and discard the upper aqueous phase. Replace this with 50 mM Tris–HCl pH 8.0. Agitate the mixture and immediately aliquot 1 to 10 ml samples in microfuge or Falcon tubes and store at −20°C.

3. Thaw aliquots at 20°C prior to use. It is the dense lower organic layer which is used for treatment of DNA. The phenol remains stable at 4°C for about a month, but storage at room temperature or exposure to light enhances oxidation and the liquid becomes pink. Such samples must be discarded because they can lead to the modification and cleavage of DNA in solution. Oxidation can be inhibited by addition of 8-hydroxyquinoline at 0.1% (w/v). This has the advantage of giving a bright yellow colour which identifies the organic layer during phenol extractions.

B. *24:1 (v/v) chloroform–isoamyl alcohol*

1. Mix 96 ml chloroform with 4 ml isoamyl alcohol; store in a capped bottle.

C. *'Phenol–chloroform'*

1. Thaw a 10 ml aliquot of phenol and transfer 5 ml of the lower phase to a universal bottle. Mix in 5 ml of 24:1 (v/v) chloroform–isoamyl alcohol and 10 ml TE pH 8.0 (*Appendix 4*) and store at 4°C. Use the lower organic layer.

DNA in aqueous solution is added to 0.5–1.0 vol. of phenol, mixed, centrifuged to separate the phases, and the upper aqueous phase removed immediately without disturbing the interface. High-molecular-weight DNA (> 10 kb) must be treated carefully to avoid shearing, so phenol extractions are usually performed by inversion or gentle rocking of the tube several times to mix the phases. It is inevitable that some DNA will be left in the phenol layer, but this poses few problems when several micrograms of DNA are manipulated. However, with less than 1 μg serious loss can be encountered because substantial proportions of the DNA partition into the phenol. Consequently, it is worth extracting the phenol once or twice with TE pH 8.0 (*Appendix 4*) to enhance recovery of particularly precious samples. Some workers favour the use of phenol–chloroform (*Protocol 1C*) because DNA is even less soluble in this medium. Samples of DNA which have been treated with phenol are routinely then extracted with 24:1 (v/v) chloroform–isoamyl alcohol (*Protocol 1B*) to precipitate remaining protein and reduce the amount of dissolved phenol which is in the aqueous phase. The use of diethyl ether to remove phenol from aqueous samples should be avoided because ether is easily oxidized and can lead to modification of the DNA sample.

DNA in aqueous solution can be precipitated by mixing with either ethanol or isopropanol in the presence of sodium or ammonium acetate and cooling to −20°C. The DNA is collected by centrifugation, except when it is in amounts of 50 μg or more in which case it can be spooled from the supernatant. Regardless of how it was precipitated the sample is rinsed with 70% (v/v) ethanol to remove any traces of salt and then dried under vacuum.

Small quantities of DNA can be assessed by electrophoresing samples in an agarose gel with known standards (see Chapter 5, Section 4.4), but samples from large-scale plasmid, genomic, or lambda phage preparations are usually determined spectrophotometrically as described in *Protocol 2*.

Protocol 2. Spectrophotometric determination of DNA

1. Transfer 10–100 μl of the DNA sample to 900–990 μl TE pH 7.5 (*Appendix 4*) in a 1 ml quartz cuvette, mix the contents thoroughly and record a spectrum from 250–320 nm. A smooth peak should be observed with an absorption maximum at around 260 nm and no noticeable shoulder at 280 nm. The 260:280 nm absorbance ratio should be more than 1.9. If the ratio is less than 1.9 then contamination with protein should be suspected.

2. Calculate the quantity of DNA using the guide that 1 ml of a solution with an A_{260} of 1.0 is equivalent to 50 μg of double-stranded or 35 μg of single-stranded DNA.

---

2. Genomic DNA

Genomic DNA is easy to isolate and characterize but will be of little use unless it is of high molecular weight, readily cleaved by a variety of restriction enzymes and clonable. Before starting to make genomic DNA, read Chapters 1–3 in Volume II of this book to be sure of your strategy, because in many cases it might be easier to follow the cDNA route to your gene. This is because the gene of interest could contain many large introns and the genome may be so large that a vast number of recombinants would have to be screened. Genomes of small to modest size (e.g. 1500 kb for *Thermoplasma acidophilum*, 35 000 kb for protozoans, and 80 000 kb for nematodes) can be handled easily via cloning in lambda because only a few thousand clones require screening.

There are many protocols for the isolation of genomic DNA but I have chosen three methods which are in common use and are appropriate, but not exclusive, for different types of starting material. DNA suitable for chromosome size measurement using pulsed field electrophoresis is not described here, but new methods are being continually developed to isolate intact chromosomes and the reader should acquaint themselves with more appropriate texts (e.g. ref. 1).

2.1 Pre-treatment of material

The techniques used in the isolation of genomic DNA depend on the physical characteristics of the starting material. Tissue culture cells and the nuclei from blood samples can be treated directly by incubation with SDS and proteinase K (Section 2.3). Other materials may require some form of pre-treatment. Bacteria will require cell wall lysis with lysozyme and EDTA (*Protocol 3B*), and some types of animal tissue will require shredding and grinding at $-180°C$ in liquid N_2 to form a powder to enable efficient proteolysis (*Protocol 3C*). Some tissues give cleaner preparations than others; for example, liver samples often contain glycogen in such quantities that it becomes difficult to separate the DNA cleanly. If small invertebrates are used it is often impractical to isolate individual tissues, but it is worthwhile to remove the gut to avoid contamination of the preparation with DNA originating from microorganisms.

DNA from plants is more difficult to obtain in a clonable form than that from animals, due to the presence of secondary metabolites and polysaccharides which interfere with the extraction procedure and co-purify with the DNA. Some of these problems can be overcome by using unexpanded leaves as starting material, these having a high cell density and containing less polysaccharide or secondary metabolites. Plant tissue is robust and requires grinding in liquid N_2 with alumina powder to enhance cell disruption (*Protocol 3D*), followed by extraction with high-salt cetyl trimethyl ammonium bromide (CTAB) buffer which dissolves the DNA (Section 2.4). If the sample is unsatisfactory subsequent samples could be prepared from isolated nuclei. This offers a clean starting material which could be treated using either the proteinase K–SDS or CTAB method.

Protocol 3. Pre-treatment of samples for DNA preparation

A. *Blood samples*

1. Mix 20 ml blood, fresh or stored at $-70°C$, with 180 ml of lysis buffer [0.32 M sucrose, 1% (v/v) Triton X-100, 5 mM $MgCl_2$, 10 mM Tris–HCl pH 7.4] at $4°C$ for 5 min.

2. Centrifuge at 15 000 g for 20 min at $4°C$. Discard the supernatant and resuspend the nuclei in 4.5 ml 75 mM NaCl, 50 mM EDTA pH 8.0. Proceed to *Protocol 5*.

B. *Bacteria*

1. Resuspend 0.5 g bacterial paste, fresh or stored at $-70°C$, in 40 ml TE pH 8.0 (*Appendix 4*). Centrifuge at 2500 g for 5 min at $4°C$. Discard the supernatant.

Protocol 3. *Continued*

2. Resuspend the pellet in 3.2 ml 50 mM Tris–HCl pH 8.0, 0.7 M sucrose then add 0.6 ml of freshly-prepared lysozyme solution (20 mg ml^{-1}) and store on ice for 5 min.

3. Add 0.6 ml 0.5 M EDTA pH 8.0 and 0.5 ml 10% (w/v) SDS, gently mix and store on ice for 5 min. Proceed to *Protocol 5*.

C. *Animal tissue*

1. Whole small invertebrates should first be dissected to remove the gut and non-contributing tissue such as wings, cuticle, and legs; animals less than 5 mm in length are used whole. Tissues such as spleen must be sliced into manageable fragments.

2. Pre-cool an unglazed ceramic mortar of at least 100 mm outside diameter by half-filling with liquid N_2. Carefully immerse the pieces of tissue in the mortar and as they freeze grind them with a suitable pestle until a powder is obtained. This will take 2–5 min for soft tissues. Replenish the N_2 as it evaporates.

3. When the tissue has powdered allow the liquid N_2 to evaporate but be careful to ensure that the pulverized sample remains cold, dry, and powdery. Proceed to *Protocol 5*. If you know the sample contains mucopolysaccharide or glycogen proceed to *Protocol 6*.

D. *Plant material*

1. Pre-cool a mortar as in step C1 above. Place 3 g unexpanded leaves into the liquid N_2, add 2 g alumina powder (Sigma type A-5).

2. Using a suitable pestle grind the sample to a fine powder (this will take 5 min), replenishing the liquid N_2 as necessary.

3. Allow the N_2 to evaporate as in step C3 above. Proceed to *Protocol 6*.

2.2 Isolation of DNA by SDS–phenol extraction

This is the simplest and fastest method for isolation of DNA and is widely used with filamentous fungi and plant material. The technique uses SDS with phenol to denature and dissolve macerated hyphae leaving the DNA intact, which is precipitated from solution with a precise volume of isopropanol (5).

Protocol 4. Isolation of DNA by SDS–phenol extraction

1. Collect hyphae by filtration of a fungal culture through a buchner funnel. Rinse with 20 mM EDTA pH 8.0, remove as much liquid as possible, freeze in liquid N_2 and lyophilize.

2. Grind the dried material in a small mortar to produce a powder. Resuspend

Protocol 4. *Continued*

50 mg in a microfuge tube by stirring in 0.5 ml extraction buffer [0.2 M Tris–HCl pH 8.5, 0.25 M NaCl, 25 mM EDTA, 0.5% (w/v) SDS].

3. Add 0.35 ml phenol (*Protocol 1A*, equilibrated with an equal volume of extraction buffer) and mix by inverting the tube several times. Add 0.15 ml chloroform and again invert to mix.

4. Centrifuge for 1 h at 15 000 g then immediately transfer the upper aqueous layer, avoiding the interface, to a tube containing 25 μl RNase A (20 mg ml^{-1}, see *Appendix 4*) and incubate for 10 min at 37°C.

5. Add an equal volume of chloroform–isoamyl alcohol (*Protocol 1B*), mix, and centrifuge for 10 min at 15 000 g. Transfer the upper aqueous layer to a sterile microfuge tube and record its volume.

6. Add a 0.54 vol. of isopropanol to the sample and invert to mix. The DNA precipitates and forms a clump in the tube.

7. Centrifuge the sample by pulsing for 5 sec and remove the liquid using a drawn-out pasteur pipette. Rinse the tube contents with 70% (v/v) ethanol, and recentrifuge to settle the pellet of DNA. Remove the liquid from the tube and dry the sample *in vacuo*.

8. Dissolve the DNA in 100 μl TE pH 8.0 (*Appendix 4*) by incubating at 4°C for several hours. Record a spectrum of a diluted sample (*Protocol 2*) to assess the results of the preparation.

2.3 Isolation of DNA by SDS–proteinase K treatment

This technique is applicable to most material and relies on proteinase K and SDS to dissolve the sample and digest the protein component without affecting the DNA. The sample is then extracted with protein denaturants (phenol, phenol–chloroform, chloroform) and the aqueous liquid containing the DNA dialysed and precipitated with ethanol or isopropanol in sodium or ammonium acetate solution. Generally, the fewer manipulations with genomic DNA the better, but some steps cannot be omitted. *Protocol 5A* is designed for samples where small quantities of DNA are required. RNase A is not used in this protocol because little if any RNA survives the various manipulations. *Protocol 5B* is designed for use with powdered tissue samples which often retain particulate material. This technique employs RNase A and includes a dialysis step to reduce the salt and detergent content prior to precipitation.

Protocol 5. Isolation of DNA with proteinase K and SDS

A. *Lysed cells, nuclei, and monolayers*

1. Adjust fluid samples from *Protocol 3* to 1% (w/v) SDS and 0.5 mg ml^{-1} proteinase K. Drained flasks containing confluent monolayers of tissue

Protocol 5. *Continued*

culture cells are incubated with 10 ml of digestion buffer [1% (w/v) SDS, 0.5 mg ml^{-1} proteinase K, 50 mM Tris–HCl pH 9.0, 0.1 M EDTA, 0.2 M NaCl]. Incubate the samples at 55°C for 3–16 h with very gentle shaking.

2. Add 1.0 vol. of phenol (*Protocol 1A*) and mix the sample very carefully for 3 h at room temperature. It is often convenient to transfer the sample to a vessel which will allow for a large contact area between the two phases so that the sample requires the minimum of agitation for adequate mixing.

3. Transfer the mixture to a Falcon tube and centrifuge at 4000 *g* for 10 min at 25°C. Remove and discard the lower phenol layer using an aspirator connected to a long pasteur pipette which passes through the aqueous phase.

4. Add 1.0 vol. of phenol–chloroform (*Protocol 1C*), gently invert to mix the liquid phases, then centrifuge at 3000 *g* for 10 min at room temperature.

5. Transfer the upper aqueous layer to a fresh tube, place on ice for 5 min then add 2.5 ml 7.5 M ammonium acetate and 10 ml ethanol at −20°C. Hook the clump of DNA from the tube using a pasteur pipette, rinse with 70% (v/v) ethanol and then dry under vacuum.

6. Place the sample in a microfuge tube with 1 ml TE pH 7.5 (*Appendix 4*) and allow to dissolve overnight. Assess yield and quality by recording a spectrum (*Protocol 2*) and running on a 0.3% (w/v) agarose gel (Chapter 5, Section 4.4).

B. *Pulverized tissue samples*

1. Transfer the frozen powder from 5–10 g tissue (*Protocol 3*) in small portions on to the surface of 20 ml buffer (50 mM Tris–HCl pH 9.0, 0.1 M EDTA, 0.2 M NaCl, 0.1 mg ml^{-1} RNase A) in a 0.5-litre beaker using a cold spatula. Swirl the solution for 10 min, then transfer the homogeneous liquid to a Falcon tube. Adjust to 1% (w/v) SDS and 0.5 mg ml^{-1} proteinase K. Incubate for 16 h at 55°C.

2. Transfer the sample to a flat-sided bottle or a 1-litre beaker and add 20 ml of phenol (*Protocol 1A*). Leave at room temperature for 3 h with maximal contact between the phases and occasionally agitate the mixture.

3. Place the sample in a Falcon tube and centrifuge at 3000 *g* for 10 min at 20°C to separate the phases.

4. Transfer the aqueous layer, avoiding the interface, to a 30-ml corex tube and centrifuge at 10 000 *g* for 20 min at 25°C to sediment any undissolved materials.

5. Dialyse the supernatant against 2 litres of TE pH 7.5 (*Appendix 4*), exchanging the buffer twice.

6. Transfer the dialysed DNA sample to a 0.1-litre beaker and add 0.1 vol. of 3 M sodium acetate pH 6.5 and 0.8 vol. of isopropanol. Mix gently with a

Protocol 5. *Continued*

pasteur pipette and hook the clumps of DNA together, rinse with 70% (v/v) ethanol and transfer to a 5 ml bijou bottle.

7. Dry the sample under vacuum for 10 min to remove the ethanol. Some water will be left with the sample, which will appear translucent. Place 2.5 ml TE pH 7.5 into the bottle and leave for several hours to dissolve the DNA. Assess the quality of the DNA as in step A6 above.

2.4 Isolation of DNA with CTAB

Isolation of DNA from plants is preferably achieved by treatment with CTAB. This detergent carries a positive charge which interacts with the negative charge of DNA in high salt concentrations and forms a soluble complex. In subsequent steps a decrease in the salt concentration causes precipitation of DNA, leaving other compounds, especially polysaccharides, in solution. The preparation of DNA from animal tissues which are rich in mucopolysaccaride could also be by this technique which yields good quality material.

Protocol 6. Isolation of DNA with CTAB

1. Transfer 2.5 g powdered tissue (*Protocol 3D*) to a Falcon tube and add 10 ml CTAB buffer at 65°C [2% (w/v) CTAB, 0.1 M Tris–HCl pH 8.0, 20 mM EDTA, 1.4 M NaCl, 1% (w/v) PVP (40 kd)]. Incubate for 15 min at 56°C with occasional gentle mixing to dissolve all nucleic acids. (PVP is polyvinylpyrrolidone.)

2. Add 10 ml chloroform–isoamyl alcohol (*Protocol 1B*), securely cap the tube and invert several times to form an emulsion.

3. Centrifuge the sample at 10 000 *g* for 10 min at 20°C. Transfer the upper aqueous layer to a Falcon tube and discard the lower chloroform layer.

4. Add 0.2 vol. of 5% (w/v) CTAB, mix and repeat the chloroform–isoamyl alcohol extraction described in steps 2 and 3.

5. Measure the volume of the aqueous phase, transfer to a 30-ml corex tube, and add an equal volume of precipitation buffer [1% (w/v) CTAB, 50 mM Tris–HCl pH 8.0, 10 mM EDTA]. Seal the tube, gently invert to mix and incubate at 4°C for 15 min; the solution will become opaque as the DNA precipitates.

6. Centrifuge at 3000 *g*, 15 min, 20°C to sediment the nucleic acid, and discard the supernatant. Add 2 ml TE pH 8.0 (*Appendix 4*) containing 1 M NaCl to redissolve the sample; this may be aided by warming to 56°C for 10 min.

7. Place the sample on ice, then add 2 vol. of ethanol pre-cooled to −20°C. Mix gently by inverting the sealed tube and centrifuge the precipitated DNA at 10 000 *g* for 10 min at 4°C, then discard the supernatant.

Protocol 6. *Continued*

8. Carefully resuspend the DNA sample in 70% (v/v) ethanol, centrifuge as in step 7, and dry *in vacuo*.

9. Dissolve the DNA in 0.2 ml TE pH 7.5 by leaving for several hours at 4°C.

10. The nucleic acid sample consists of genomic DNA and RNA contaminants, but the RNA can be removed by incubation with RNase A for 20 min at 37°C (see *Appendix 4*) and the DNA retrieved by repeating steps 7–9.

2.5 Assessment of quality

A problem encountered during the manipulations described above is that the DNA is unavoidably sheared, which is not surprising considering that chromosomes, which are the source of the DNA, range in size from 0.3 to 200 megabase pairs. Careful sample handling is required throughout, transfer by pipette should be with wide diameter tubes only, any mixing should be gentle, and rotamixers should be avoided. When DNA is precipitated with ethanol the solvent must be at −20°C to avoid shearing. The precipitated DNA can then be spooled from the solution using a glass rod; in theory, collection by centrifugation could cause contaminants to co-sediment with the sample.

The size of the DNA molecules recovered can be estimated by electrophoresis in a 0.3% (w/v) agarose gel with lambda (49 kb) or phage T4 (150 kb) as a size marker (Chapter 5, Section 4.4). The sample on the gel should appear as a discrete band migrating slightly slower than the marker and with no discernable trail in the direction of migration, which would indicate extensive shearing. Such a sample could be used for Southern hybridization and cloning in lambda vectors, where inserts of around 20 kb are required. No guarantee could be offered as to the suitability for cloning of these DNA samples in yeast artificial chromosomes, nor should it be used for preparation of cosmid libraries or cloning in phage P1 unless you are assured that the fragment size is 250 kb or more, as measured by orthogonal field agarose gel electrophoresis (Chapter 5, Section 6). Genomic DNA should be digested cleanly with a selection of restriction enzymes and be visible as a smear of restriction fragments from a few hundred base-pairs to several thousand after electrophoresis. Depending on the source of the DNA you should expect to see several intense discrete bands, which arise from repetitive sequences within the genome, superimposed on the smear. If the sample cannot be digested with restriction enzymes it is possible to clean it up substantially by CsCl centrifugation (*Protocols 15* and *16*); however, the proportion of sheared material will increase considerably.

3. Phage cloning vectors

Two types of phage are commonly used in recombinant DNA work. Lambda phage is used for preparing gene banks in viable phage and can also be used to package cosmid DNA. M13 phage is used to prepare single-stranded DNA for chain termination sequencing.

3.1 Lambda phage

Phage lambda is used to prepare libraries of genomic DNA with insert sizes of about 20 kb in replacement vectors such as λEMBL3. Insertion vectors, such as λgt10, can be used for the cloning of smaller inserts, usually those derived from synthesis of cDNA up to 9 kb in size. A lambda library would be maintained as a suspension of phage particles in a storage buffer from which individual clones containing the gene of interest would be identified and isolated. An individual homogeneous clone would be stored and used as a stock for preparation of DNA in bulk. The DNA insert would then be isolated and subcloned into other more appropriate vectors for restriction mapping and sequencing. The relevant techniques are described elsewhere in Volumes I and II of this book.

Phage lambda will replicate in *E. coli* which lyse during growth and release millions of phage particles. This can be achieved in liquid cultures or on lawns of bacteria (Chapter 2, Section 5), either route providing a stock of phage with high titre which is suitable for long-term storage or preparation of DNA. Growth on lawns allows unimpeded individual phage replication and is also used for screening and amplification of libraries.

Typical yields of DNA isolated from the lambda phage contained on four 90-mm diameter confluent lawns is 200 to 500 μg, whereas cultures yield, for example, 2–5 mg λgt10 DNA from 1 litre of *E. coli* C600. The latter method is time-consuming, but the provision of several mg of DNA is useful when several attempts at subcloning are required; bear in mind that λgt10 with a 1 kb cDNA insert will at most yield 1 μg of fragment from 40 μg starting material.

Phage yield can be optimized by including maltose in the growth medium of the cells prior to infection, as this induces the lambda receptor protein and ensures efficient adsorption of phage particles (Chapter 2, Section 5.2). Control of the multiplicity of infection (m) is crucial and varies depending on the host-vector system but can be optimized in small-scale cultures. However, this may be unnecessary, since a poor yield still provides sufficient DNA for most purposes. The most critical problem with lambda phage is its sensitivity to detergent, which reduces its infectivity; to reduce this problem all glassware which will come into contact with phage should first be rinsed in 5% (v/v) acetic acid and then oven-baked at 180°C.

The following protocols describe how to obtain quantities of DNA from lambda phages grown on agar plates or in broth cultures. Additional details on the growth and purification of lambda phage are given in Chapter 2, *Protocols 10–14*.

Protocol 7. Lambda phage preparation from agar plates

1. Prepare a 50 ml overnight culture of an appropriate *E. coli* host strain in YT

Protocol 7. *Continued*

supplemented with 0.2% (w/v) maltose and 10 mM $MgSO_4$ (see *Appendix 4*). Centrifuge the sample in two sterile Falcon tubes at 3000 g for 5 min. Discard the supernatants and resuspend the cell pellets in approximately 10 ml 10 mM $MgSO_4$ to obtain an OD of 2.0 at 650 nm, corresponding to about 10^9 cells ml^{-1}.

2. Mix 100 μl cells with 10 μl phage stock [titre 10^7 plaque forming units (p.f.u.) ml^{-1}] in 5 ml clean, sterile glass tubes. Leave at ambient temperature for 15 min. As a control incubate cells in the absence of phage.

3. Add 3 ml YTS agar (*Appendix 4*) at 45°C to the tubes, mix by rolling between the palms of the hands and pour immediately on to pre-warmed YT agar in 9 cm Petri dishes. Leave to set and dry slightly for 20 min, with the lids ajar in a sterile hood. Incubate overnight at 37°C.

4. The control plate will be covered with an opaque lawn of bacteria. On the test plates the lawn will have disappeared, due to confluent lysis of the cells. Remove the soft agar overlay from up to four plates and place in a Falcon tube, add 5 ml 10 mM Tris–HCl pH 7.5, 10 mM $MgCl_2$ and shake gently to dissipate the pieces of agar.

5. Add 0.2 ml chloroform (<u>not</u> chloroform–isoamyl alcohol) and rotamix gently. Centrifuge at 2000 g for 15 min at 4°C. Collect the supernatant without disturbing the agar pellet, which may be re-extracted with fresh buffer.

6. The phage stock (approximately 10^{11} p.f.u. in total) may be stored almost indefinitely after addition of a few drops of chloroform at 4°C, or can be purified further by CsCl centrifugation (*Protocol 9*).

Protocol 8. Lambda phage preparation from liquid culture

1. Prepare an overnight culture of an appropriate *E. coli* host in YT supplemented with 0.2% (w/v) maltose (see *Appendix 4*).

2. Transfer 0.9 ml culture and 0.1 ml of 10 mM $CaCl_2$, 10 mM $MgCl_2$ solution to a Falcon tube. Add 100 μl phage suspension containing approximately 10^8 viable phage and incubate at 37°C for 15 min. Phage stock can be titred by making serial dilutions of phage in SM (*Appendix 4*) and counting plaques on lawns as in *Protocol 7*. See also Chapter 2, *Protocol 10*.

3. Make up the cell/phage suspension to 10 ml with pre-warmed YT, mix, transfer to 500 ml pre-warmed (37°C) YT containing 10 mM $MgCl_2$, and incubate for 10 h with vigorous shaking.

4. Add 2.5 ml chloroform (<u>not</u> chloroform–isoamyl alcohol) to the flask and shake for 15 min, then centrifuge the sample for 30 min at 6000 g to remove cell debris. (Chloroform lyses residual cells.)

Protocol 8. *Continued*

5. Decant the supernatant into a large conical flask, add 35 g NaCl and shake gently until completely dissolved. Add 80 g PEG 6000, place the flask in a cold room and stir with a magnetic stirrer for 1 h, or until the PEG is dissolved. Leave the mixture overnight in the cold room to enable the phage to precipitate.

6. Centrifuge the sample for 30 min at 3000 g, 4°C. Discard the supernatant, dry the inside of the tube and resuspend the phage pellet in 10 ml SM (*Appendix 4*). Leave the sample overnight at 4°C to allow the phage particles to dissipate.

7. Store the sample with a few drops of chloroform to inhibit growth of microorganisms, or further purify it by CsCl centrifugation.

Suspensions of lambda phage can be used directly for the preparation of DNA but the product is usually difficult or impossible to digest with restriction enzymes. This can be remedied to some extent by further purification through ion-exchange columns, but lambda DNA is prone to shear so it is preferable to purify the phage by CsCl density gradient centrifugation, and then digest away the phage coat proteins with SDS and proteinase K before precipitating the DNA from solution.

Protocol 9. Purification of phage lambda by CsCl centrifugation

1. Add DNase I at 100 μg ml^{-1} final concentration to the phage suspension and incubate at 37°C for 30 min.

2. Add an equal volume of chloroform to the phage and mix well for 10 min. Separate the phases by centrifugation at 1000 g in a bench-top centrifuge.

3. Remove the upper aqueous layer, containing the phage, and record its volume. Add 0.77 g CsCl per millilitre of solution and shake gently until dissolved.

4. Depending on the volume of the sample, fill either a small or large Beckman Quick-Seal tube and centrifuge for 16 h at 300 000 g in the Vti 65 or Ti50 rotor.

5. The phage will be present about half-way down the tube as a thin bluish-grey band (similar to the DNA bands shown in *Figure 2*). Puncture the top of the tube with a needle to admit air and remove the phage into a syringe by puncturing the tube with a 19-gauge needle.

6. Either dialyse the sample against 2 litres of 10 mM Tris–HCl pH 7.5, 10 mM MgSO$_4$ twice, or pass the sample through a small gel filtration column, such as Pharmacia PD10.

Protocol 10. Preparation of DNA from lambda phage by proteinase K–SDS treatment

1. Transfer the phage suspension to a Falcon tube after measuring its volume, and adjust to 10 mM EDTA (pH 8.0) and 0.2% (w/v) SDS using 0.5 M and 20% (w/v) stock solutions respectively. Mix, add 1 mg proteinase K, and incubate at 37°C for 1 h.

2. Add an equal volume of phenol–chloroform (*Protocol 1C*), place the capped tube on its side and rock gently on a table shaker for 10 min at room temperature. Centrifuge at 1000 *g* for 5 min.

3. Transfer the upper aqueous phase with a wide-bore pasteur pipette to a fresh tube and repeat step 2 but using 24:1 (v/v) chloroform–isoamyl alcohol (*Protocol 1B*).

4. Transfer the upper aqueous phase to a 30-ml corex tube, add 0.1 vol. of 3 M sodium acetate (pH 5.5), and place on ice for 5 min.

5. Add 2.5 vol. of ethanol ($-20°C$) and gently rock the tube to mix the contents. The precipitated phage DNA molecules will clump together as a white mass (*Figure 1*).

Figure 1. Lambda DNA after precipitation with ethanol (see *Protocol 10*).

Protocol 10. *Continued*

6. Hook the DNA on to a pasteur pipette and transfer the pipette and DNA to a glass test-tube and dry in vacuum for 30 min.

7. Add 1 ml TE pH 7.5 (*Appendix 4*) to the tube and place at 4°C overnight. The DNA will dissolve completely, or if several milligrams are present remain as a gelatinous lump. In this case add more TE and leave several hours more for dissolution.

8. Check the yield and quality of the DNA by recording a spectrum (*Protocol 2*) and electrophoresing 2–10 μl in a 0.3% (w/v) agarose gel (Chapter 5, Section 4.4).

3.2 M13 phage DNA

The single-stranded DNA present in the filamentous phage M13 is predominantly used for chain termination sequencing. Recombinant M13 is stable with inserts of up to 2 kb. Larger fragments can also be accommodated but are prone to deletion during growth, forming a mini-circle phage which is of no use.

M13 DNA is prepared from small culture volumes which are well-aerated during growth at 37°C (*Protocol 11*). This provides sufficient material (2–5 μg) for several sequencing attempts or for other uses such as oligonucleotide site-directed mutagenesis. If larger quantities of phage DNA are required the protocol can be scaled up but it is preferable to duplicate the method described because of the quality of the material which results. The procedure also provides a stock of M13 phage particles which can be stored and used for re-infection and isolation of DNA at a later date.

Protocol 12 describes a modified procedure that is suitable for use with phage obtained from phagemids (6). In fact, the convenience of these vectors has proved so popular that M13 is being surpassed as the most widely-used vector for sequencing of DNA (see Chapter 8, Section 3.2).

Protocol 11. Preparation of single-stranded DNA from M13 phage

1. Transfect an appropriate strain of competent *E. coli* with 50 ng of M13 double-stranded DNA into which is ligated the DNA of interest (see Chapter 8, Section 1.1). Plate out and incubate as described in Chapter 8, Section 2.3.

2. Take the agar plate from the incubator and pick clear plaques (see Chapter 8, Section 2.3) with a toothpick into 3 ml pre-warmed DYT (*Appendix 4*) in culture tubes (180 × 14 mm) and incubate with vigorous shaking for 10 h at 37°C. Control incubations should be set up with a blue, non-recombinant plaque, and from an area on the plate devoid of plaques.

3. Transfer 1.5 ml of culture to a microfuge tube and centrifuge at 15 000 *g* for 15 min; ensure that all the cells are pelleted.

Protocol 11. *Continued*

4. Remove 0.9 ml supernatant, being careful to avoid disturbing the pellet, and place in a fresh tube containing 0.15 ml 2.5 M NaCl, 20% (w/v) PEG 6000. Mix and leave at ambient temperature for 15 min.

5. Centrifuge the phage–PEG suspension at 15 000 g for 10 min. A small white pellet of phage should be visible at the bottom of the tube from cultures infected with phage. The control culture of bacteria alone should have no discernible pellet.

6. Aspirate and discard the supernatants using a drawn-out pasteur pipette, without dislodging the pellet.

7. Recentrifuge the tubes for 10 sec to collect any remaining liquid and aspirate this. Ensure no liquid remains in the tube by probing with a tapered piece of absorbent tissue.

8. Add 100 μl TE pH 7.5 (*Appendix 4*) to the sample and mix thoroughly to bring the pellet into suspension. Add 40 μl phenol (*Protocol 1A*) and rotamix frequently until all particulate material is dissipated. Keep the suspension mixed for a further 10 min, then place on ice for 5 min.

9. Centrifuge the sample at 15 000 g for 5 min and remove the clear upper aqueous phase into a fresh tube, avoiding any precipitate at the interface. This procedure must be carried out immediately after centrifugation to avoid the liquid layers diffusing into each other.

10. Add 0.1 vol. of 3 M NaAc pH 5.5 and 2.5 vol. of ethanol, mix, and incubate at $-70°$C for 10 min or $-20°$C for 30 min. Centrifuge at 15 000 g for 30 min.

11. Pour off and discard the supernatant, rinse the tube with 0.5 ml 70% (v/v) ethanol without disturbing the contents, and dry in vacuum.

12. The tube should now appear empty, with no odour of phenol. Add 40 μl TE pH 7.5 to dissolve the sample and estimate recovery by electrophoresing 5 μl of sample in a 0.8% (w/v) agarose gel (see Chapter 5, Section 4.4).

Protocol 12. Phage preparation from phagemid vectors

1. Prepare an overnight culture of cells harbouring the phagemid in DYT (*Appendix 4*) with appropriate antibiotic selection.

2. Inoculate 3 ml DYT containing the appropriate antibiotic with 60 μl of overnight culture. Incubate at 37°C with vigorous shaking for 30 min.

3. Add 20 μl helper phage stock (see Chapter 8, Section 3.2) containing about 10^5 viable phage particles and continue incubation for 10 h. The culture can now be treated as an M13 phage preparation and single-stranded DNA isolated.

Protocol 12. *Continued*

4. If the yield of DNA is poor it can be improved by a modification to step 3. Helper phage carry antibiotic-resistance genes (e.g. kanamycin resistance with M13KO7; see Chapter 8, Section 3.2) consequently it is worthwhile adding antibiotic to the culture 60 min after addition of helper phage. This ensures that surviving cells carry both the phagemid and the helper phage.

---

4. Plasmid vector DNA

It is quite straightforward to isolate plasmid DNA from *E. coli* and this can be performed on either a small or 'mini-prep' scale to obtain sufficient material for cursory analysis, or on a large, 'maxi-prep' scale to provide a stock of plasmid for longer-term use. The isolation of plasmids is performed in essentially three stages. The bacterial cell wall is first weakened by the action of lysozyme, and the cells then lysed by EDTA and a detergent at high pH. Finally, the insoluble cell debris consisting of genomic DNA and protein is precipitated with high salt and centrifuged down, leaving the plasmid in solution. In the context used here the term 'plasmid' also refers to the double-stranded replicative form (RF) of M13 and to cosmids (Chapter 8, Section 3.1).

4.1 Small-scale plasmid preparation

There are numerous ways of isolating plasmids on a small scale but one particular method (7) is universally popular because it is fast, reproducible, and provides several micrograms of plasmid which is readily cleaved by all restriction enzymes. The preparation will be contaminated with RNA, which co-migrates on electrophoresis gels with DNA fragments of 200–500 bp, but generally this is not a problem unless the RNA masks the identification of DNA fragments generated by restriction digest of the plasmid. In this case, RNase A can be added to the restriction digest, or the whole of the sample can be treated and the plasmid retrieved by precipitation with ethanol and sodium acetate. The attraction of this small-scale preparation is that several clones can be processed simultaneously. Furthermore, the sample is often suitable for double-stranded DNA sequencing, although to ensure reproducibility with this technique a Qiagen Tip 20 (Hybaid UK) should be used during the purification.

Protocol 13. Small-scale isolation of plasmid DNA

1. Prepare an overnight culture of *E. coli* containing plasmid in 3 ml DYT (*Appendix 4*) with the appropriate antibiotic.

2. Transfer 1.5 ml culture to a microfuge tube on ice and centrifuge for 5 min at 15 000 *g*.

Protocol 13. *Continued*

3. Discard the supernatant and resuspend the cell pellet in 0.1 ml of buffer (50 mM glucose, 10 mM EDTA, 25 mM Tris–HCl pH 8.0 containing 2 mg ml^{-1} freshly-weighed lysozyme). Incubate on ice for 10 min.

4. Add 200 μl lysis buffer and mix gently [lysis buffer: 0.2 M NaOH, 1% (w/v) SDS; prepared fresh from stocks of 5 M NaOH and 10% (w/v) SDS]. Incubate on ice for 10 min.

5. Add 150 μl potassium acetate and mix gently (potassium acetate is prepared by adding glacial acetic acid to 5 M potassium acetate until pH 4.8 is obtained). Centrifuge at 15 000 g for 15 min.

6. Transfer the supernatant to a microfuge tube containing an equal volume of phenol–chloroform (*Protocol 1C*), mix and centrifuge at 15 000 g for 5 min.

7. Remove and retain the upper aqueous layer in a microfuge tube and add 1 ml cold ($-20°$C) ethanol. Incubate on ice for 10 min.

8. Centrifuge for 30 min at 15 000 g. Discard the supernatant and rinse the tube contents carefully with 0.5 ml 70% (v/v) ethanol.

9. Centrifuge at 15 000 g for 10 min. Pour off and discard the supernatant.

10. Rinse the pellet in the tube with cold 70% (v/v) ethanol and dry the sample in vacuum.

11. Resuspend the sample in 50 μl water and electrophorese 5 μl in a 0.8% (w/v) agarose gel to estimate yield (see Chapter 5, Section 4.4).

12. Remove contaminating RNA by adding 1 μl of RNase A (1 mg/ml) to 10 μl aliquots of sample along with the restriction endonuclease, and incubate for 60 min at 37°C. Alternatively, RNase A can be used alone and the DNA retrieved via ethanol precipitation. See *Appendix 4* for preparation of RNase A.

4.2 Large-scale plasmid preparation

The large-scale purification technique is more time-consuming and uses relatively large volumes of media, but results in large quantities of very pure plasmid. The rationale of the method is similar to that described for the small-scale preparation, but the crucial purification step entails centrifugation of the plasmid in caesium chloride solution. This exploits the slight differences in density between covalently-closed supercoiled plasmid molecules, other nucleic acids, and protein. The yields are generally very high and provide an excellent source of DNA which is suitable for all restriction enzymes and double-strand sequencing.

There are many high-copy number plasmids which yield between 400 and 3000 μg supercoiled DNA from 0.5 litre media. The higher yields can be optimized by adding more antibiotic to the culture during growth although this is inappropriate with the replicative form of M13 phage, which has no antibiotic resistance genes. Low-copy number plasmids may well yield only 200 to 400 μg DNA. Yields can be increased by amplification with chloramphenicol (grow the culture until the OD at 550 nm is 0.8–1.0, then add solid chloramphenicol to a final concentration of 150 μg ml^{-1} and continue incubating at 37°C overnight), or in the case of M13 by carefully controlling conditions such as the time of phage infection. However, it is often simplest to increase the culture volume to provide more plasmid.

Protocol 14. Large-scale isolation of plasmid DNA

1. Prepare a 0.5-litre culture of cells harbouring plasmid in a 2.5-litre conical flask. Incubate with vigorous shaking for at least 10 h at 37°C: cultures grown overnight are usually convenient. The RF of M13 is prepared by inoculating 0.5 litres of DYT media (*Appendix 4*) with 5 ml of an overnight culture of cells. Incubate for 2 h at 37°C, add 2 ml of phage supernatant (*Protocol 11*, step 4) and continue incubation for a further 8–10 h.

2. Centrifuge the cells in large bottles at 2000 *g* for 30 min at 5°C. Do not use excessive gravitational force at this stage as it will make the pellets difficult to resuspend in the next step. Dispose of the supernatant and dry the inside of the centrifuge bottles with tissue.

3. Gently resuspend the cell pellet in 8 ml of buffer (25 mM Tris–HCl pH 8.0, 50 mM glucose, 10 mM EDTA containing 5 mg ml^{-1} fresh lysozyme). Draw the mixture in and out of a pipette a few times to disperse any clumps of cells.

4. Transfer the suspension to a 30-ml corex tube and stand at room temperature for 10 min, during which time a slight change in hue should be discernable due to the action of lysozyme on the cell walls.

5. Add 16 ml of lysis buffer (see *Protocol 13*, step 4). Seal the tube with parafilm and gently invert a few times until the tube contents become a transparent, straw-coloured, moderately viscous solution. Incubate on ice for 10 min.

6. Divide the sample equally into two 30-ml corex tubes and add 5 ml of potassium acetate solution (see *Protocol 13*, step 5) to each tube. Seal the tubes with parafilm and invert gently to ensure complete mixing. A dense white granular precipitate coagulates which is composed of genomic DNA, protein and other cellular debris. Incubate the tubes for 10 min on ice.

7. Centrifuge the tubes for 30 min at 25 000 *g* at 4°C to pellet the debris.

8. Carefully pour off and retain the clear supernatants in a 25-ml cylinder and note the volume. Add 0.6 vol. of isopropanol, mix well and leave at room temperature for 15 min. Divide the turbid solution into two 30-ml corex tubes and centrifuge at 20 000 *g* at 4°C for 30 min.

Protocol 14. *Continued*

9. Pour off and discard the supernatant, and carefully rinse the tube walls and pellets with ice-cold 70% (v/v) ethanol, being careful not to dislodge any material. If the pellets become dislodged it is as well to resuspend them and centrifuge at 10 000 *g* at 4°C for 10 min.

10. Warm the tubes at 37°C for a few minutes, place perforated aluminium foil over their tops and dry in vacuum.

11. Dissolve the dried pellets in 10 ml of TE pH 7.5 (*Appendix 4*). This may take some time and can be encouraged by slight warming or by gentle disruption using a loose-fitting homogenizer.

The large-scale plasmid preparation, described above, may be performed with culture media of 0.1–0.5 litre. Smaller volumes allow the plasmid to be redissolved in less than 10 ml TE which then allows a choice of tube size for CsCl centrifugation of the whole sample. Plasmid preparations at this stage consist of covalently-closed- and open-circles of plasmid, sheared genomic DNA fragments from the host, RNA, and protein. CsCl centrifugation (*Protocol 15*) permits the discrete separation of covalently-closed-circular plasmids because they are marginally denser than other forms of DNA; in addition RNA sediments and protein forms a pellicle within the tube.

The centrifugation is preferable in a Quick-Seal tube in an angle ròtor, since slight variations in the density of the CsCl solution can be tolerated: the plasmid band becomes slightly shifted within the tube but can still be identified and recovered. Quick-Seal tubes can also be centrifuged in vertical rotors, in which the gradient forms across the width of the tube and re-orientates along with the plasmid band as the centrifuge slows down. Utmost accuracy is required in preparing CsCl solutions for plasmid separation when vertical rotors are used because variations in density will result in the plasmid banding near to the tube wall, which means that upon re-orientation it will be present as a diffuse region at one end of the tube.

Plasmid DNA is recovered from Quick-Seal tubes with a syringe (*Protocol 16*). The utmost care should be exercised to avoid personal injury. The contaminating ethidium bromide can easily be removed from the sample by extraction with an organic solvent, and the CsCl can be removed by dialysis, passage through a gel filtration column, or by precipitation with ethanol after suitable dilution of the CsCl solution to avoid it precipitating.

Protocol 15. Caesium chloride centrifugation of plasmid preparations

A: Samples for angle rotors

1. Measure accurately the volume of solution available, adjust to exactly 10 ml, and transfer to a 30-ml corex tube.

Protocol 15. *Continued*

2. Add precisely 10.00 g CsCl and mix gently to dissolve.

3. Add 0.3 ml ethidium bromide solution (10 mg ml^{-1}).

4. If the sample becomes opaque due to a lipid-like material precipitating from the solution, centrifuge the tube at 2000 g for 5 min at ambient temperature to collect the debris on the surface of the solution.

5. Transfer the clear pink solution, avoiding the lipid layer, to a Beckman Quick-Seal tube (16 × 76 mm) with pasteur pipettes, one acting as a funnel and the other for transfer. There should be enough liquid to fill the tube to its neck.

6. Balance the tubes in pairs using a topping-up solution (10.00 g CsCl per 10.30 ml water). Remove any drops of liquid from the mouth of the tube. Seal using the heating block provided by Beckman.

7. Place the tubes with their collars into an angle rotor and centrifuge for at least 36 h at 250 000 g or 16 h at 400 000 g at 15°C.

B: Samples for vertical rotors

1. This protocol is based on small maxi-prep cultures for plasmids from 50–100 ml culture, and varies slightly from the protocol above.

2. Measure accurately and adjust the volume of plasmid solution to precisely 4.65 ml.

3. Add 4.60 g CsCl and mix gently to dissolve.

4. Add 0.2 ml 0.5 M EDTA pH 8.0 and 0.2 ml ethidium bromide solution (10 mg ml^{-1}). Remove any insoluble material by centrifugation.

5. Transfer to a Beckman Quick-Seal tube (13 × 51 mm). Balance with CsCl solution (4.6 g per 5.05 ml water) and seal.

6. Place the tubes with their collars into the vertical rotor (Vti 65) and centrifuge for 16 h, 300 000 g at 15°C.

Protocol 16. Recovery of plasmid DNA from Quick-Seal tubes

1. Carefully remove the Quick-Seal tube from the rotor and place in a suitable stand.

2. Visualize the bands of DNA using a UV radiation source. The covalently-closed circular plasmid is the most dense and is nearest to the base of the tube. It is quite often visible without the aid of a UV source (see *Figure 2*).

3. Clamp the tube in a stand and place a beaker below it. Pierce the top of the tube with a 19-gauge needle; with a second needle attached to a 2-ml syringe pierce the side of the tube 1 to 2 mm below the plasmid band so that the needle

Protocol 16. *Continued*

Figure 2. The appearance of a large-scale plasmid preparation after centrifugation in CsCl. In each tube two bands can clearly be seen, the lower of which is the supercoiled plasmid.

point is well into the liquid. Carefully angle the needle into the centre of the band and draw liquid into the syringe while changing its angle to accommodate the relative shift of the band. Withdraw the syringe containing 0.4 to 1.0 ml of plasmid solution, and discard the Quick-Seal tube and its contents.

4. Transfer the liquid from the syringe barrel to a 15 ml corex tube and extract the ethidium bromide from the solution by mixing with an equilibrated organic solvent. This can be isopropanol, or any of the butyl or amyl alcohols to hand, which should be stored equilibrated with a saturated aqueous CsCl solution from which the upper organic layer is used. Mix the tube contents thoroughly, and centrifuge at 2000 *g* for 2 min at ambient temperature. Remove and discard the upper pink organic layer, then repeat the extractions until the organic layer is colourless.

5. To remove the CsCl, adjust the volume of plasmid DNA to 2.5 ml and pass through a gel filtration column (PD10, Pharmacia). Alternatively, dilute the sample with 2 vol. of water and add 6 vol. of cold (4°C) ethanol, mix, incubate at 4°C for 30 min then centrifuge at 15 000 *g* for 30 min at 4°C, discard the supernatant and dry the pellet in vacuum. A third method is to dialyse the sample against 2 litres TE pH 7.4 (*Appendix 4*), which should be replaced after 12 h.

6. Whichever method was followed in the previous step, the sample in solution, or after redissolving in TE, should be re-precipitated with ethanol and rinsed in 70% (v/v) ethanol to remove residual CsCl.

Protocol 16. *Continued*

7. Dissolve the dried sample in 1 ml TE pH 7.4 and transfer 10 µl to 990 µl TE pH 7.4 in a 1-ml quartz cuvette, mix, and record a spectrum from 250–320 nm (*Protocol 2*).

Acknowledgements

I am greatly indebted to all of the staff and students in the School of Biological Sciences at Bath who have tried and tested so many of the techniques which I have described here, and to Dr P.D. Harris for his careful reading of the manuscript.

References

1. Sambrook, J., Fritsch, E.F., and Maniatis, T. (1989). *Molecular Cloning, A Laboratory Manual*, 2nd edn. Cold Spring Harbor Laboratory, Cold Spring Harbor, New York.
2. Perbal, B. (1988). *A Practical Guide to Molecular Cloning*, 2nd edn. Wiley Interscience, New York.
3. Berger, S.L. and Kimmel, A.R. (ed.) (1987). *Methods in Enzymology*, Vol. 152. Academic Press, London.
4. Rodgers, S.O. and Bendich, A.J. (1988). In *Plant Molecular Biology Manual* (ed. S.B. Gelvin and R.A. Schilperoort), pp. A6/1–A6/11.
5. Raeder, U. and Broda, P.M.A. (1985). *Letters in Applied Microbiology*, **1**, 17.
6. Vieira, J. and Messing, J. (1987). In *Methods in Enzymology*, Vol. 153 (ed. R. Wu and L. Grossman), pp. 3–11. Academic Press, London.
7. Birnboim, H.C. and Doly, J. (1979). *Nucleic Acids Research*, **7**, 1513.

4

Purification of RNA

MILES WILKINSON

1. Introduction

The most important consideration in the preparation of RNA is to rapidly and efficiently inhibit the endogenous ribonucleases which are present in virtually all living cells. There are several classes of ribonucleases which have been well characterized in eukaryotic and prokaryotic organisms, including both endonucleases and exonucleases. Methods designed for the preparation of RNA depend on speed and/or efficiency of neutralization of ribonucleases for their success. Most methods stipulate the use of ribonuclease inhibitors such as RNasin, vanadyl–ribonucleoside complexes, guanidinium hydrochloride, guanidinium isothiocyanate, heparin, and dextran sulphate, to name a few. Proteinase K is sometimes used to degrade ribonucleases, and organic solvents such as phenol and chloroform are used to remove ribonucleases by extraction procedures. Another important concern in the preparation of RNA is to avoid accidental introduction of trace amounts of ribonucleases from hands, glassware, and solutions. Lastly, it is desirable that the methods used for the preparation of RNA are relatively rapid and simple. Literally hundreds of procedures for RNA preparation exist, many of which are time-consuming and labour-intensive. The methods described here are relatively simple to perform and provide intact RNA suitable for cDNA cloning, RNA (Northern) blotting, or *in vitro* translation. Techniques for the preparation of total cellular RNA from animal and plant cells, lower eukaryotes, and prokaryotic cells are provided. Several alternative methods are given which allow for the purification of cytoplasmic RNA, nuclear RNA, and poly(A)$^+$ RNA from eukaryotic cells. A rapid mini-prep method for the isolation of cytoplasmic RNA from small numbers of eukaryotic cells is also provided. Techniques to prepare polysomal RNA and immunoprecipitate specific polysomal mRNAs are described elsewhere (1–3).

2. Ribonuclease-free conditions

It is worth devoting time towards the task of making a 'ribonuclease-free environment' in part of your laboratory. This includes setting aside an area of the laboratory for reagents and utensils to be used specifically for RNA work. It is desirable, for example, to use a particular colour tape to label ribonuclease-free solutions. Autoclaving effectively destroys many ribonucleases which may be present in laboratory solutions, but autoclaving does not inactivate some classes of ribonucleases (e.g. RNase A is commonly *prepared* by boiling; see *Appendix 4*). Most reagents used for RNA work can be made ribonuclease-free by treating them with diethyl pyrocarbonate (DEP), a strong, but not absolute inactivator of most ribonucleases (4). Because DEP can modify RNA molecules (by carboxymethylation) and efficiently inactivate most enzymes, it is critical to deplete all traces of DEP before using a treated solution. DEP is removed by high temperatures which converts it to ethanol and CO_2. The following is a list of precautions for preparing ribonuclease-free laboratory materials:

(a) Use gloves for all work involving RNA.

(b) Bake glass bottles and all other glassware for at least 4 h at 250°C. Incubate the bottle tops in 0.1% (v/v) DEP-water for at least 4 h at 37°C, replace the tops on the bottles and autoclave for 15 min to inactivate the DEP. **Warning:** open the DEP container in a fume hood.

(c) Most plasticware, including conical tubes and pipettes, are relatively ribonuclease-free and hence can be used without any special preparation.

(d) Autoclave pipette tips in their original containers before use. Microfuge tubes should be transferred to ribonuclease-free containers (without directly handling the tubes) and autoclaved. The tubes are removed for use with forceps and carefully handled so that the top of the tube does not touch the bench or gloved hands.

(e) Ribonuclease-free water is prepared by adding DEP to a final concentration of 0.1% to double distilled water (or equivalent). Autoclave for 1 h, open the autoclave door briefly to release steam, and leave overnight in the autoclave (turned off, but hot) to decompose residual DEP. Alternatively, boil the water for at least 30 min on a heating pad. After the water has cooled down, smell it to be sure there are no traces of the sweet-smelling odour characteristic of DEP. If some DEP remains, autoclave further or boil the water for 15 min or more.

(f) Ribonuclease-free solutions are prepared, whenever possible, by treating with DEP as described above.

(g) Solutions containing the buffer Tris cannot effectively be DEP-treated since Tris inactivates DEP. Sucrose cannot be DEP-treated since autoclaving sucrose causes it to caramelize. Since ammonium acetate is relatively volatile, it also cannot be autoclaved. For reagents in this category, it is recommended to prepare them using ribonuclease-free water. Remove the reagents from their original containers by tapping gently into a plastic weighing boat. Avoid using spatulas, magnetic stirrers, or beakers unless they have been baked at 250°C for at least 4 h to destroy ribonucleases. Autoclave the solutions for at least 30 min. Autoclaving alone will effectively inactivate many classes of ribonucleases which may contaminate reagents in trace amounts.

3. Quantitation of RNA

RNA is accurately quantitated by measuring its absorbance in a spectrophotometer. The optical density (OD) of RNA is measured at its maximum absorbance wavelength of 260 nm. One OD unit is equivalent to 40 μg/ml of RNA. Typically, a small aliquot of a RNA sample is used for quantitation and then discarded. Use a dilution of the RNA sample which will give an OD value of between 0.1–0.5 (2–10 μg in a 0.5-ml cuvette). In cases where only small amounts of RNA are being assessed, it may be more appropriate to save the RNA after OD determination. In this case, set aside a cuvette which will only be used for nucleic acid (preferably only RNA) quantitation. Before use, fill the cuvette with freshly prepared 0.1% DEP-water and incubate at least 10 min at 37°C, followed by three rinses with ribonuclease-free water.

Contamination of RNA preparations with substantial amounts of protein is demonstrated by measuring the OD at a wavelength of 280 nm, since proteins typically have a maximum absorbance at this wavelength. A 260/280 absorbance ratio of 1.8–2.0 is appropriate for pure RNA. Contaminating protein will result in lower values of the 260/280 ratio (values below 1.6 indicate serious protein contamination). Significant quantities of DNA are rarely present in RNA preparations made by the methods described in this chapter. However, if DNA contamination is suspected (for example, if the sample is extremely viscous and 'stringy') this can be assessed by electrophoresing the sample on a denaturing agarose gel and staining with ethidium bromide as described in Chapter 5, Section 4.4.6. DNA cannot be distinguished from RNA by spectrophotometry since both exhibit a maximum absorbance at a wavelength of 260 nm.

4. Precipitation and storage of RNA

Like DNA, RNA is precipitated in the presence of alcohol and salt. A standard method to precipitate RNA is to add a one-tenth volume of 3 M sodium acetate (pH 5.2), 2.0–2.5 vol. of ethanol, followed by vigorous mixing, and precipitation at -20°C. Studies with DNA have indicated that precipitation of nucleic acids occurs most efficiently between -20°C and room temperature (5), and not at -70°C as is commonly believed. The length of time for precipitation and for centrifugation depends on the amount of RNA to be precipitated. For RNA samples at a concentration in ethanol of greater than 10 μg/ml, precipitate for at least 20 min at -20°C, and centrifuge for 10 min at room temperature. For RNA samples at a concentration of less than 10 μg/ml, centrifuge for 30 min. Trace amounts of RNA (<100 ng/ml) should be precipitated overnight to improve recovery.

Generally, RNA is centrifuged in 1.5 ml tubes in a microfuge at maximum speed. If samples larger than 1.5 ml are to be centrifuged, several options are available:

- split the sample into multiple 1.5 ml microfuge tubes;
- for RNA samples at a concentration of greater than 10 μg/ml, pellet in 15 or 50 ml polypropylene conical tubes at 2500 g;
- for RNA samples at low concentration (<10 μg/ml) it may be necessary to centrifuge at 10000 g in sturdy tubes (e.g. 15 ml Falcon 2059 tubes or 40 ml 'oakridge' tubes).

Free nucleotides (e.g. radioactively labelled NTPs) are efficiently ($>90\%$) removed by precipitation in the presence of 2.5 M ammonium acetate (0.5 vol. of 7.5 M ammonium acetate) and 2.0–2.5 vol. of ethanol. Precipitate at room temperature for 15 min, and centrifuge for 10–30 min, depending on the amount of RNA being precipitated (see above).

After centrifuging the sample, examine the white RNA pellet: it should be visible if more than 3 μg of RNA is present. Carefully remove the supernatant without disturbing the pellet, leaving 5–20 μl of the supernatant behind. Add 1 ml 80% (v/v) ethanol (for samples in microfuge tubes), centrifuge for 1 min, and carefully remove the supernatant. The 80% ethanol wash removes salt and trace amounts of organic solvents which may have been present in your RNA sample. Dry the RNA sample by air-drying for at least 15 min. Drying in a speed-vac is more rapid, but be very careful that the dry RNA pellet does not become airborne and vacates the tube.

RNA stored at $-70°$C in the presence of ethanol and salt is stable for several years. Storage at $-20°$C is appropriate for short-term storage, but is not recommended for periods of greater than 1 month. Because RNA in this form is a precipitate, it is important to vigorously mix the contents of the stock tube before taking an aliquot to centrifuge. For convenience, RNA can also be frozen at $-70°$C in water alone. If the RNA sample is absolutely free of ribonucleases, it can be stored for years in this form without detectable degradation.

5. Total cellular RNA

Three methods for the preparation of total cellular RNA are described below. The first method, the 'guanidinium–CsCl method' depends on the potent chaotropic agent guanidinium isothiocyanate to both lyse the cells and rapidly inactivate ribonucleases (6, 7). Since RNA has a higher buoyant density than DNA and most proteins, it is purified from the lysate by ultracentrifugation through a dense caesium chloride cushion. The advantage of this method is that samples can be quickly lysed and then stored for several days before ultracentrifugation. For example, this method is useful for kinetic studies where samples must be taken at several different time points. Guanidinium isothiocyanate is such a powerful denaturant of ribonucleases that this method can be effectively used for preparation of RNA from ribonuclease-rich tissue such as pancreas (6). The disadvantage of this method is that it requires an overnight ultracentrifugation step. However, this allows for the purification of genomic DNA, in addition to RNA. The second technique, the 'guanidinium–LiCl method' also depends on guanidinium isothiocyanate as a denaturing agent, but does not require ultracentrifugation over caesium choride (8). Both of these methods can be used for preparation of RNA from most mammalian cells, and they can be used for lower eukaryotic cells and prokaryotic cells by substituting an appropriate wash buffer in the first step. The third protocol, the 'hot phenol method', is specifically for prokaryotic cells. This method depends on the organic solvent phenol to extract and purify prokaryotic RNA (9). The hot phenol method provides high yields of RNA within hours, without an ultracentrifugation step. However, this method requires speed on the part of the worker to avoid RNA degradation.

Protocol 1. Guanidinium–CsCl method for preparation of total cellular RNA

Reagents

- Tris-saline (25 mM Tris, 0.13 M NaCl, 5 mM KCl). For a 1 litre solution, add 3 g Tris-base, 8 g NaCl and 0.36 g KCl. Adjust the pH to 7.2–7.4.

- Guanidinium lysis buffer (4 M guanidinium isothiocyanate, 30 mM sodium acetate, 1 M β-mercaptoethanol). Dissolve 47 g guanidinium isothiocyanate (high purity, enzyme grade) in about 50 ml water by heating to about 60°C. Add 1 ml 3 M NaAc (pH 5.2). Add water to 93 ml and filter through a 0.45–0.80 μm pore size membrane or Whatman filter No. 1. Add 7 ml β-mercaptoethanol after filtration.

- CsCl–EDTA (5.7 M CsCl, 5 mM EDTA). Dissolve 96 g of CsCl in 80 ml water, add 1 ml 0.5 M EDTA (pH 7.0) stock, bring the volume exactly to 100 ml with water, and filter through a 0.20–0.45 μm pore size filter.

1. Prepare the cells or tissues as follows:

    (a) <u>Adherent cells</u>: wash twice with ice-cold Tris-saline. Remove the cells either with 1 mM EDTA in Tris-saline, trypsin/EDTA in a physiological buffer, or with a rubber policeman. Pellet the cells by centrifugation, decant off the supernatant, blot most of the residual liquid on a paper towel, vigorously tap the bottom of the tube with a finger to resuspend the pellet, and put on ice. Alternatively, if a small number of dishes are harvested, it is feasible to lyse the cells (step 2) while still attached to the dish.

    (b) <u>Suspension cells</u>: wash once with ice-cold Tris-saline by centrifugation, decant off the supernatant, blot most of the residual liquid on a paper towel, vigorously tap the bottom of the tube with a finger to resuspend the pellet, and put on ice.

Protocol 1. *Continued*

(c) <u>Tissues</u>: One method to pulverize tissues is to grind the tissue with a cold mortar and pestle in the presence of lysis buffer and either dry-ice pellets or liquid nitrogen (step 2). A better approach is to homogenize the tissue in the presence of lysis buffer with a high speed tissue solubilizer such as an omnimixer or polytron.

2. Add guanidinium lysis buffer to the single cell suspension,[a] and quickly vortex or tap the tube vigorously to completely lyse the cells. If it is desired to also prepare high-molecular-weight genomic DNA from the cells, it is best to avoid excessive mechanical stress (e.g. vortexing). The volume of guanidinium lysis buffer depends both on the number of cells and the size of the ultracentrifuge tubes to be used. In general, add at least 10 vol. of lysis buffer to a tissue fragment or cell pellet, or about 1 ml of lysis buffer per 3×10^7 dispersed mammalian cells. The volume of the cell lysate should be two-thirds the volume of the ultracentrifuge tube used in step 4.

3. Proceed to the ultracentrifugation step immediately, or store the cell lysate at $-70°C$ or $-20°C$. Lysates can usually be stored for several weeks in this form without detectable deterioration of the RNA.

4. Rinse the ultracentrifuge tubes with DEP-treated water to remove possible debris from the tube. Add CsCl–EDTA to the ultracentrifuge tube so that it takes up one-third the volume of the tube.

5. Layer the cell lysate on top of the CsCl. Add more lysis buffer, if required, to bring the volume to within 2 mm from the top of the tube, and to balance the buckets.

6. Spin in the ultracentrifuge for at least 12 h at $18°C$ (lower temperatures may precipitate the CsCl). For most ultracentrifuge rotors (e.g. Beckman SW50.1 and SW60), 35 000–40 000 r.p.m. is sufficient to pellet the RNA. The TLS-55 rotor used with table-top Beckman ultracentrifuges must be spun a 55 000 r.p.m. for at least 3 h to pellet the RNA (10).

7. After the centrifugation, examine the tube. An RNA pellet may not be visible unless there is more than 100 μg of RNA. A DNA band is typically observed in the lower third of the tube if more than 10^7 mammalian cells were used to prepare the lysate. The DNA can be purified by standard procedures (11). If only RNA is desired, aspirate the supernatant, being careful to remove all of the DNA. Leave the tube inverted so that the proteins and ribonucleases present in the tube will drain away from the RNA pellet.

8. Cut off the bottom 1 cm of the ultracentrifuge tube with a razor blade and quickly resuspend the RNA pellet (visible or not) in 100 μl water or RNA resuspension buffer [0.4% (w/v) SDS, 5 mM EDTA]. Transfer the RNA to a microfuge tube containing 0.4 ml chloroform:butan-1-ol (4:1).[b]

9. Add another 100 μl of water or buffer to the ultracentrifuge tube, resuspend the residual pellet and transfer to the same tube containing chloroform:bu-

Protocol 1. *Continued*

tanol. Repeat this operation a total of four times so that the RNA pellet is resuspended in a total volume of 400 μl water or RNA resuspension buffer. Vortex briefly, centrifuge for 2 min. Take the aqueous phase (upper) and transfer to a second microfuge tube containing chloroform:butanol. Vortex, and repeat the centrifugation.

10. To determine the yield of RNA, take 10 μl of the upper phase and quantitate the RNA using a spectrophotometer (see Section 3).

11. Ethanol precipitate the RNA (remaining upper phase) with 0.1 vol. of 3 M NaAc (pH 5.2) and 2.0–2.5 vol. of ethanol (see Section 4).

[a] In some cases, particularly for tissue lysates, it may be desirable to add 0.4 g solid CsCl per millilitre of lysis buffer. This normally restricts the migration of cellular debris during ultracentrifugation to the upper half of the tube.

[b] For preparation of RNA from tissues, it is advisable to extract twice with phenol:chloroform:isoamyl alcohol (25:24:1) prior to chloroform:butanol extraction in step 8.

Protocol 2 describes a modification of the method originally described by Cathala *et al.* (8). An important advantage of this method is that the steps prior to overnight LiCl precipitation can be done relatively rapidly. Hence, it is a useful method under circumstances where many RNA samples must be harvested, particularly if they are to be harvested at different times. The critical step in this method is the precipitation of RNA by LiCl. If the number of cells in the cell lysate is too few, RNA may not be efficiently precipitated. On the other hand, if the density of cells is too great, contaminants such as DNA may be co-precipitated along with the RNA. A similar method is described by Chomczynski and Sacchi (12) where RNA is prepared in a period of hours by guanidinium isothiocyanate lysis and phenol–chloroform extraction.

Protocol 2. Guanidinium–LiCl method for preparation of total cellular RNA

The following instructions are intended for processing moderate to large numbers of cells ($> 3 \times 10^6$ mammalian cells) in either 15 ml Falcon 2059 or 40 ml 'oakridge' tubes. A mini-prep version (for 1 ml of cell lysate or less) can be performed in microfuge tubes; for the extraction steps, centrifuge at full speed for 2 min.

Reagents

- GTEM lysis buffer [5 M guanidinium isothiocyanate, 50 mM Tris (pH 7.5), 10 mM EDTA, 8% (v/v) β-mercaptoethanol]. Dissolve 59 g guanidinium isothiocyanate (high purity, enzyme grade) into about 50 ml of water by

Protocol 2. *Continued*

heating to approximately 60°C. Add 5 ml of 1 M Tris and 2 ml of 0.5 M EDTA, bring to a volume of 92 ml with water, filter, and store at 4°C. Add β-mercaptoethanol to a concentration of 8% (v/v) just before use.

- 6 M LiCl.
- PK buffer [50 mM Tris (pH 7.5), 5 mM EDTA, 0.5% (w/v) SDS].
- Proteinase K (20 mg/ml stock made in water and stored at −70°C).

Method

1. Refer to *Protocol 1*, steps 1–2, for instruction how to wash and lyse the cells (substitute the GTEM lysis buffer for the guanidinium lysis buffer). Use at least 7 ml of GTEM lysis buffer for every gram of tissue. For cultured eukaryotic cells, lyse about 3×10^6 cells per ml of lysis buffer (the actual value for a given cell type is empirical).

2. Shear the DNA with an 18 gauge needle or with a tissue solubilizer (e.g. polytron at medium speed).

3. For tissues, add an equal volume of chloroform:isoamyl alcohol (24:1), vortex for 15 seconds, centrifuge at 2500 *g* for 10 min, and transfer the upper aqueous phase to another tube.

4. Add 1.4 vol. of 6 M LiCl and precipitate for at least 15 h at 4°C.[a]

5. Centrifuge for 30 min at 10 000 *g*. Aspirate as much of the supernatant as possible.

6. Add a volume of PK buffer which is one-half of the original lysis volume.

7. Add proteinase K to a final concentration of 200 μg/ml (add 10 μl per ml), resuspend the pellet, and incubate for 30 min at 45°C.[b]

8. Add NaCl to a final concentration of 0.3 M. Extract with an equal volume of phenol, followed by one or more extractions with phenol:chloroform:isoamyl alcohol (25:24:1) until no protein is evident at the interface. Extract once with chloroform:isoamyl alcohol (24:1). For each extraction, centrifuge for 10 min at 2500 *g* in conical tubes.

9. Ethanol precipitate with 2.0–2.5 vol. of ethanol (Section 4).

10. Pellet the RNA by centrifugation, resuspend the RNA in 400 μl water, take at least 5 μl for quantitation (see Section 3), transfer to a microfuge tube and ethanol precipitate the RNA with 0.1 vol. of 3 M sodium acetate (pH 5.2) and 2.0–2.5 vol. of ethanol (see Section 4).

[a] Samples can normally be stored in LiCl for several days without detectable RNA degradation.
[b] If there is difficulty resuspending the pellet, alternate proteinase K incubations with pipetting or vortexing.

Protocol 3 provides a procedure developed for obtaining RNA from prokaryotic cells (9). In this method, the bacteria are lysed in a hot SDS solution, followed by extraction with phenol. The bacterial DNA remains associated with the cell debris at the interface of the tube during extraction. This method is not appropriate for preparation of RNA from most eukaryotic cells since the DNA is less likely to be trapped at the interface of the tube during extraction. However, to prepare RNA from non-bacterial cells, the DNA can be sheared with an 18-gauge needle during step 3.

Protocol 3. Hot phenol method for preparation of total cellular RNA

Reagents

- Phenol equilibrated with 0.1 M sodium acetate (pH 5.2).
- RNA lysis solution [0.15 M sucrose, 10 mM sodium acetate (pH 5.2), 1% (w/v) SDS].

Method

1. Add 4 ml equilibrated phenol to 40 ml 'oakridge tubes'. Warm these tubes to 65°C in a water bath. Also, warm the RNA lysis solution to 65°C.

2. Add the RNA lysis solution to a dispersed bacterial cell pellet. Use at least ten volumes of RNA lysis solution to process one volume of cells. Quickly transfer the cell lysate to a tube containing hot phenol.

3. Gently invert the tube several times and incubate the tube for 10 min at 65°C. Invert the tube twice during the 10-min incubation period.

4. Centrifuge the tubes at 10 000 g for 5 min at 4°C.

5. Transfer the upper phase to a new tube containing phenol, avoiding the white interface layer which contains protein and DNA. Gently invert the tube several times and incubate at 65°C and centrifuge as before.

6. Transfer the upper phase to a tube containing phenol:chloroform:isoamyl alcohol (25:24:1), vortex briefly, and centrifuge as before.

7. Transfer the upper phase to a tube containing chloroform:isoamyl alcohol (24:1), vortex, and centrifuge as before.

8. Ethanol precipitate the RNA: Transfer the upper phase to a tube containing 0.1 vol. of 3 M sodium acetate (pH 5.2; final concentration is 0.3 M), add 2.0–2.5 vol. of ethanol, and precipitate at −20°C (Section 4).

9. Pellet the RNA by centrifugation, resuspend the RNA in 400 μl water, take at least 5 μl for quantitation (Section 3), transfer to a microfuge tube and ethanol precipitate the RNA (Section 4).

Note: Although this method provides high yields of RNA, in some cases the RNA prepared may not be entirely free of contaminating proteins. In this case, two options may be pursued. After ethanol precipitation, the sample can be resuspended in 400 μl water and extracted one time with phenol:chloroform:isoamyl alcohol (25:24:1) and one time with chloroform:isoamyl alcohol (24:1),

Protocol 3. *Continued*

followed by another ethanol precipitation. Alternatively, the pellet can be resuspended in PK buffer [50 mM Tris (pH 7.5), 5 mM EDTA, 0.5% (w/v) SDS], and incubated with 200 μg/ml proteinase K for 30 min, followed by extraction, and ethanol precipitation (see Section 4).

---

6. Cytoplasmic and nuclear RNA

6.1 Cytoplasmic RNA

6.1.1 Cell lysis and extraction

The procedure described in *Protocol 4* can be used to prepare cytoplasmic RNA from either tissues or cell lines (7). A step which is critical for this method is the 'lysis step' where the cell membrane is disrupted, allowing the release of relatively intact nuclei. The method described here utilizes NP-40 under isotonic conditions (physiological salt concentrations) to achieve cell membrane rupture. This method works for many cell types. However, the conditions may require alteration to suit some cell lines and tissues. Cells which are not efficiently lysed will not release all of their cytoplasmic RNA, and thus a low RNA yield will result. On the other hand, sensitive cells have fragile nuclei which are disrupted by high concentrations of detergent, liberating DNA. The nuclear stain toluidine blue-O (Sigma Corp.) can be used to assess the state of the nuclei. Add 0.1 vol. of a toluidine blue-O stock solution (1% dissolved in DMSO) to the nuclei and examine under a light microscope. The ratio of NDD buffer to Tris-saline can be altered to achieve appropriate lysis and release of nuclei. Alternatively, the concentration of NP-40 and DOC can be independently altered. In addition, some cell types (e.g. fibroblasts) may require Dounce homogenization and/or a 5–10 min incubation prior to centrifugation.

The number of cells processed by this procedure is critical. Because of the large amount of protein present in most cells, only moderate numbers of cells can be processed by phenol–chloroform extraction. Generally, the maximum number of cells which can be processed per tube is 5×10^7 tissue culture cells. If the cells possess a high nuclear to cytoplasmic volume ratio it may be possible to process 2×10^8 cells (e.g. quiescent lymphocytes). On the other hand, typically only 2×10^7 adherent fibroblasts can be processed per tube. One can *not* scale up in a single tube; the procedure usually does not work well with large numbers of cells in large tubes.

Protocol 4. Isolation of cytoplasmic RNA

Reagents

- Tris-saline (see *Protocol 1*).
- NDD lysis buffer: to 90 ml of Tris-saline, add 10 ml 10% (v/v) Nonidet P-40 (NP-40), 0.5 g sodium deoxycholate (DOC) and 10 mg dextran sulphate.

Protocol 4. *Continued*

Method

1. Prepare the cells or tissues as follows:

   (a) <u>Adherent cells</u>: wash twice with Tris-saline, remove the cells with either 1 mM EDTA in Tris-saline, trypsin/EDTA in a physiological buffer, or with a rubber policeman. Wash the cells twice with Tris-saline by centrifugation, decant off the supernatant, and put the cells on ice.

   (b) <u>Suspension cells</u>: wash twice with ice-cold Tris-saline by centrifugation, pour off the residual supernatant, and put the cells on ice.

   (c) <u>Tissues</u>: prepare tissues as a single cell suspension. The method used depends on the tissue type. The use of proteolytic enzymes such as collagenase and/or Dounce homogenization is often required.

2. Resuspend up to 5×10^7 cells with 2.5 ml of ice-cold Tris-saline in a 15 ml polypropylene conical tube. Either use a pipette or resuspend the cells prior to adding the Tris-saline by sharply tapping the bottom of the tube with a finger.

3. Add 2.5 ml of NDD lysis buffer, invert the tube 10 times (or Dounce homogenize for some cell lines and tissues; see Section 6.1.1), and centrifuge for 3 min at 2500 g at 4°C. This step is the most important for achieving a good RNA preparation. It is imperative that this step be done as quickly as possible to decrease ribonuclease degradation. A centrifuge with an efficient braking system is desirable.

4. Carefully remove all but the last 0.1 ml of supernatant (cytosolic fraction) and quickly transfer to a 15-ml conical tube containing 5 ml of ice-cold phenol:chloroform:isoamyl alcohol (25:24:1). Add 0.25 ml of 20% (w/v) SDS (final concentration: 1%) and 150 μl of 5 M NaCl (0.3 M final concentration, including NaCl contributed by the Tris-saline). Vortex the tube for 30 sec and centrifuge for 10 min at 2500 g.

5. Discard the nuclear pellet, or prepare nuclear RNA (see Section 6.2). If the nuclear pellet is 'loose' with viscous DNA emanating from it, see Section 6.1.1 for measures to prevent nuclear lysis.

6. Once the cytosolic fraction has completed its centrifugation, examine the sample. The upper phase should be relatively clear and the interface should contain a white proteinaceous precipitate. If the upper phase is extremely turbid, it is likely that too many cells were processed. The turbidity problem is normally alleviated by increasing the volumes used for extraction.

7. Transfer the upper aqueous phase (leaving behind the protein at the interface) to a fresh 15 ml polypropylene conical tube containing 5 ml of phenol:chloroform:isoamyl alcohol (25:24:1), vortex briefly, and centrifuge as before. Repeat this step until no more protein is visible at the interface.

Protocol 4. *Continued*

8. Extract as above with chloroform:isoamyl alcohol (24:1).

9. Ethanol precipitate the RNA by transferring to a 15 ml polypropylene conical tube containing 10 ml of ethanol (2.0–2.5 vol.). Precipitate at −20°C for at least 20 min (see Section 4).

10. Pellet the RNA by centrifugation at 2500 *g* for 20 min. Resuspend the RNA in 400 μl of water and quantitate by spectrophotometry (see Section 3). If the RNA in solution is not clear, it may contain trace amounts of lipids and/or protein. In this case, extract with phenol:chloroform:isoamyl alcohol (25:24:1) in a microfuge tube (2 min centrifugation) until the interface is clean, then extract once with chloroform:isoamyl alcohol (24:1). Ethanol precipitate the RNA by the addition of 0.1 vol. of 3 M sodium acetate (pH 5.2) and 2.0–2.5 vol. of ethanol (see Section 4).

6.1.2 Ribonuclease degradation

Protocol 4 typically provides intact RNA suitable for Northern blots (13). The dextran sulphate present in the lysis buffer and the ionic detergent SDS added before extraction are ribonuclease inhibitors, albeit incomplete inhibitors. The cell lysis step is performed as quickly as possible at cold temperatures to decrease the possibility of RNA degradation. The extraction step depends on phenol and chloroform to act as denaturing agents which will inactivate endogenous ribonucleases. Although these precautions have been made to inhibit ribonuclease attack, some cell types possess ribonuclease activity which is sufficiently high (e.g. pancreatic exocrine cells) that intact RNA cannot be prepared by this method. One alternative is to use the strong denaturing agent guanidinium isothiocyanate to prepare cytoplasmic RNA (see Section 6.1.4).

Another alternative is to use other ribonuclease inhibitors during the preparation of cytoplasmic RNA. For example, vanadyl–ribonucleoside complexes (obtained from Gibco-BRL) inhibit many of the known ribonucleases, but do not interfere with most enzymatic reactions (14). Vanadyl ribonucleoside complexes should be used at a concentration of 10 mM during the cell lysis step.

Some protocols dictate adding urea and EDTA during extraction with phenol–chloroform (15). Urea is a strong denaturing agent and EDTA chelates Mg^{++} ions required for the activity of some ribonucleases. The use of these agents is suggested for cells with a high content of ribonucleases. *Protocol 4* can be modified as follows: for step 4, instead of adding SDS and NaCl, add an equal volume of a 2× extraction buffer: 7 M urea, 450 mM NaCl, 10 mM EDTA, 10 mM Tris–HCl (pH 7.4) and 1% (w/v) SDS (prepare fresh before use).

6.1.3 Rapid mini-prep method

Protocol 5 provides a simple and efficient method of purification of cytoplasmic

RNA (16) from small numbers of eukaryotic cells (as few as 10^5–10^6 cells). Using this mini-prep method, RNA can be prepared from ten different cell samples in considerably less than 2 h. The mini-prep method will allow purification from as many as 2×10^6 to 2×10^7 cells (depending on the cell type). The RNA yield from the mini-prep method is about 10 μg per 10^6 cells, although this value can vary considerably depending on the cell type. The method is essentially a scaled-down version of *Protocol 4*. From most cell types, the quality of RNA isolated by this mini-prep method is suitable for Northern analysis or the polymerase chain reaction (see Volume II, Chapters 5 and 7). If problems with RNA degradation or cell lysis are encountered, see Sections 6.1.1 and 6.1.2 for possible solutions.

Protocol 5. Rapid mini-prep of cytoplasmic RNA

1. Pellet the cells, resuspend in 1 ml of ice-cold Tris-saline (see *Protocol 1*), transfer to a 1.5 ml microfuge tube and centrifuge at low speed (3000–6000 g) at 4°C for 30 sec. If only a high-speed microfuge is available, centrifuge at 10 000 g for 10 sec, but this may seriously disrupt the cells.

2. Discard the supernatant, resuspend the cells in 250 μl ice-cold Tris-saline. Add 250 μl ice-cold NDD lysis buffer (see *Protocol 4*), invert the tube ten times and centrifuge at 4°C for 30 sec. Preferably, centrifuge at low speed (3000–6000 g).

3. Carefully transfer the supernatant (being careful to avoid disturbing the nuclear pellet) to a microfuge tube containing 0.5 ml of phenol:chloroform:isoamyl alcohol (25:24:1). Quickly add 25 μl of 20% (w/v) SDS and 15 μl of 5 M NaCl to the tube, vortex or shake for 15 sec, and microfuge at maximum speed at 4°C for 2 min. Discard the nuclear pellet or prepare nuclear RNA as described in Section 6.4.

4. Transfer the upper aqueous phase to another microfuge tube containing phenol:chloroform:isoamyl alcohol, vortex, and centrifuge at room temperature. Repeat this extraction procedure until there is no visible protein at the interface (normally two or three extractions steps, total). If the extraction is difficult because of excess turbidity or viscous DNA contamination see Section 6.1.1.

5. Extract once with chloroform:isoamyl alcohol (24:1) to remove residual phenol.

6. Transfer the upper aqueous phase to a microfuge tube containing 1 ml of ethanol and precipitate at -20°C for at least 20 min (see Section 4).

6.1.4 Guanidinium method

In most circumstances, *Protocols 4* and *5* provide high yields of intact cytoplasmic RNA. However, for cell types with high levels of ribonuclease activity or transcripts which are highly unstable, these methods may not be appropriate. The following method provides high quality cytoplasmic RNA from moderate numbers of cells:

Protocol 6. Guanidinium method for cytoplasmic RNA

1. Lyse the cells as described in *Protocol 5*, steps 1 and 2.

2. Transfer the cytoplasmic supernatant (0.5 ml) to a tube containing 2.5 ml of guanidinium lysis buffer (see *Protocol 1*).

3. Ultracentrifuge the sample over CsCl and process the RNA as described in *Protocol 1*, steps 2–11.

6.2 Nuclear RNA

Nuclear RNA is useful for studies examining precursor transcripts and RNA splicing intermediates. *Protocol 7* is an effective method for obtaining nuclear RNA (7). To determine whether the RNA prepared is enriched for nuclear RNA, it is best to examine the RNA by gel electrophoresis. One approach is to stain a suitable gel (Chapter 5, Section 4.4.6) with acridine orange to visualize the rRNA precursor transcripts. *Figure 1* shows that eukaryotic nuclear RNA prepared by this method is enriched for 32S and 45S rRNA precursor transcripts which are absent in cytoplasmic RNA. Both cytoplasmic and nuclear RNA contain mature

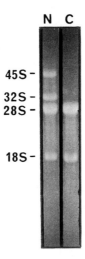

Figure 1. 10 μg of cytoplasmic (C) and nuclear (N) RNA from SL12.4 T lymphoma cells were prepared as described in *Protocols 4* and *7*, respectively, and electrophoresed in a 1% denaturing agarose gel containing formaldehyde (13). After electrophoresis, the gel was stained for 2 min in 10 μg/ml acridine orange, 10 mM sodium phosphate (1:1 ratio of monobasic and dibasic) and 1.1 M formaldehyde, followed by destaining for 20 min in the same buffer lacking acridine orange.

18S and 28S rRNA. These transcripts can also be observed directly on nylon-based transfers (see Volume II, Chapter 5) by staining with 0.03% (w/v) methylene blue and 0.3 M sodium acetate (pH 5.2) for 45 sec, followed by destaining with water for 1–2 min. The transfers must be baked or UV cross-linked prior to staining.

Protocol 7. Preparation of nuclear RNA

1. Release the nuclei from large numbers of cells (>1–5×10^7) as described in *Protocol 4*, steps 1–4. Smaller numbers of cells can be lysed as described in *Protocol 5*, steps 1–3.

2. Discard the excess supernatant from above the pellet with a P200 pipetter. Vigorously tap the bottom of the tube to resuspend the pellet. Perform this step rapidly so the pellet does not warm up to room temperature.

3. Add 1 ml of ice-cold guanidinium lysis buffer (see *Protocol 1*) per 3×10^7 dispersed mammalian cells or add a ten times volume of lysis buffer as compared with the cell pellet, and immediately agitate the tube so the pellet dissolves and forms a clear solution. If preparing DNA, do not vortex, instead vigorously tap the bottom of the tube with a finger to dissolve the pellet.

4. Add an appropriate amount of guanidinium lysis buffer to the nuclear lysate so that its final volume is equal to two-thirds of the volume of the ultracentrifuge tube to be used. Ultracentrifuge the sample through a CsCl cushion and process the RNA as described in *Protocol 1*, steps 2–11.

Note: An optional step prior to nuclei lysis is to wash the nuclei in ice-cold TNM buffer [0.1 M Tris-HCl (pH 7.5), 10 mM NaCl, 3 mM $MgCl_2$]. Add TNM buffer carefully so as not to resuspend the nuclear pellet and centrifuge at 1000–2000 g for 3 min at 4°C. This wash step removes most of the residual cytoplasmic RNA present. However, the extra time required for this step could potentially lead to some degradation of the nuclear RNA.

---

7. Poly(A)$^+$ RNA

Ribosomal RNA (rRNA) is, by far, the most abundant RNA species in eukaryotic cells, representing 80–90% of the total cellular RNA. The two most abundant rRNA transcripts, 18S and 28S, have sizes of approximately 2 kb and 5 kb, respectively. In contrast, messenger RNA (mRNA) represents only 1–5% of the total cellular RNA, and is heterogenous in size, ranging from less than 0.5 kb to over 10 kb. Most of the mRNA transcripts present in mammalian cells are polyadenylated at the 3'-end. The tail of adenylate residues typically extends 50–200 bases. This poly(A)-tail allows mRNA to be purified by affinity chromatography on oligo(dT)-cellulose. Described below are two methods which are based on oligo(dT)-cellulose chromatography. The first technique allows up to 10 mg of RNA to be processed per ml of oligo(dT)-cellulose (17). The principle of the method is that the RNA is bound in the presence of high concentrations of salt, washed in intermediate salt concentrations, and eluted in a low salt buffer. Yields are increased by recycling the RNA over the column twice. The second method is a rapid method of preparation of poly(A)$^+$ RNA directly from cell lysates (18). The advantage of this method is that poly(A)$^+$ RNA can be prepared from cells in about 4 h. A potential disadvantage to this approach is that it is not recommended for cells which possess a high content of ribonucleases.

Protocol 8. Enrichment for poly(A)$^+$ RNA from nuclear, cytoplasmic, or total RNA

Reagents

- 1 × loading buffer [50 mM sodium citrate (pH 7.5), 0.5 M LiCl, 1 mM EDTA, 0.1% (w/v) SDS].
- 2 × loading buffer [100 mM sodium citrate (pH 7.5), 1 M LiCl, 2 mM EDTA, 0.2% (w/v) SDS].
- Wash buffer [50 mM sodium citrate (pH 7.5), 0.1 M LiCl, 1 mM EDTA, 0.1% (w/v) SDS].
- Elution buffer [10 mM sodium citrate (pH 7.5), 1 mM EDTA, 0.05% (w/v) SDS].
- Regeneration solution (0.1 M NaOH, 5 mM EDTA).

Method

1. Prepare a ribonuclease-free chromatography column (e.g. Bio-Rad Econo-Pac polypropylene column) by rinsing with freshly prepared 0.1% (v/v) DEP-water, followed by autoclaving.
2. Suspend at least 25 mg of oligo(dT)-cellulose in regeneration solution for every milligram of RNA to be processed. Transfer the oligo(dT)-cellulose to the column. The packed bed volume is 0.1 ml per 25 mg of oligo(dT)-cellulose.
3. Wash the column with three column volumes of water. Wash with more water if the column effluent has a pH of greater than 8.0.
4. Equilibriate the oligo(dT)-cellulose with five column volumes of 1 × loading buffer.
5. Dissolve the RNA in water to a final concentration of 5 mg/ml, heat to 65°C for 5 min, and cool to room temperature. Warming the RNA helps to remove secondary structure which may interfere with poly(A)$^+$ RNA binding to the oligo(dT).
6. Add an equal volume of 2 × loading buffer to the RNA, load on the column, wash with one column volume of 1 × loading buffer, and collect the combined effluent.
7. Pass the RNA effluent over the column a second time to improve the yield: heat the effluent to 65°C for 5 min, cool to room temperature, and load on the column.

Protocol 8. *Continued*

8. Wash with ten column volumes of $1 \times$ loading buffer.

9. Wash with five column volumes of wash buffer.

10. *Optional:* measure the OD of fractions collected after the addition of wash buffer. If the final wash volumes are devoid of measurable RNA, proceed to the next step. If significant amounts of RNA are being eluted, wash the column with more wash buffer.

11. Add three column volumes of elution buffer to elute the poly(A)$^+$ RNA. The RNA which is eluted is typically 20-fold enriched for poly(A)$^+$ RNA and is suitable for many applications, but still possesses over 50% rRNA contamination. If additional purification is required, proceed to the next step. Otherwise, ethanol precipitate the RNA as described in step 14.

12. Equilibrate the column with five volumes of $1 \times$ loading buffer. Incubate the eluate at 65°C for 5 min, let cool to room temperature, and repeat steps 6–11.

13. Take an aliquot of the eluted RNA for quantification (Section 3). Typically, 1–3% of the input RNA is recovered after oligo(dT)-chromatography.

14. Ethanol precipitate the RNA by the addition of sodium acetate (pH 5.2) to a final concentration of 0.3 M, and 2.0–2.5 vol. of ethanol (see Section 4).

15. Regenerate the column by sequential washing with three column volumes of regeneration solution, water and $1 \times$ loading buffer.

16. Store the oligo(dT)-cellulose at 4°C in $1 \times$ loading buffer containing 0.05% (w/v) sodium azide.

Protocol 9. Poly(A)$^+$ RNA prepared directly from cell lysates

Reagents

- Poly(A)$^+$ lysis buffer [0.2 M NaCl, 0.2 M Tris–HCl (pH 7.5), 2% (w/v) SDS, 0.15 mM $MgCl_2$, 200 μg/ml proteinase K]. To prepare 10 ml of buffer, add the following to 6.5 ml of water: 0.4 ml 5 M NaCl, 2 ml 1 M Tris–HCl (pH 7.5), 1 ml 20% (w/v) SDS, 1.5 μl 1 M $MgCl_2$ and add 100 μl of 20 mg/ml proteinase K *immediately* before use.

- Binding buffer [0.5 M NaCl, 10 mM Tris (pH 7.5)].

Method

1. Prepare the cells or tissues as follows:

   (a) Cultured cells: wash as described in *Protocol 1*, step 1(a) or 1(b). Add 10 ml of lysis buffer per 10^8 cells and quickly homogenize the cell suspension in a high-speed tissue solubilizer (e.g. polytron) or by shearing the DNA through an 18-gauge needle until no longer viscous.

Protocol 9. *Continued*

    (b) <u>Tissue</u>: Quick freeze in liquid nitrogen. Grind up the tissue in a chilled mortar and pestle, transfer the slurry to a 50 ml conical tube, and add 10 ml of lysis buffer per gram of tissue. Vortex until no tissue fragments are visible. Alternatively, pulverize the tissue in the presence of the lysis buffer with a high-speed tissue solubilizer.

2. Incubate the cell lysate for at least 1 h at 45°C. Either provide constant agitation in a shaking water bath, or agitate by hand every 10 min.

3. Hydrate the oligo(dT)-cellulose with water in a 50 ml conical tube. Prepare 0.2 g of oligo(dT)-cellulose per 10^9 cells or 10 g of tissue. Pellet by brief centrifugation and equilibrate the cellulose in 20-fold excess of binding buffer. Pellet, remove all but 1 ml of the binding buffer.

4. Adjust the NaCl concentration of the cell lysate to that of the binding buffer by adding 60 μl of 5 M NaCl per millilitre of lysate. Add the oligo(dT)-cellulose to the lysate, mix well, and incubate at room temperature for 30 min with constant agitation.

5. Pellet the oligo(dT) at room temperature (lower temperatures may precipitate the SDS).

6. Aspirate or pour off the supernatant and resuspend the pellet in an equal volume of binding buffer. Repeat until the supernatant appears clear.

7. Transfer the washed oligo(dT)-cellulose to a column (e.g. Bio-Rad Econo-Pac polypropylene column) and continue washing with three volumes of binding buffer or until the UV absorbance (260 nm) of the output gives a reading of less than 0.05 (Section 3).[a]

8. Elute the poly(A)$^+$ RNA with three column volumes of water.

9. Precipitate the RNA by adding 0.1 vol. 3 M sodium acetate (pH 5.2) and 2.0–2.5 vol. of ethanol (see Section 4).

10. The yield of poly(A)$^+$ RNA from 10^8 cells ranges from 5–100 μg.

11. The oligo(dT)-cellulose can be regenerated as described in *Protocol 8*, step 15.

   [a] If the column is running slowly due to the presence of high molecular weight genomic DNA, resuspend the contents of the column with a pipette.

Acknowledgements

I wish to thank Carol MacLeod, Patricia Salinas, Livia Theodor, Randy McCoy, and Jim Garret who introduced me to some of the methods described herein. I am also grateful to Thomas Herrick and Elisa Burgess who contributed data and provided ideas concerning these methods. I am indebted to Scott Landfear for his helpful hints and valuable discussion.

References

1. Schutz, G., Kieval, S., Groner, B., Sippel, A.E., Kurtz, D.T., and Feigelson, P. (1977). *Nucleic Acids Research*, **4**, 71.
2. Gough, N.M. and Adams, J.M. (1978). *Biochemistry*, **17**, 5560.
3. Shapiro, S.Z. and Young, J.R. (1981). *Journal of Biological Chemistry*, **256**, 1495.
4. Kumar, A. and Lindberg, U. (1972). *Proceedings of the National Academy of Sciences of the USA*, **69**, 681.
5. Zeugin, J.A. and Hartley, J.L. (1985). *Focus*, **7** (1), 1.
6. Chirgwin, J.M., Przybyla, A.E., MacDonald, R.J., and Rutter, W.J. (1979). *Biochemistry*, **18**, 5294.
7. Wilkinson, M. (1988). *Nucleic Acids Research*, **16**, 10934.
8. Cathala, G., Savouret, J., Mendez, B., West, B.L., Karin, M., Martial, J.A., and Baxter, J.D. (1983). *DNA*, **2**, 329.
9. Von Gabain, A., Belasco, J.G., Schottel, J.L., Chang, A.C.Y., and Cohen, S.N. (1983). *Proceedings of the National Academy of Sciences of the USA*, **80**, 653.
10. Verma, M. (1988). *Biotechniques*, **6**, 848.
11. Iverson, P.L., Mata, J.E., and Hines, R.N. (1987). *Biotechniques*, **5**, 521.
12. Chomczynski, P. and Sacchi, N. (1987). *Analytical Biochemistry*, **162**, 156.
13. Wilkinson, M. and MacLeod, C.L. (1988). *EMBO Journal*, **7**, 101.
14. Sambrook, J., Fritsch, E.F., and Maniatis, T. (1989). *Molecular Cloning, A Laboratory Manual*, 2nd edn. Cold Spring Harbor Laboratory, Cold Spring Harbor, New York.
15. Pearse, M., Gallagher, P., Wilson, A., Wu, L., Fisicaro, N., Miller, J.F.A.P., Scollay, R., and Shortman, K. (1988). *Proceedings of the National Academy of Sciences of the USA*, **85**, 6082.
16. Wilkinson, M. (1988). *Nucleic Acids Research*, **16**, 10933.
17. Aviv, H. and Leder, P. (1972). *Proceedings of the National Academy of Sciences of the USA*, **69**, 1408.
18. Badley, J.E., Bishop, G.A., St. John, T., and Frelinger, J.A. (1988). *Biotechniques*, **6**, 114.

5

Electrophoresis of nucleic acids

ANTHONY T. ANDREWS

1. Introduction

The modern research worker in the life sciences area is often faced with the problem of having to separate large biomolecules such as nucleic acids, proteins and complex lipids or carbohydrates either for the qualitative or quantitative analysis of mixtures or for the preparation of individual components. In most cases it is also desirable to cause as little damage as possible to the molecules so that their properties are not changed significantly, and nowhere is this more important than in the handling of large DNA molecules which are particularly susceptible to cleavage by chemical or enzymic processes and even by physical stress. Present-day methods for the separation of large biomolecules usually rely therefore on physical processes which cause the minimum disturbance to the structure of the molecule and which result in the maximum retention of any biological activity.

A large group of separation methods are based on some specific biological or chemical property or interaction of the molecule under investigation (for example, affinity chromatography, immuno-adsorption, precipitation procedures, differential centrifugation) but these fall outside the scope of this chapter. Apart from these, most other separation methods are based on differences in molecular charge or molecular size, or sometimes a combination of the two. Electrophoretic methods can be designed to exploit either or both of these parameters, although size differences are much the most important in the case of nucleic acid separations.

Historically, the first form of electrophoresis was moving-boundary electrophoresis in which the components to be analysed are in free solution. Density gradients (such as sucrose gradients) were used to eliminate convection mixing and to reduce diffusion so that separated components remained as sharp, well-defined zones. The next developments were the use of paper and then cellulose acetate strips as anti-convective supporting media, but in none of these do size differences play a role, so separations are charge-based, which gives reasonable separations for proteins but is of little use for nucleic acids. Paper and cellulose acetate matrices are also somewhat heterogeneous which can interfere seriously with resolution capability. The first, much more homogeneous, gel matrix used was starch but although mechanical hindrance to the passage of molecules through the pores of the gel introduced a size-sieving effect in addition to charge difference making it useful for nucleic acid separations, starch is a natural product and somewhat variable and gel properties are difficult to control. Modern methods rely almost exclusively on agarose and on polyacrylamide gels, which as well as being very homogeneous and reproducible have the added benefit that it is possible to control and manipulate the gel properties at will to suit experimental requirements.

2. Basic factors influencing electrophoretic mobility

During electrophoresis the driving force on a particle is determined by the charge on the particle and the potential gradient (voltage). At constant velocity this is balanced by the frictional resistance of the medium and in free solution Stokes' Law is obeyed. However, in a gel medium Stokes' Law is not obeyed strictly and the frictional resistance is then influenced by a number of factors including not just solvent viscosity but also the gel density and pore size and the size and shape of the particle. For a large biological macromolecule it is the net charge on the molecule which is important, so for a molecule such as a protein which possesses many different ionizable groups with differing pK values, charge and hence electrophoretic mobility are strongly pH dependent, but this is less true for nucleic acids where most ionizable groups have very similar pK values. The ionic strength of the medium determines the electrokinetic potential which reduces the net charge on the molecule to the effective charge and in practice it is found that the mobility of the molecule is approximately inversely proportional to the square root of the ionic strength. Low ionic strength results in high electrophoretic mobility but sharper zones are obtained in high ionic strength buffer (1). The higher the ionic strength the greater the conductivity and since the amount of heat generated is proportional to the current flowing, unless the apparatus is constantly cooled, the temperature will rise. This in turn increases the rates of diffusion, causes an increase in ionic mobility (amounting to about 2.4% per degree Centigrade) and results in a fall in the viscosity of the medium.

Thus there are a great number of factors which influence electrophoretic mobility in gel media and it is the ability to manipulate and exploit these effects which gives electrophoretic methods unrivalled versatility for separating biological macromolecules of all types.

In the case of nucleic acids size is the principal factor which determines the composition chosen for the gel matrix, the pore size of the gel being of great importance. Small oligonucleotides, restriction fragments and small species of DNA and RNA are usually examined on polyacrylamide gels while agarose gels are universally employed for large nucleic acids. Composite gels of agarose and polyacrylamide are intermediate in pore size and were introduced for the analysis of relatively small nucleic acids and large proteins, but they are more complicated to prepare and have few real advantages in most cases so are only used infrequently.

3. Theory of nucleic acid separations

The theory of the behaviour of RNA during gel electrophoresis in terms of the influence of factors such as $\%T$, $\%C$ (both defined in Section 4.1), sample load, sample volume, buffer concentration, joule heating and gel properties has been discussed by Richards and Lecanidou (2) and would apply equally well to DNA separations.

Both diffusion and electrophoretic motion involve movement of sample molecules through the gel so the frictional resistance offered by the matrix would be the same. This means that the diffusion coefficient D_0 of sample molecules in free solution and the electrophoretic mobility u_0 in buffer in the absence of gel (i.e. free solution) are diminished in the presence of gel in the same proportion, so that

$$\frac{D}{D_0} = \frac{u}{u_0} = \alpha \qquad \alpha \equiv \text{retardation coefficient}$$

where D and u are the diffusion coefficient and electrophoretic mobility respectively in gel. The retardation coefficient α depends upon the properties of the gel (especially $\%T$ and $\%C$, see Section 4.1) and upon the size and shape of the sample molecule.

The distance d moved by a sample zone during gel electrophoresis is proportional to time (t) and the voltage gradient g (in V cm^{-1}) so that

$$d = ugt = \frac{uit}{k}$$

where i is the electrical current density (A cm^{-2}) and k is the specific conductivity (Ω^{-1} cm^{-1}). This enables u to be calculated so if u_0 for the particular sample species is known, α can then be calculated. If D_0 is known or measured then D can also be calculated. It is found that for RNA and denatured DNA u_0 is independent of molecular weight and is related to the viscosity of the medium while D_0 varies inversely with the square root of the molecular weight if the sample molecules are assumed to behave as random coils. Richards and Lecanidou (2) reported that $u_0 = 31 \times 10^{-5}$ cm^2 V^{-1} s^{-1} in 0.05 M buffer at 25°C and found that for RNA molecules ranging in molecular weight from 10^4 to 10^6 kd D_0 varied from 1.3×10^{-6} to 1.4×10^{-7} cm^2 sec^{-1}.

During electrophoretic migration the shape of a sample zone is given by the variation in concentration c with the displacement x along the axis of motion. With a Gaussian distribution of sample molecules, the shape is characterized by c_m, the maximum concentration and the zone width $(2w)$ at half-height (i.e., $c = c_m/2$). The load volume v applied per unit cross-sectional area of the gel is equal to the height of the column of sample solution containing the load (m g of

RNA of charge equivalent weight M^0), so that the integrated area of the zone is $m/\pi r^2 M^0$.

If the potential gradient in the sample solution and gel was the same then at zero time the starting zone width $2w_0$ is given by

$$2w_0 = \alpha v,$$

so if α and v are known or calculated $2w_0$ can be found. In practice, the starting zone is often smaller than this because the potential gradient in the sample layer is often greater than in the gel. Indeed, this is the basis for diluting or dialysing sample solutions to give an ionic strength lower than that of the gel buffer, which was introduced by Hjertén *et al.* (3) as a useful practical means of producing narrow starting zones and improving the resolution in subsequent separations.

If the starting zone is rectangular and diffusion was the only zone-widening effect as it migrates through the gel then the width of the starting zone would have a negligible effect on that of the final zone after separation providing that $2w_0 < (2DT)^{\frac{1}{2}}$ or $v < (2D_0t/\alpha)^{\frac{1}{2}}$. When $\alpha = 0.1$, during typical runs load volumes up to 2 mm cause no loss of resolution and even those of 10 mm result in only a 25% increase in zone width so the volume of sample applied to the gel is seldom a major factor in determining resolution. In practice, diffusion is not the only zone-broadening factor however and zone width increases with the mass of RNA within the zone. Ignoring diffusional effects, this is described by

$$2w = \left(\frac{2000 \; gu \; Amt}{\mu \pi r^2 \; M^0}\right)^{\frac{1}{2}}$$

where A depends on the mobilities of the RNA and buffer ions. This concentration zone-broadening is comparable to that caused by diffusion when

$$m = \frac{16 D_0 \mu \pi r^2 \; M^0 \; \ln 2}{2000 \; gu_0 A}$$

For loads greater than this the concentration effect is the most important factor determining zone width while for small ones diffusion is more important. With small gel rods of about 5 mm diameter and in 0.05 M buffers this change-over point corresponded to only about 0.5 μg of RNA (2), but it would be higher with buffers of higher ionic strength. A further consequence of these effects is that zones become asymmetric during the separation with the front edge becoming sharper and moving faster than the trailing edge which becomes more diffuse. If the loading in a particular zone is too high it may even overtake and obscure the zone of a more minor component moving just ahead of it.

Richards and Lecanidou (2) also arrived at two further conclusions, namely that there should be a linear relationship between sedimentation coefficient and electrophoretic mobility, and second, that mobility should vary linearly with $\sqrt{T}$, with resolution being better the higher the value of T (discussed in Section 4.1). In general these conclusions have been verified experimentally by many workers and there is an empirical linear relationship between mobility and log(molecular weight). While this can give a practical method of measuring the molecular weights of RNA and denatured DNA it assumes a flexible random molecular structure which is not necessarily true and it ignores any conformational effects on mobility. In fact, nucleic acid conformers are often separable by electrophoresis, and it is precisely to avoid such conformational effects that denaturing agents such as formamide or 8 M urea are added and gels are run warm (e.g. 50–80°C) for nucleic acid sequencing so that all sample components are fully denatured and in the same conformational state (4).

While a simplistic treatment such as that given above holds generally true for small nucleic acids run most usually in polyacrylamide gels, there are many situations even within this restricted range where the relationships are not strictly adhered to. In recent years there has been much work on the migration of nucleic acids through both polyacrylamide and agarose gels and it is now obvious that many factors are involved and the passage of nucleic acid molecules through gel matrices is very complicated, with several aspects of the observed electrophoretic behaviour still remaining to be resolved. Many models employ a reptation mechanism (primary, secondary, biased, etc.) and seek to explain anomalous forms of migration behaviour, nucleic acid compositional effects, nucleic acid–gel interactions, electric field effects, and so on. An extensive treatment of theoretical aspects of nucleic acid separations is outside the scope of this chapter and the reader is referred to recent papers and reviews (5–15).

4. Gel composition, apparatus, and electrophoresis

It has already been mentioned that small- and medium-sized nucleic acids are best separated using polyacrylamide gel electrophoresis while larger molecules are separated on larger-pore composite gels of polyacrylamide mixed with agarose, or more usually on gels of agarose alone which have the largest pore size and are therefore most suitable for very large molecules.

4.1 Polyacrylamide gels

Polyacrylamide gels are formed by the vinyl polymerization of acrylamide monomers ($CH_2\!=\!CH\!-\!CO\!-\!NH_2$) into long random chains of polyacrylamide which are cross-linked by the inclusion into the mixture of small amounts of an appropriate bifunctional co-monomer, usually N,N'-methylene-bis-acrylamide ($CH_2\!=\!CH\!-\!CO\!-\!NH\!-\!CH_2\!-\!NH\!-\!CO\!-\!CH\!=\!CH_2$) commonly known as 'Bis'. The resulting cross-linked chains form a gel structure (*Figure 1*), the pore size of which is determined by the initial concentrations of both acrylamide and the cross-linker. The nomenclature introduced by Hjertén (16) is now widely used to describe gel composition, the term T being the total monomer concentration (acrylamide + Bis) in grams per 100 ml (that is, weight per volume per cent) and C being the percentage (by weight) of total monomer T which is contributed by the cross-linker (Bis).

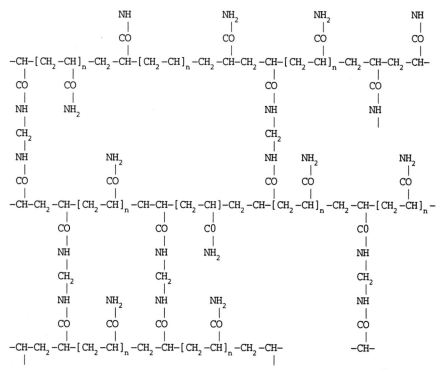

Figure 1. Diagrammatic structure of a polyacrylamide gel matrix formed by the co-polymerization of acrylamide monomer and a small proportion of the cross-linking agent, Bis.

It will be apparent from this that the pore size of the gel can be altered in an easy and controllable fashion simply by changing the concentrations of the two monomers. If both are altered together, keeping C constant, pore size decreases as T is increased. If C is varied then for any given value of T the pore size is a minimum when C is about 5% (17), although at high values of T (above about 15%) this no longer holds and the value of C required for minimum pore size is influenced by T (18, 19). For a typical $T=5\%$, $C=5\%$ gel the pore size is about 20 nm, but as the proportion of cross-linker is increased the polymerization reaction kinetics slow down and the gel formed becomes progressively less homogeneous. Clumps of polymerized fibres form with much less dense areas in between and when C is increased to a level of 30–50% the effective pore size may be as high as 500–600 nm (20). The temperature at which polymerization occurs also affects pore size, with both low (0–4°C) and high (>50°C) temperatures leading to unsatisfactory gel structures (21). The optimum polymerization temperature is 25–30°C. Experimentally it is found that both the absolute and relative mobilities of molecules are influenced by many factors which influence the length of the polyacrylamide chains. These include not only the acrylamide and Bis levels and the temperature but also the concentration of catalysts used to initiate the polymerization reaction and the time taken for gelation to occur (22). The polymerization proceeds by a free-radical mechanism and the most common way of initiating this is to use ammonium persulphate which produces oxygen free-radicals via a base-catalysed mechanism, the bases used generally being tertiary aliphatic amines such as N,N,N',N'-tetramethylethylenediamine (TEMED) or 3-dimethylaminopropionitrile (DMAPN). Alternatively, free radicals can be generated photochemically using small amounts of riboflavin in place of the persulphate, but still with TEMED. For both chemical and photochemical reactions the bases must be in the free-base form so at acid pH values either higher levels of catalysts must be used to avoid unacceptably long gelation times or different catalyst systems must be used (23). With both catalyst systems, oxygen at above trace levels acts as an inhibitor and consequently many workers advocate the de-aeration of solutions. While this does have the advantage that catalyst levels can be substantially reduced, unless it is done carefully, it may be difficult to get the catalyst levels correct for a suitable reproducible gelation time. Whatever the conditions chosen polyacrylamide solutions should always gel between about 10 and 30 min after catalyst addition. Mixtures which gel in less than 10 min or which have still not gelled after 60 min should be discarded because uneven polymerization will occur resulting in non-homogeneous gels and poor separations. Gelation times are most easily adjusted by altering the amounts of catalysts, and levels in the range 1–10 mM are usually appropriate, with equimolar amounts of ammonium persulphate and TEMED generally being best.

4.1.1 Apparatus for polyacrylamide gels

Polyacrylamide gels are nowadays nearly always run in the vertical slab gel format and the earlier arrangement of preparing gels in glass tubes, sometimes mistakenly referred to as disc electrophoresis (mistaken because the term 'disc' is in fact an abbreviation for discontinuous and applies to the type of buffer system employed; it is unrelated to the shape of the gel or of separated sample zones) is no longer to be recommended. This is because slab gels have the advantages that gel composition and electrical conditions are uniform across the slab so that comparison between different sample zones is far more accurate, a large number of samples can be run on a single gel, gel removal from glass tubes is avoided (it can be difficult and requires practice) and staining and autoradiography are simplified and more reproducible. All the major manufacturers of electrophoresis equipment (Pharmacia-LKB, Bio-Rad, Hoefer, Shandon, Desaga, Buckler, Sartorius, Camag, and many others) make vertical slab gel apparatus that are intended for protein separations in polyacrylamide gels and which are also perfectly suitable for nucleic acid separations. Unlike proteins, however, it is common for nucleic acid separations to be run in gels with only air cooling and in this case a more simple apparatus is used.

4.1.2 Gel formulations

Gel formulations required for nucleic acid analysis are more simple and straightforward than for proteins because all nucleic acid molecules have similar charge densities so that separations during electrophoresis are dependent upon size differences and, to a lesser extent, upon any conformational differences. The concentration of sample components into sharp zones in a stacking gel phase before entering the main separation gel, which necessitates discontinuous buffer systems and gels made in a number of stages, is very widely exploited for protein separations but is not effective for nucleic acids. Nucleic acids are retarded and concentrated when entering the top of the gel anyway, so elaborate stacking procedures to give sharp zones are unnecessary. Thus slabs of uniform composition and homogeneous buffer systems are much more appropriate. Since the nucleic acids are negatively charged, basic buffer systems such as Tris–HCl, Tris–phosphate, sodium borate, or veronal (sodium diethyl-barbiturate) at pH values in the range 7.0–9.5 are usually best, with buffer molarity usually being 0.025–0.05 M (*Table 1*).

Table 2 shows typical concentration ranges (%*T* values) of polyacrylamide gels suitable for separating RNA and DNA molecules of various sizes. Higher molecular weight species (i.e. 10^6 or more) are better examined with agarose gels (see Section 4.4) since at low concentrations (below $T = 3\%$) polyacrylamide gels become mechanically very weak and difficult to handle. Gels with T greater than

---

Table 1. Buffers for polyacrylamide and agarose gel electrophoresis

Buffer	Recipe for 1 litre (10 × stock)
TBE	108.0 g Tris-base, 55.0 g boric acid, 9.3 g EDTA. Final pH 8.2–8.4.
TAE	48.4 g Tris-base, 11.4 ml glacial acetic acid, 20.0 ml 0.5 M EDTA (pH 8.0). Final pH 7.6.
TPE	108.0 g Tris-base, 15.5 ml 85% phosphoric acid, 40.0 ml 0.5 M EDTA (pH 8.0).

Table 2. Optimal range of resolution for different polyacrylamide gel concentrations

Acrylamide (%)	Range of resolution dsDNA (bp)	ssDNA (n)	Dye mobilities (bp dsDNA) XC	BPB
3.5	100–1000	750–2000	450	100
5.0	75–500	200–1000	250	65
8.0	50–400	50–400	150	45
12.0	35–250		70	20
15.0	20–150		60	15
20.0	5–100		45	12

about 25% become rather brittle and are likewise less easy to handle. The ratio of Bis to acrylamide is important in determining the physical properties of the gel and if C is much greater than about 10% gels become opaque and brittle. However, if a very porous gel is needed a low value of T may be combined with values for C of 20% or more. Gelation does not occur readily if T is less than about 2.5%, so rather than reducing T to very low values porous gels can be more readily obtained by keeping T at 3.0–3.5% and increasing C to 20–25%. At the other extreme, if C is less than about 1.0% the resulting gels are glue-like and mechanically poor. These are rather sweeping statements, however, as the best level for C depends upon T and Davis (24) found that optimum handling characteristics are obtained if C is decreased as T increases. Suitable values may be $C = 5$–7% at $T = 5$%, falling to $C = 1.0$ to 1.5% when $T = 20$%.

Typical recipes for the composition of polyacrylamide gels are shown in *Table 3* and the procedure for their preparation is outlined in *Protocol 1*.

Table 3. Composition of polyacrylamide gels (per 100 ml)

Constituents	Gel concentration (%T)					
	3.5	**5**	**7.5**	**10**	**15**	**20**
Acrylamide (g)	3.24	4.7	7.13	9.6	14.55	19.5
Bis (g)	0.26	0.3	0.37	0.4	0.45	0.5
TEMED (ml)	0.1	0.1	0.1	0.1	0.1	0.1
Buffer, stock 10 × concentrated (ml)[a]	10	10	10	10	10	10
Water	86.4	84.9	82.4	79.9	74.9	69.9
Ammonium persulphate 10% (ml)[b]	1.0	1.0	1.0	1.0	1.0	1.0

[a] See *Table 1*.
[b] Prepared fresh daily. Add last as ammonium persulphate initiates polymerization. For a denaturing gel add urea to give a final concentration of 7 M.

Protocol 1. Preparation of polyacrylamide gels

1. Assemble the gel mould (the precise procedure will depend upon the particular design of apparatus).
2. Calculate the desired quantities of reagents from *Table 3* needed to make up sufficient solution to fill the gel mould.
3. Dissolve the calculated amounts of acrylamide and Bis (**caution: highly toxic,** so wear gloves when handling solutions together with a dust-mask when weighing solids) in the water and buffer (*Table 1*) and add the TEMED.
4. Immediately before use add the ammonium persulphate solution and mix in well.

Protocol 1. *Continued*

5. Using a syringe with a long needle or a peristaltic pump, quickly fill the mould with solution taking care not to trap any air bubbles in the mould.

6. Apply a sample well-forming comb to the top of the gel mould.

7. Flush out the needle or tubing with water before the gel polymerizes in them and blocks them!

8. When the gel has polymerized (10–30 min) remove the sample well-comb and place the gel in the electrophoresis apparatus.

Acrylamide and Bis can be weighed out (using gloves and a dust-mask, as both Bis and acrylamide are toxic by skin adsorption and inhalation of dust particles), mixed and kept as a stock solution if desired, but as acrylamide is very slowly hydrolysed to acrylic acid, a small amount of a mixed bed ion-exchange resin such as Amberlite MB-1 should be placed in the storage vessel to remove it. TEMED would also be removed by the resin so cannot be added to the stock solutions. Persulphate actually initiates the polymerization reaction and so is only added immediately before use.

To avoid the hazards associated with weighing out acrylamide and Bis powders, stabilized stock solutions of these compounds are now commercially available from a number of suppliers. These are usually in the form of concentrated aqueous solutions of the individual components, appropriate volumes of which are mixed with buffer and polymerization catalysts, or of pre-packaged mixtures to which only the TEMED and persulphate catalysts need to be added.

4.1.3 Sample application and electrophoresis

In most designs of vertical gel apparatus samples are inserted into sample wells pre-formed in the top of the gel slab with the aid of a well-forming comb during the polymerization process. The apparatus is then filled with buffer (the same as that used in the gels) and the samples injected down through 1–2 cm of buffer into the wells. In order to do this the sample density must be increased by adding to them a small amount (about 5%) of sucrose, sorbitol or glycerol (see *Table 4* for loading buffer recipes). A small amount of a tracking dye, such as bromophenol blue, can be added conveniently at the same time. This dye serves not only as a visual indication that the samples have been loaded correctly, but also enables the progress of the electrophoretic separation to be followed. This is, in fact, the easiest way to monitor progress because the migration of zones depends upon so many different factors that it is otherwise difficult to generalize about the electrophoretic conditions. If neutral or slightly basic buffers are employed, nucleic acids will migrate towards the anode which consequently should be in the bottom chamber. Voltages applied across the gel are usually in the region of 10–30 V cm^{-1} and electrophoresis is continued until the tracking dye has almost

Table 4. Loading buffers for polyacrylamide and agarose gel electrophoresis

Buffer	Recipe (6 × working concentration)
SBX	40% (w/v) sucrose, 0.25% (w/v) bromophenol blue, 0.25% (w/v) xylene cyanol. Store 4°C.
FBX	15% (w/v) Ficoll 400, 0.25% (w/v) bromophenol blue, 0.25% (w/v) xylene cyanol.
GBX	30% (v/v) glycerol, 0.25% (w/v) bromophenol blue, 0.25% (w/v) xylene cyanol. Store 4°C.
PAGE denaturing[a]	0.3% (w/v) bromophenol blue, 0.3% (w/v) xylene cyanol, 10 mM EDTA (pH 8.0) in 80% (v/v) formamide.

[a] With a denaturing polyacrylamide gel, heat the sample to 100°C for 5 min immediately before loading on to the gel.

reached the bottom of the gel slab. With a 17 cm gel run at 500 V this will probably take 80–90 min. The current is then switched off and the gel removed from the apparatus for zone detection (Section 5).

4.2 Polyacrylamide concentration gradient gels

While polyacrylamide slab gels of uniform composition and %T are very frequently employed, the use of gels with a gradient of increasing concentration has gained in popularity in recent years. In a gel with %T increasing linearly the migration velocity of a sample band is inversely related to time and varies exponentially with the distance travelled. This has the advantage that small DNA or RNA components in a mixture rapidly enter a region of decreasing gel pore size and are slowed relative to larger, more slowly-moving components. It is thus possible to extend running times to give a good separation of large components without losing the smaller components off the end of the gel slab. Gradient gels may also give a higher resolution than gels of constant T because throughout the run the leading edge of any band of sample is moving into a more concentrated region of gel than the trailing edge, hence encountering greater resistance, so that there is a band-sharpening effect (25). A further advantage is that since migration rates slow down as the separation progresses electrophoretic conditions do not need to be controlled so precisely as in constant T gels and simple unstabilized direct current power supplies can be used, while it may be possible to run gels for 3–8 h during the day or overnight as convenient. Diffusion is also reduced in more concentrated gels, so they can be kept unfixed or unstained for quite long periods without loss of resolution, which gives time, if necessary, to test portions of the gel, or duplicate gels, before deciding upon how to proceed with material separated in the main gel. Gels can be prepared with any shape of gradient to suit the particular separation but linear or simple concave gradients are usual. Since the gel pore size is a function of $1/\sqrt{T}$, in a linear gradient the pore size changes more gradually at high values of T than at low values, but this is overcome if concave gradients are used.

4.2.1 Gradient gel preparation and running

Most devices for preparing solution density gradients can also be used for preparing concentration gradient gels and some simple designs are described by Andrews (23). The simplest merely consists of two identical flasks joined by a siphon bridge so that as dense (concentrated) solution either flows from or is pumped from one flask into the gel mould it is replaced with half its volume of light (dilute) solution from the other flask. If the first flask is placed on a magnetic stirrer, so that the light solution is mixed in well as it flows from the siphon bridge, such an arrangement will yield a linear concentration gradient. Attaching a third flask filled with light solution joined by another siphon bridge to the second one will give a concave gradient. In this example the gel mould must be filled from the top so that the more concentrated solution flows in first, but a similar arrangement can be drawn up for bottom-filling moulds where the light solution is pumped in first. There are, of course, also many designs of commercial gradient gel makers available. With many designs the most important precaution is to avoid convection currents caused by the heat generated during polymerization. To minimize these it is usual to arrange for a gradient in TEMED as well, so that polymerization occurs first in the upper (most dilute) part of the gel gradient and proceeds downwards.

The upper and lower limits chosen for the gel concentration depend upon the size range of nucleic acids being studied. Gels with T less than about 3.0–3.5% are quite difficult to handle and while using relatively high C values (10–20%) can improve the mechanical properties to a limited extent it is really much better if the gels are supported on a plastic backing film, such as GelBond (FMC Corp.), or on a glass plate. If this is done then gels with T as little as 2% may be feasible. Caton and Goldstein (26) have suggested that gradient gels with $T = 2$ to 16% or 2.5 to 12% are suitable for nucleic acids up to 30S, and Jeppesen (27) has used 2.5 to 7.5% or 3.5 to 7.5% gels for DNA fragments within the mol. wt. range 7×10^4 to 1.4×10^7.

Because nucleic acid zones gradually slow down as they reach the more concentrated regions of gel, gradient gels generally take a little longer to run than slabs of homogeneous gel of the same size. Depending upon the molecular size and hence mobility of the fastest-moving zone it may be most advantageous to the separation of larger components for electrophoresis to be continued for a while even after the tracking dye has reached the end of the gel slab. For 8×8 cm square gels, running times of 2–6 h at 50–100 V may be suitable, but this will depend upon many factors, not least the concentration range of the gel, and separation conditions will need to be established for each type of gel and sample combination.

4.3 Polyacrylamide–agarose composite gels

These composite gels were introduced with the idea that agarose, dilute solutions of which give quite firm gels, could be added to increase the mechanical strength of dilute (low $\%T$) polyacrylamide gels. The resulting large-pore gels would be useful for studying high molecular weight nucleic acids, viruses or lipoproteins, which do not enter polyacrylamide gels of about $T = 4\%$ or more. The approach enables T to be reduced to 1% or less, which would be totally unusable with polyacrylamide alone. Except for very large molecules the agarose does not contribute to molecular sieving (size separation) effects which remain proportional to the concentration of acrylamide.

Both possible approaches to preparing composite gels have been tried, Uriel and Berges (28, 29) mixing all the necessary ingredients and keeping them above 35°C until the acrylamide had gelled and then cooling to solidify the agarose, while Peacock and Dingman (30) cooled the mixture to 20°C or less immediately after adding the acrylamide polymerization catalysts so that the agarose gelled first (quite a low level of persulphate was used to ensure that the acrylamide polymerization was delayed sufficiently). According to Peacock and Dingman it is important that when T is less than 3% the agarose component gels first, although at higher values of T the order of gelling has little effect on mechanical properties. These authors prepared gels containing 0.5–2.0% agarose and 1.0–3.5% acrylamide, over which range agarose addition is helpful, but Uriel and Berges (28, 29) employed 0.8% agarose and T varying from 2.5 to 9.0%. Above about $T = 3-4\%$ the mechanical and separation properties of gels containing polyacrylamide alone are perfectly satisfactory and they are more straightforward to prepare, so there is little purpose in using composite gels. Indeed, Serwer (31) has pointed out that concentrated agarose gels can have similar properties and resolution to large-pore polyacrylamide gels with either low $\%T$ or with high cross-linking ($\%C$) and the agarose gels are much easier to make, so there may not be any need to resort to the use of composite gels at all.

4.4 Agarose gels

Agarose gels are nowadays much the most popular medium for the electrophoretic separation of medium- and large-sized nucleic acids. Particles with average hydrodynamic radii of as much as 30 nm and duplex DNAs with molecular weights up to 10^8 can be studied on very dilute gels of as little as 0.03% agarose (32). However, more typical agarose concentrations fall within the range 0.3–2.5%, the concentration used depending upon the sizes of nucleic acids to be separated. An approximate guide is shown in *Table 5*, but the separation properties depend on the type of agarose and you should follow the recommendations in the supplier's product sheets. Although agarose gels lack mechanical strength, especially with the more dilute gels, handling is facilitated by the use of horizontal gels which can then be supported on glass plates or plastic sheets. The vertical gel format which used to be commonplace, but which demanded a relatively high gel strength so that gels of less than about 0.5–0.6% were difficult to deal with, has now greatly diminished in popularity. If vertical gel

Table 5. Agarose concentrations required to separate nucleic acids of various sizes

Agarose (%)	Linear nucleic acid size (kb)
0.3	1.0–70
0.5	0.7–45
0.8	0.4–20
1.0	0.3–10
1.2	0.2–8
1.5	0.2–6
2.0	0.1–5

slabs are to be used however apparatus such as that described by Sugden *et al.* (33) is suitable.

4.4.1 Gel preparation

Regardless of the size or orientation (vertical or horizontal) of the gel, agarose gels are always prepared in the same very simple manner. For most applications only a single gel component is needed (agarose) and no polymerization catalysts are required so they are quick and easy to prepare, with far less chance of error than with polyacrylamide gels. These factors coupled with the complete lack of toxicity (unless toxic buffers are used) are largely responsible for their popularity.

Agarose gels for DNA separations are prepared as outlined in *Protocol 2*. Various agarose preparations with different gelation temperatures are commercially available but most set to a gel on cooling to 40–45°C. If it is necessary to handle the gel extensively after preparation, or through staining and destaining steps after electrophoresis, it may be convenient to cut a sheet of hydrophilic plastic support, such as GelBond (FMC Corp.), to the required size and place it in the bottom of the gel mould, hydrophilic side (readily wettable) up, before pouring in the warm agarose solution. Once set, the mould can be dismantled and the gel on the backing GelBond sheet lifted out and transferred to the electrophoresis apparatus.

Protocol 2. Preparation of agarose gels for DNA separations

1. Assemble the gel mould.

2. Add the weighed amount of agarose to the volume of buffer (*Table 1*) needed to fill the mould.

3. Heat on a hot (boiling) water bath until the agarose has dissolved. Alternatively use a microwave oven.

4. Add a denaturing agent if desired (e.g. urea to 6–8 M).

5. Use a small portion of the agarose solution in a pasteur pipette to seal the edges of the mould.

Protocol 2. *Continued*

6. Cool the rest of the agarose solution to about 50°C, pour into the mould and immediately place the sample well-forming comb in position.

7. When the gel has completely cooled and set (30–40 min), remove the comb and place the gel in the electrophoresis apparatus.

8. Add sufficient buffer (*Table 1*) to fill the electrode chambers and cover the gel with a depth of about 1 mm.

4.4.2 Gel buffers

Almost any buffer system could theoretically be used, but the ionic strength should preferably be between 0.03 and 0.1 and since all nucleic acids have very similar charge properties regardless of size, similar buffers can be used for all separations. The most widely used are shown in *Table 1*. Concentrated stock solutions (10- to 50-fold concentrated) are often made up and diluted as needed. Since the use of discontinuous buffer systems does not improve the sharpness of nucleic acid zones simple homogeneous buffers (i.e. the same buffer in both gel and apparatus) are perfectly satisfactory and usually preferred on grounds of convenience.

4.4.3 Sample application

The loading buffer for sample application should contain 0.05% bromophenol blue and 0.05% xylene cyanol as tracking dyes and 5% sucrose, glycerol, or Ficoll to increase sample solution density. These percentages are final values in the sample to be applied, so for a tenfold concentrated stock buffer they should be 10-fold higher than this. Suitable recipes are given in *Table 4*. Small amounts (typically 5–20 μl) of sample containing 0.2–1.0 μg of DNA are then applied with a microsyringe (or Gilson or Eppendorf-type pipette) by injecting them down through the thin layer of buffer covering the gel into the sample wells. When samples contain a complex mixture of nucleic acids, which will be distributed over a large number of electrophoretic zones, loadings as high as 5–10 μg per well may be acceptable.

4.4.4 Running conditions

Because in *Protocol 2* the gels are actually covered by a thin (1 mm deep) layer of buffer the technique is often called submarine gel electrophoresis. The electrical resistance of the gel is almost identical to that of the buffer so a substantial proportion of the current passes through the gel, but of course the deeper the overlaying buffer layer the smaller this proportion is and the less efficient the process becomes. The principal advantages of the submarine technique are that the buffer layer prevents drying out of the gel and also provides some degree of cooling. If no buffer overlay is used and electrical contact between electrode buffer chambers and the gel is made using buffer-moistened filter paper or cloth wicks, which used to be the general procedure, it is necessary to place a sheet of polythene over the surface of the gel and wicks in order to prevent the gel from drying out during the run.

It is generally recommended that for maximum resolution voltages applied to the gels should not exceed 5 V cm^{-1} as voltages higher than this tend to preferentially increase the migration rate of higher molecular weight DNA and reduce the useful fractionation range that can be studied. Indeed, overnight separations with voltages of about 1 V cm^{-1} are quite frequently used (31). Relative mobilities of nucleic acids of different sizes are little affected by temperature so gels are usually run at room temperature, but very dilute gels gain useful mechanical strength if they are run in the cold (e.g. $4°C$).

4.4.5 Modifying the voltage gradient

In a homogeneous potential gradient two different nucleic acid zones will move with their migration being in a fixed proportion to each other. Thus inter-zone separations become larger the further the sample zones have migrated and when a sample containing nucleic acid species of widely-differing sizes is examined the result is that the smaller, fast-moving components will have reached the end of the gel before the larger components have become adequately resolved. When polyacrylamide gels are employed this is easily countered by using a gel concentration gradient (see Section 4.2) but with agarose gels a similar approach is much less satisfactory. In a gel slab of homogeneous composition, when a voltage is applied across it the current passing through each element of gel is constant throughout the slab but the voltage drop varies inversely as the resistance. Usually, of course, this also is homogeneous so the potential gradient (V cm^{-1}) is constant all the way along the gel, but if the gel varies in thickness or the concentration of ions in some parts of the gel is greater than in others, then the resistance will vary and so too will the potential gradient. It is quite practical to prepare gradient gels with a buffer concentration increasing towards the lower part of the gel, to give decreasing resistance and hence a shallower potential gradient so that the faster-moving nucleic acid zones become subject to a lower driving force than large, slow-moving zones. However, more frequently the same result is achieved by running wedge-shaped gels. Once an appropriate mould has been prepared, or purchased (a suitable one is marketed by Pharmacia-LKB Ltd), making the gels is just as simple as preparing ordinary agarose gels. The usual gel mixture is merely poured into the mould in the usual way. When set, the mould is disassembled and the resulting gel is approximately twice as thick at one end as the other (for instance, 2 mm reducing to 1 mm). Samples are applied to wells in the thinner end in the usual way but since in a homogeneous gel mixture electrical resistance is proportional to cross-sectional area, during the run zones will move progressively into regions of lower electrical resistance and hence of reducing potential gradient.

Any approach resulting in the preferential slowing-down of faster moving nucleic acid zones will not only enable greater resolution of slowly-moving zones by permitting longer run times without losing fast zones, but will also extend the size range of the nucleic acids that can be separated on a single gel and give a better distribution of zones along it.

4.4.6 Separation of RNA by agarose gel electrophoresis

Denaturing agarose gels are used to analyse single-stranded DNA molecules and for separations involving RNA. Several methods have been developed but the most popular are alkaline agarose gels for DNA (*Protocol 3*) and the glyoxal system for RNA and DNA (*Protocol 4*).

Protocol 3. Alkaline agarose gel electrophoresis

1. Prepare an agarose gel in a neutral buffer comprising 50 mM NaCl, 1 mM EDTA (pH 8.0).

2. When the gel is set, soak it in running buffer for 30 min. Running buffer = 30 mM NaOH, 1 mM EDTA.

3. Ethanol precipitate the DNA samples and take each one up in 10 μl of loading buffer [50 mM NaOH, 1 mM EDTA, 3% (w/v) Ficoll 400, 0.025% (w/v) bromocresol green, 0.025% (w/v) xylene cyanol].

4. Run the gel as normal. Usually the DNA will be labelled and so will be detected by autoradiography, If the DNA is unlabelled, soak the gel in 1 M Tris–HCl pH 7.6, 1.5 M NaCl for 45 min then stain with ethidium bromide (Section 5.2.3).

Protocol 4. Glyoxal denaturation for gel electrophoresis of single-stranded DNA and RNA

1. Deionize a 6 M 40% (w/v) glyoxal solution by stirring with a mixed-bed ion-exchange resin until neutral.

2. Denature up to 10 μg of DNA or RNA in 8 μl of 1 M glyoxal, 50% (v/v) dimethyl sulphoxide, 10 mM sodium phosphate pH 7.0. Denaturation will take 1 h at 50°C.

3. Cool on ice, then add 2 μl of loading buffer. Loading buffer = 10 mM sodium phosphate pH 7.0, 50% (v/v) glycerol, 0.4% (w/v) bromophenol blue.

4. Load on to a standard agarose gel prepared in 10 mM sodium phosphate pH 7.0 and run in the same buffer.

5. Autoradiograph or stain with ethidium bromide (Section 5.2.3).

4.4.7 Size markers

To determine the sizes of nucleic acid molecules in test samples a standard is normally separated in an adjacent lane. The nature of the standard depends on the application. Possibilities are:

- restriction digests of lambda or other DNA molecules, which will give fragments of known sizes (see *Appendix 5*).
- oligonucleotide 'ladders', which comprise fragments that are all multiples of the basic oligonucleotide (e.g. 121, 242, 363 bp etc.).
- unrestricted DNA molecules (e.g. λ DNA, 49.5 kb).
- for RNA gels, rRNA molecules, or commercial preparations of known fragment sizes.

5. Detection methods

5.1 Unstained gels

Nucleic acids have relatively high absorbances at 250–260 nm, so their quantitative estimation (or qualitative detection) by the direct densitometry of unstained gels is more practical than it is with proteins. When polyacrylamide gels are used, both the acrylamide and bisacrylamide monomers should be of high purity, because many of the impurities frequently found absorb strongly at this wavelength, but with care as little as 0.05 μg of nucleic acid can be detected by this method (34). Below about 250 nm background absorption becomes excessive, even with purified reagents. One of the major factors influencing opacity is the proportion of the cross-linking agent (Bis) and the extent of this light-scattering varies approximately exponentially with Bis concentration, so it is best if C is 3% or less. It is also helpful if gels can be pre-run before samples are analysed to remove polymerization catalysts or any other impurities, which may absorb UV radiation.

5.2 Gel staining methods

5.2.1 Stains-All

There are various methods for staining gels for RNA and DNA detection, some of which are summarized elsewhere (23). One of the most popular of the earlier staining techniques employed the dye Stains-All which, as its name implies, stains nucleic acids blueish-purple, phosphoproteins blue, proteins red, and glycos-aminoglycans and glycoproteins various colours from blue to brown. In a typical method for nucleic acids (35), gels are soaked overnight in 0.005% (w/v) Stains-All in 50% (v/v) formamide in a dish wrapped in aluminium foil and then washed free of excess stain with running tap water. The dye is somewhat photosensitive, so band patterns fade if exposed to bright light. It is not a particularly sensitive nucleic acid stain, but makes up in convenience for what it lacks in sensitivity. The most recent and sensitive methods, however, are silver staining techniques and those using the fluorescent intercalating dye ethidium bromide.

5.2.2 Silver staining

Although there are variants of the silver-staining protocol specifically developed for application to agarose gels (36) they are optimized for protein detection, and the agarose matrix itself can give rise to unacceptably high background coloration unless care is taken. It is more suitable for use with polyacrylamide gels in which the sensitivity of nucleic acid detection is at least as great as with ethidium bromide. The procedure of Goldman and Merril (37) is outlined in *Protocol 5* and is capable of detecting down to 1 ng of nucleic acid.

Protocol 5. Silver staining of polyacrylamide gels for nucleic acid detection

1. Fix gels for 20 min in 50% (v/v) methanol, 12% (v/v) acetic acid.
2. Wash in three changes of 10% (v/v) methanol, 5% (v/v) acetic acid; 10 min per wash.
3. Soak gels for 5 min in 3.4 mM potassium dichromate, 3.2 mM HNO_3.
4. Transfer to 12 mM $AgNO_3$ for 20 min.
5. Develop with agitation for 20–25 min in 0.28 M Na_2CO_3 containing 0.5 ml per litre of 37% (v/v) formaldehyde. Change the solution once or twice to prevent precipitation of silver salts on the gel surface (brown precipitate).
6. A yellowish background should appear. Now stop development by 5 min in 20% (v/v) acetic acid.
7. Wash gels twice for 10 min in water.
8. Store in water or soak in 3% (v/v) glycerol for 10 min and vacuum dry.

Merril *et al.* (38) state that, unlike with proteins, in many cases no fixing step is necessary and it is sufficient to wash the gels with water only, in order to remove buffer salts. If required, silver staining can also be applied to ethidium bromide stained gels. *Protocol 5* is optimized for $T = 5\%$ polyacrylamide gels, 0.8 mm thick, and for thinner or thicker gels times may need alteration (thicker gels require longer to reach equilibrium at each step).

5.2.3 Ethidium bromide staining

Much the most popular method for detecting nucleic acid zones is ethidium bromide staining. This can be used with either polyacrylamide or agarose gels in one of two ways. Either a low level (0.5 μg ml^{-1}) of ethidium bromide is added directly to the buffer used for making up the gel, and also to the buffer in the electrophoresis tank, or it can be used as a post-separation stain. The former has the advantages that the progress of the separation can be followed at any stage by illuminating the gel with UV radiation and, of course, no separate gel staining step is needed, but the disadvantage is that the mobility of linear duplex DNA is reduced by the presence of the dye by about 15%. As a post-separation stain, the gel, run in dye-free buffer, is removed from the apparatus and immersed in a staining bath of ethidium bromide (0.5 μg ml^{-1}) in the same buffer or in water for 15–60 min at room temperature. Under these circumstances the gel matrix itself only picks up small amounts of the dye and background fluorescence is quite low, so nucleic acid zones show up strongly, but if the background is troublesome (for example, with very low levels of nucleic acid), it can be reduced by washing out the unbound ethidium bromide by soaking the gel in 1 mM MgSO$_4$ for 1 h.

Either long or short wavelength UV illumination (260, 300, or 360 nm) can be used as the excitation wavelength and the emitted light is at 590 nm, so the zones show up as an orange-red fluorescence. The usual limit of detection is about 1–2 ng of double-stranded DNA or RNA, but since ethidium bromide is an intercalating dye, its binding to single-stranded nucleic acids is much weaker and the fluorescence yield (and hence sensitivity) is considerably lower.

A further disadvantage of using ethidium bromide in either buffers or gels is that it is a powerful mutagen (see *Appendix 2*). It is **essential**, therefore, that gloves are always worn when making up solutions or handling gels containing it. The ethidium bromide itself should always be measured out in a fume cupboard and, of course, any spillages should be cleared up immediately to avoid endangering the safety of yourself and others. Likewise, solutions and materials used for cleaning up spillages must be disposed of safely in accordance with your local safety regulations.

5.2.4 Recording and quantification

Quantification of band patterns can be achieved by photographing them and then scanning the photographic negatives with a densitometer: most conventional densitometers cannot be used for the direct scanning of ethidium bromide fluorescent bands. Photography with standard black and white film (such as Kodak Tri-X, Ilford FP) or with high resolution film (e.g. Kodak AHU) of the gel transilluminated by placing it on a light box is satisfactory for Stains-All or silver-stained gels. Transillumination with a UV lamp at about 300 nm and Polaroid film (Type 55 or 665) and a Wratten 9 or 23A filter is often used with ethidium bromide stained gels, although standard black and white film could probably be used in this case also. Problems with this technique are to ensure even illumination of the whole gel and to be certain that all parts of the gel are the same distance from the camera lens, so that some bands are not viewed obliquely. This can often be largely overcome by bending the gel slightly towards the camera lens (assuming it is not mounted on a glass plate, of course!).

There are many forms of data handling, varying from simple manual estimation of peak areas to fully computerized systems, depending largely upon the densitometer used for scanning the negatives, but in all cases quantification is achieved by comparing the unknown samples with a standard mixture containing known amounts of nucleic acid, run on the gel at the same time as the unknowns, to provide a calibration plot. Good examples of the technique can be found in the literature (37, 39). No matter how sophisticated the system, however, because it introduces additional factors which can influence both resolution and quantitative accuracy, the scanning of photographic negatives can never be expected to equal the performance of direct scanning of stained or fluorescent band patterns, but it is perfectly adequate for many purposes and has the advantage of also providing a permanent record.

A recently introduced alternative to photography is the electronic imaging of fluorescent band patterns with a modified TV camera equipped with a charge-coupled device (CCD) detector (40, 41). Compared to conventional photography this has the advantages of a linear response, a greater dynamic range and immediate availability of the image in digital form for computerized data analysis.

5.3 Autoradiography

As mentioned above, the most popular method for detecting nucleic acid zones on gel slabs is using the fluorescent dye ethidium bromide, but running at a very close second must come radiolabelling followed by autoradiography. Ethidium bromide staining has the advantages of being quick, convenient and relatively simple to perform, but both the reagent and gels stained with it are hazardous to handle and strong, preferably short wavelength UV illumination is needed, which can also be hazardous unless due care is taken (safety goggles, screens and such like; see *Appendix 2*). Radiolabelling and autoradiography can be more sensitive, have a greater usable range (autoradiographic exposures can be short to detect only the most intense zones or long for maximum sensitivity) and a permanent record is obtained at the same time and this is in a convenient form for data handling and quantification. Against these advantages must be balanced the need to pre-label the samples with radioactive isotopes, which must also be handled with appropriate care and often preferably in a purpose-built laboratory or at least in a suitable safety chamber (precise requirements are determined by legal regulations and will depend on both the quantities and identities of isotopes being used; see *Appendix 2*). These considerations may add considerably to the complexities of the procedure, but generally once the facilities are set up, radiolabelling is quite straightforward. Nevertheless, the very necessity to pre-label samples and then set up the autoradiography as separate operations in addition to the electrophoresis itself means that it is considerably slower than ethidium bromide (or any other) staining.

Details of methods for radiolabelling nucleic acids and for their subsequent detection by autoradiography are given in Volume II, Chapter 4, and will not be discussed further here.

6. Pulsed field gel electrophoresis methods

All the discussion so far in this chapter has assumed that nucleic acid samples are applied to a slab of gel and a voltage then applied so that in a neutral or basic buffer system the nucleic acids have a net negative charge and are, therefore, attracted by the anode towards which they migrate in an orderly fashion under conditions of more or less constant electrical field strength and direction. As indicated in *Table 2*, when polyacrylamide gels are used, the largest nucleic acids that can be separated are only about 1 kb long. Even with the most dilute agarose gels this can only be extended to 50–100 kb at most (*Table 5*). Thus when yeast chromosomes, for example, contain DNA of a few megabase pairs (Mb) in length and human chromosomes are typically 50–200 Mb, it is obvious that constant-field electrophoresis has very limited capabilities in genetic studies. Techniques such as 'chromosome walking' can only proceed by a large number of small (< 50 kb) steps, which is time-consuming and not always practical. Fortunately a great leap forward was made in 1982 when Schwartz *et al.* (42) introduced the first pulsed field gel electrophoresis method. This and the large number of related variations on it developed since have increased the size limit for nucleic acid separations by two orders of magnitude or more to at least 12 Mb (43). The very fact that there have been such a large number of variations described, as well as a considerable number of both theoretical and application papers in such a short time, testifies to the dramatic advance that pulsed field methods represent. They have made possible the separation of intact yeast chromosomal DNAs (44, 45), the mapping of the Duchenne muscular dystrophy gene (46), of the genome of *Escherichia coli* (47) and of human chromosome 21 (48). Such maps are a prelude to the sequencing of genomes and eventually to whole chromosomes, which is a relatively formidable goal for even simple organisms but one which is rapidly becoming a practical proposition with the aid of new technologies, including pulsed field methods.

6.1 Theory of pulsed field methods

During migration through an agarose gel under the influence of a constant low electric field strength, DNA molecules smaller than about 12 kb are separated in accordance with their molecular weights almost entirely on the basis of a sieving mechanism as the molecules pass through the gel matrix (49). This can be described by the Ogston theory of pore size distribution if extrapolations to zero field strength are made, but in practice the conformation of DNA molecules is distorted by any finite electric field and the theory no longer holds. Thus the theory of nucleic acid separations on gels differs from that of proteins which are not distorted in this way at the sort of field strengths which are usually used. The passage of nucleic acids through a gel is better described by theories of end-to-end migration, which is known as the reptation theory. Simplistically, this can be envisaged as the nucleic acids snaking endwise on through the pores in the gel, but in fact the precise mechanism is by no means fully understood and it is still the subject of intensive investigation (12–14).

In such a situation the 'body' of the snake has limited scope for lateral motion and the ends of the DNA molecules are more readily oriented by the direction of the electric field applied (biased reptation model). In a unidirectional field the prediction is that the mobility of very large DNA molecules would become independent of molecular weight (10, 50). However, when the direction of the field applied to the gel alternates, the DNA molecules are forced to continually orient and re-orient themselves. If the field on and off times are chosen to coincide with the orientation times of at least some of the DNA molecules in the sample, then the finite times required for the whole molecular realignment process into the new field direction leads to much improved molecular weight separations. As mentioned above, there have been various theories to describe the centre-of-mass velocity and the reptation dynamics of DNA molecules in pulsed electric fields of various durations and geometries. The situation is in fact even more complicated than it might appear, since it has now been shown (14) that very short electrical pulses of high amplitude (e.g. 1–10 kV cm^{-1} for 10–1000 μsec duration) can orient individual agarose chains or bundles of chains of the gel matrix itself. Smaller applied fields with much longer pulse times (such as 10–100 V cm^{-1} for 0.5 to 2 sec) result in slow time-dependent effects that indicate that domains in the agarose matrix become oriented and by reversing the direction of the applied field the domains change their direction of orientation from parallel to perpendicular (or *vice versa*). Orientation and re-orientation of microdomains of the matrix under the influence of alternating pulsed electric fields would increase the fluidity of the matrix making it easier for very large DNA molecules to migrate through the gel. Thus, effects of alternating fields on both the DNA molecules themselves (reptation and biased reptation effects) and on the gel matrix have to be considered.

An alternative hypothesis to standard reptation theory for describing the mechanism of DNA fractionation during pulsed field methods has been put forward by Serwer (51). This theory, termed the gel hysteresis hypothesis, has much in common with standard reptation theory but also includes the influences of DNA–gel interactions, and it certainly goes some way towards explaining some of the poorly understood aspects of using pulsed fields.

Experimentally, when the electric field is reversed in polarity, very little change is seen in the DNA orientation if sufficient time has elapsed for the DNA molecules to have become completely stretched and oriented in the direction of that field before it is reversed in direction. All that then happens is that the DNA molecules simply reverse their direction of migration with little loss of orientation. If, however, they have not had time to become fully stretched or to have reached their equilibrium orientation, field reversal causes substantial disorientation, this forced rate of disorientation being in fact greater than that which occurs in the absence of an applied electric field, and complicated re-orientation patterns are observed.

111

When Schwartz and Cantor first described the pulsed field approach, they showed that DNA mobility depended upon the field switching interval (pulse time), but others (e.g. ref. 52) have since shown that the contributions of other physical factors are also vital. Changes which altered the velocity of DNA migration (temperature, agarose concentration, voltage, ionic strength, and many others) also invariably changed the relative mobilities of different zones and did not act uniformly on the mobilities of all sizes of DNA being studied. It was found that all these factors are interdependent, so that while changes in one parameter might adversely affect resolution, this can be restored by making compensatory changes in another factor. As examples, shorter switching intervals are needed to maintain resolution as the temperature is raised, the switching interval required varies inversely with the field strength, and decreased separation rates in more concentrated agarose gels can be compensated for by raising the temperature. Manipulating these factors permits separation parameters to be adjusted over a wide range, which may be convenient and beneficial, but greatly complicates choosing the best, as opposed to merely adequate, conditions. At the present time the choice is often largely empirical and many trial experiments are often needed. Despite recent attempts (e.g. ref. 53) to describe the effects of such parameters and provide mobility surfaces from which optimum conditions can be selected, these are still limited in their applicability and there are several aspects of pulsed field electrophoresis which remain poorly understood. For example, it is not known why low field strengths are required for separating large molecules, or why they cannot be separated by two fields offset by 90 degrees but can be resolved by using multiple fields and re-orientation angles of 90 degrees or less, or why some DNA molecules do not follow the inverse monotonic relationship between size and mobility and why expanded regions of separation exist, in which molecules in a particular size range show a selective increase in mobility and a heightened sensitivity to fluctuations in the electric field compared to those in other size ranges also being separated at the same time. Likewise it is not really known why the mobility of a particular DNA zone does not increase monotonically with pulse time but instead shows a minimum at a particular pulse time, the value of which depends upon molecular weight (54). Thus, to achieve a separation over a broad range of DNA sizes, such as occur in chromosomal DNA or most other natural samples, it is common practice to gradually increase the pulse time through the experiments (termed 'ramping') to move this 'window' of good separation from that best for low molecular weight species to that more suitable for large DNA (53). This ensures that all components are subjected to good separation conditions for at least part of the time.

6.2 Gels for pulsed field methods

The whole purpose of the pulsed field approach is to give improved resolution and separation of medium- and large-sized DNA and RNA compared to conventional agarose gel electrophoresis. As the name implies, the only real difference is in the way the electrical field is applied, so the gels used are the same in terms of the buffers employed. Likewise, the migration of large nucleic acids proceeds faster the more dilute the agarose gel, but since the separation factors are rather different to those operating in conventional gels, it is not necessary to use extremely dilute gels, even for the largest nucleic acids. The most widely used gel concentration is 1%. For nucleic acids larger than about 5 Mb, 0.4–0.5% gels may be preferred, but the sharpness of bands decreases as the agarose concentration is reduced, so 0.7–0.8% gels are better for nucleic acids that are nearly as large (e.g. 1–5 Mb). The size and shape of gels is generally dictated by the design of apparatus used, and hence the particular separation method. Indeed, this may also determine other aspects of the gel and separation.

As in conventional constant field agarose gel electrophoresis, ethidium bromide can be added to the gels (at a level of 0.1 μg ml^{-1}) when they are made up, before sample application and gel running, or it can be used in a post-separation staining step. The former has the disadvantage that gels are then highly toxic and must be handled with gloves and all due precautions but the advantage that the progress of separations can be followed easily. In spite of this, post-separation staining is much the more popular procedure, however. Samples are applied to wells formed in the gel in the usual way. Electrical conditions and separation times depend upon the method used and upon the size range of nucleic acids being studied, but typically vary from a few hours for medium sized DNA to a week or more for very large DNA species (3–10 Mb). Relatively low field strengths (2–10 V cm^{-1}) are usually found to be best, especially for very large DNA molecules. While it is tempting to speed up separations by increasing the voltage gradient and/or reducing agarose concentration this often leads to reduced resolution.

6.3 Pulsed field methods

Pulsed field techniques fall into two broad categories, those where the orientation of the field is constant and those which include a change in field orientation. The former is represented by unidirectional pulsed field gel electrophoresis (UPFGE) and field inversion gel electrophoresis (FIGE). Because there are many possible ways of arranging the geometry there are many variations of the latter, which are described below. All pulsed field methods employ horizontal agarose gels in the submerged (submarine) mode so there are no problems of wicks or making electrical contacts between gel and buffer chambers. Buffers which are identical to those used in ordinary (non-pulsed field) agarose gel electrophoresis of nucleic acids (see *Table 1*) are usually circulated through a cooling bath during the run to maintain the temperature at 10–15°C.

6.3.1 Unidirectional pulsed field gel electrophoresis (UPFGE)

This method (41, 55) is the simplest of all the pulsed field techniques, but also the one that represents the smallest advance in performance over conventional agarose gel electrophoresis of nucleic acids in which a constant static electrical field is used.

Experimentally, static fields are found to be incapable of separating double-stranded DNA molecules longer than about 50 kb. This upper limit can however be extended to about 400 kb by applying the electric field in short pulses (typically a few tenths of a second) with long pauses (for instance, 10 sec) in between. Since there is no change in field angle or direction this is known as unidirectional pulsed field gel electrophoresis (UPFGE).

Sutherland *et al.* (55) were interested in measuring the number of radiation and chemically-induced single-strand breaks in DNA but since the mobilities of experimental standards would need to be compared with DNA length standards run in an adjacent sample lane on the same gel, they could not use the earlier forms of PFGE (Section 6.3.3) or OFAGE (Section 6.3.4) because in those the applied fields are not uniform and samples placed in different lanes often migrate different distances. DNA sizes were often too large for static field electrophoresis, so they applied UPFGE to submarine 0.4% agarose gels run at 10–15 V cm^{-1} in either alkaline (30 mM NaOH, 2 mM EDTA) or neutral (50 mM sodium phosphate pH 7.4, 1 mM EDTA) buffers. The apparatus was very simple, consisting of a conventional submarine gel apparatus with a relay-based gating circuit controlled by a pulse generator interposed between the power supply and the electrophoresis cell. Typical pulse times were 0.1–0.2 sec with a 5–10 sec pause between pulses. With this proportion of off-time, runs can be quite slow and may take 4 to 5 days even for a small 10 cm gel, although overnight will often give an adequate separation for many purposes. For molecules longer than the effective pore size of the agarose gel, there is no net electrophoretic migration before the molecule becomes oriented in the field direction, so to achieve the best separation the time between pulses must be long enough for the longest molecule present to return to an unoriented conformation. This can be speeded up by briefly reversing the field, as in FIGE (Section 6.3.2). However, UPFGE does have some advantages. First, Sutherland *et al.* (55) found that molecular length was a monotonic function of mobility; second, analytical expressions of the same form as those used in static field electrophoresis can still be used to characterize the relationship between length and mobility (so the same computer programmes can be used); third, the average power applied is lower which is helpful when high conductivity buffers, such as their alkaline buffer, are used; and finally, electrode wear is lower than in FIGE.

6.3.2 Field inversion gel electrophoresis (FIGE)

While attempting to optimize separations of large DNA by OFAGE (Section 6.3.4) Carle *et al.* (54) found that the separation could be made strongly dependent upon the size of the DNA simply by periodically reversing the electric field while retaining a straightforward one dimensional format. In order to achieve a separation it is, of course, necessary for electrophoresis in the forward direction to be longer than in the reverse direction and in their earliest work Carle *et al.* (54) used a 2:1 ratio with 0.5 sec forward and 0.25 sec in reverse. When used with 1% gels in TBE buffer (*Table 1*) and a constant voltage gradient of 10.5 V cm^{-1} this was found to give a 'window' for good DNA size fractionation between 15 and 30 kb while larger DNA molecules still co-migrated as they did in conventional agarose gel electrophoresis. This window of good fractionation could be shifted up in size range by lengthening the switching cycle, so that 3 sec forward and 1 sec in reverse was best for the 50 to 125 kb region. Similarly, maintaining the switching cycle but using a higher voltage in the forward direction also shifted the separation window.

With this uncomplicated arrangement it was found that the mobility of different sizes of DNA passes through a minimum, with both larger and smaller molecules migrating in accordance with size differences. The presence of a minimum was ascribed to the adoption of directional conformations by DNA molecules migrating in a steady state. Simplistically, when the field is reversed a certain time is taken for the molecules to rearrange themselves into a new conformation and until this is achieved they remain in a low-mobility intermediate state. The mobility pattern is actually rather more complex, but that is essentially the net result confirmed by others (53). The position of this minimum in the size range depends upon the separation parameters but particularly upon the switching pattern. By altering this pattern molecules previously immobile will move and take part in the fractionation process while those of a different size range come to a halt. Thus all molecules in a mixture can be separated if the switching time is continuously varied, or 'ramped'. In their original paper, Carle *et al.* (54) used a constant 3:1 forward to reverse interval but varied the switching time from 9 sec at the start to 60 sec after 18 h. By ramping the switching time it is possible to separate some of the molecules at high resolution or all of them at a lower resolution. The roles of molecular weight, pulse time and field strength on the mobility of linear DNA have been explored in detail by Crater *et al.* (53) who derived equations delineating the best separation conditions.

The ability of FIGE to give high resolution over particular size ranges has led to its application to the analysis of restriction fragment length polymorphisms (RFLPs) (56, 57). RFLPs provide some of the most important genetic markers available and enable the construction of restriction and linkage maps which are vital in human clinical genetics, forensic science and population studies. The interesting size ranges of RFLPs are relatively small [for example, Craig *et al.* (56) studied the range 0.03–5 kb and Gejman *et al.* (57) 0.1–23.1 kb] and while these can be separated quite well by conventional gel electrophoresis (as described in ref. 9) resolution may be a limiting factor in such studies. Both groups found that FIGE was capable of giving substantial improvements in resolution.

FIGE has a number of advantages and disadvantages by comparison with other pulsed field methods. Perhaps its biggest advantage is that it uses conventional electrophoresis chambers and any one of several different commercially-available designs can be used, so no new apparatus is required which reduces costs. All that is needed is a power controller unit. This should be able to supply a defined voltage and to reverse the polarity in a rapid and accurately timed manner. The apparatus is often quite compact and gels of many different sizes can be handled, including very large ones. Because the separation is in a single dimension there are no lateral forces on zones and DNA band patterns are straight and free from distortion (56). Compared to other pulsed field methods FIGE is quite fast and typical run times are 4–24 h. On the negative side, in theory separations are limited in size range to about 2 Mb, but in practice the limit is nearer 800 kb, so it cannot handle such large nucleic acids as some of the other methods. There can also be quite serious electrode corrosion and repairs can be expensive. This is not a true corrosion but rather damage caused by micro-explosions resulting from the alternate release at the electrode surface of H_2 and O_2 gases which then react violently with one another.

6.3.3 Pulsed field gradient gel electrophoresis (PFGE)

Chronologically this was the first pulsed field method (42, 44) and in this, not only is the field pulsed, but the direction is varied as well. Like all methods in which the field orientation is altered during the run PFGE employs apparatus specifically designed for it. As with all the other methods, the gel is in the submarine format and the square gel is mounted on a platform centred between removable, vertical, diode-isolated, platinum electrode arrays (58). The diodes eliminate interaction between active and inactive electrodes. Electronic control consists of three parts; timing circuitry, switching relays and power supply. The field on the y-axis consists of a field gradient established with an array of negative electrodes and a simple point electrode on one side as the positive electrode (*Figure 2*). The orthogonal x-axis field is uniform with arrays of electrodes along both sides. Typically, the y-field might be 20 V cm^{-1} and the x-field 10 V cm^{-1} with a constant alternating switching cycle between them of 40 to 60 sec. When this is applied to a 10×10 cm square 1.5% agarose gel in TBE buffer (*Table 1*) a typical run time would be about 16 h.

Unfortunately, with this original switching pattern (44), the gradient y-axis field introduced a vector in the direction of the x-axis, so bands were skewed off to

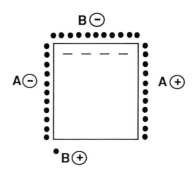

Figure 2. The electrode configuration for PFGE.

one side, which reduced the number of sample slots on the gel which could be used successfully and also led to some loss of band resolution. Later, much more sophisticated switching patterns (58) led to much improved results, at the cost of some further complexity, and it was found that the size of molecules resolved was a sensitive function of switching time; short switching times giving better separations of small molecules but longer times being better for larger molecules. At longer times band separations were reduced and McPeek *et al.* (58) suggested that at any given run time there was a zone of resolution within which DNA molecules were separated. Changes in switching time did not alter the extent of this zone but did affect the size range of DNA molecules separated within it. On the basis that large DNA molecules cannot move in the gel until they are correctly oriented and aligned with a pore in the gel matrix, switching the field direction forces the molecules to re-orient themselves into the new direction. The re-orientation time is a function of molecular size, so molecules with re-orientation times shorter than the switching time between pulses are able to migrate and will be separated while those with longer re-orientation times will not. Thus switching time is crucial to the resolution of different sized molecules. PFGE is able to give good separations of DNA molecules in the 20 kb to 2 Mb size range, but even with sophisticated field switching patterns (58) sample lanes are not entirely straight.

6.3.4 Orthogonal field alternation gel electrophoresis (OFAGE)

OFAGE was historically the next pulsed field method, introduced by Carle and Olsen in 1984 (45). In concept it is identical to PFGE, and indeed the two methods are often bracketed together, and likewise it applies alternating pulses of electrical field at different angles to the gel so that DNA molecules are forced to follow a zig-zag path through the gel.

As in PFGE, separation is on the basis of the times taken for the molecules to re-orient themselves preparatory to migration in the new direction. Thus, the fractionation range (about 20 kb to 2 Mb) is similar to that in PFGE. As in PFGE, equipment specifically designed for the method must be used. The main difference here is that instead of being along the side of the gel slab, in OFAGE the electrodes are placed at an angle across the corners of the gel (*Figure 3*). Each pair of electrodes consists of a long and a short straight length of platinum wire, so that each pair of electrodes gives a slightly non-uniform field gradient. Although DNA sample lanes down the gel are relatively straight and with a similar degree of resolution across the gel, non-uniformity in the field results in lanes being closer together in the centre than at the end of the gel, a sort of hour-glass appearance. Comparison of band mobilities between lanes is easier than in PFGE.

Typically, quite large (20×20 cm) gels are used, so the whole apparatus is quite bulky and in their work Carle and Olson (59) used submarine 1.5% agarose gels run at 300 V and 13°C with a running time of 18 h. Switching intervals between the two electric fields were usually about 50 sec but were varied to suit

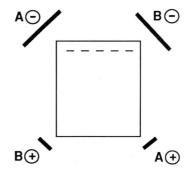

Figure 3. The electrode configuration for OFAGE.

the sample fractionation range required. Hightower and Santi (60) likewise used 1.5% agarose gels in TBE buffer (*Table 1*), in this case run at 12°C and 21 V cm^{-1} for 22 h with a 120-sec pulse time to separate chromosomal DNA. The angle between the sets of electrodes is crucial and below 90 degrees no separation occurs. The usual range is 90–150 degrees but for most runs about 120 degrees was found to be the optimum. The angle can be varied for different runs but is fixed during the run itself.

6.3.5 Transverse alternating field electrophoresis (TAFE)

This approach is rather different in concept to all the other pulsed field methods in that while they generate fields that lie in the plane of the gel, TAFE is a three-dimensional design with the gel placed vertically between two sets of wire electrodes placed parallel to the gel. The fields generated are therefore transverse to the gel plane (*Figure 4*). This arrangement has the advantage that all sample lanes are exposed to the same strength and direction of field at all times. In the apparatus described by Gardiner and Patterson (48), a gel of 7.5 × 10 cm is positioned in the centre of a Plexiglass box. The gel is supported on each side by thin (0.3 × 0.3 × 11 cm) Plexiglass strips in such a way that almost all of both sides of the gel slab is exposed to the electrophoresis buffer. This, of course, means that since the gel is largely unsupported it must have considerable mechanical strength so it is not possible to use dilute gels. The minimum reported (48) was

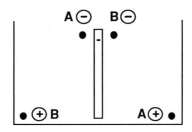

Figure 4. The electrode configuration for TAFE.

0.55% with ordinary agarose or 0.7% with low-melting-point agarose, but for this it was essential to fill the apparatus with buffer to provide some support before positioning the gel. Clearly, some manual dexterity and care is called for with setting up this technique. It is also absolutely vital that the electrodes are all accurately and symmetrically positioned relative to the centre of the gel because the fields generated pass through the plane of the gel, not parallel to it, which means that the DNA molecules follow a zig-zag path downwards through the thickness of the gel. Any loss of symmetry in the field would mean that the DNA would pass out of one face or the other of the gel slab into the surrounding buffer and be lost. Small asymmetries are overcome by running gels at constant current rather than constant voltage, but this cannot correct for any substantial imbalance. A further disadvantage of TAFE is that the field angle is fixed. It is determined by the height of the gel slab, and hence of the negative electrodes and the distances of the two lower positive electrodes from the gel. In the first model of TAFE apparatus (61) an angle of 90 degrees was used, but later (48) an angle of 115 degrees was used and permitted the separation of rather larger DNA molecules.

Of course, these are the initial field angles, and as zones migrate down the gel both the field strength and the angle change. Since both of these parameters are important in the resolution of large DNA molecules it is not possible to maintain optimal separation conditions throughout the run. The angle at any stage can only be changed by rebuilding the apparatus. Likewise any one set of apparatus needs gels of a particular size. Changing size involves repositioning the electrodes and altering the field angle or alternatively scaling the proportions of the apparatus precisely up or down to maintain the angle. The actual running conditions are simple, however, which may offset some of these disadvantages, and single voltage and pulse time conditions are used throughout, with no complex field switching, ramping, or computer control of voltages or pulse times needed.

6.3.6 Contour clamped homogeneous electric fields (CHEF)

In this system (62, 63) the gel is surrounded by multiple electrodes arranged along a polygonal contour and clamped to calculated electrical potentials. In theory an infinite number of electrodes so arranged can give a totally homogeneous electrical field, but in practice as long as they are not too close to the gel a much smaller number will suffice; for example, four along each side of a square or hexagon. Switching the field between electrodes on opposite sides of the square apparatus gives a fixed re-orientation angle of 90 degrees while the hexagonal arrangement (*Figure 5*) can give 60 or 120 degree angles. There are small deviations from ideal potential clamping of the electrodes due to buffer conductivity which are greater the higher the buffer ionic strength, but the effects are very small and DNA sample lanes run very straight down the gel. In Chu's work (62, 63) gels varying from 0.7–1.0% agarose in 0.125 to 0.5 × TBE buffer (*Table 1*) were used and as many as forty samples could be analysed

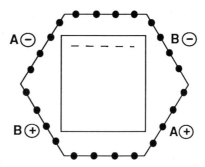

Figure 5. The electrode configuration for CHEF/PACE.

simultaneously on a single gel slab which could be up to 15×15 cm square in size. One disadvantage is that, because of the spacing between gel and electrodes, the apparatus is quite large. Good results were obtained with chromosomal DNA up to at least 7 Mb in size, although for such large molecules very low field strengths (1.5 V cm^{-1}) and long pulse times (60 min) were needed so runs were slow (130 h). For smaller DNA up to 2 Mb, 8 V cm^{-1} could be used with 80 sec pulse times and runs were complete in 24 h. These conditions are very similar to those required in other pulsed field separations.

DNA molecules of the same size can have different conformations (including supercoiled, nicked, bent, branched and cruciform) and these do not respond in the same way as linear DNA to changes in field strength. Thus, when mixtures of DNA were run (63) on gels with varying field strengths and conditions arranged so that the product of field strength and pulse time remained constant, the linear DNA migrated along an arc while supercoiled and nicked DNA moved to outside the line of the arc. The resolution of different DNA conformations was dependent upon field re-orientation angle, with 120 degrees being better than 90 degrees. Although in the CHEF apparatus there is limited scope for varying this angle further, it is clear that this method does have value in separating at least linear from other DNA conformations, although it is not yet established whether the latter can be separated from one another by CHEF.

6.3.7 Programmable, autonomously-controlled electrode (PACE) system

The apparatus used for PACE, a method introduced by Hood and co-workers (52, 64), is the same as that used for CHEF (see *Figure 5*) and consists of 24 electrodes arranged in a hexagonal array around a central gel. The difference is that while in CHEF voltages are clamped to pre-determined values by fixed resistors, PACE employs an electronic controlling unit and the voltage at each electrode is independently regulated through 24 separate amplifier circuits regulated by a microcomputer. This enables the voltage at any electrode to be altered at any stage during the run. Any number of electric fields with any magnitude, orientation or duration can be generated, so that fields are not limited by the geometry of the apparatus and homogeneous fields can be set up for each orientation. This enormous flexibility enabled Birren *et al.* (52) to mimic the conditions of pulsed field methods, including UPFGE, FIGE, and 120-degree pulsed field methods (65) for separating relatively small (1–50 kb) DNA using the single apparatus. The computer control also permitted very sophisticated switching regimes to be employed, so that it is possible to progressively change the switching interval during runs, a procedure called 'switch interval ramping'. When pulsed fields with a single switching interval are used there is a region in which zones are compressed and poorly resolved by comparison with both larger and smaller DNA molecules (66), termed 'central zone 2' by Vollrath and Davis (67). By ramping the switching time, this is eliminated, and separations are then a linear function of DNA size over the whole gel (64). By combining multiple fields of different strength, orientation and duration, conditions can be chosen that separate any size range of DNA with higher resolution than with other pulsed field methods limited to only two alternating electric fields.

The close control over the electric fields also enabled Birren *et al.* (52, 68) to explore the roles of various factors controlling DNA velocity and resolution. These factors included not only voltage, switch time, temperature, and gel concentration (68) but also re-orientation angle, buffer strength, agarose type and the potential for high voltages in separating very large (3–6 Mb) chromosomal DNA (52). Interactions between these factors were also examined and the data obtained permitted excellent separations of DNA over a very wide size range. Resolution in a particular size range could often be improved by incorporating a field reversal cycle (as with FIGE) in between the forward orthogonal cycles. The gels used for all this work (52, 68) were submarine 13 cm square 1% agarose gels made up in TBE, TAE, or $0.5 \times$ TBE buffer (*Table 1*) run at 13°C, with the apparatus buffer cooled by circulating through an external cooling bath.

6.3.8 Rotating gel electrophoresis (RGE)

Several of the methods described so far involve electronic switching to alter the direction of the applied field, but an alternative is to have a fixed electrode configuration and to rotate the gel through a set angle between pulses. This is rotating gel electrophoresis (RGE) and several designs of suitable apparatus have been described (66, 69–71). The electrodes are generally mounted in large buffer compartments at each end of the electrophoresis chamber with the gel mounted on a rotating table in the centre (*Figure 6*). The table is usually circular, so it is common practice also to cast circular gels to mount on it, although sample wells are placed in a straight line and lanes of sample components should also run straight. In most designs the gel table is rotated by an electric motor between pulses, but Sutherland *et al.* (71) have described a pneumatically-driven table which may be both cheaper and safer, since there are no electrical components under or in the vicinity of the gel. Since the pneumatic actuator gives a fixed rotation, different-sized pulley wheels mut be used to alter the field angle. This is

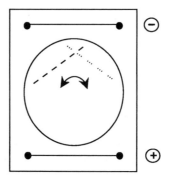

Figure 6. The electrode configuration for RGE.

simple to do but means that unlike PACE the field angle cannot be changed during the actual run. Rotating the gel with electric motors (66) is more flexible in this respect. Gel rotation takes a finite time, even though in the apparatus of Sutherland *et al.* (71) it is less than 0.5 sec, but even when the equipment is designed to give relatively gentle angular accelerations and decelerations at the beginning and end of the rotation there is some stress on the gel. For this reason high gel strengths are needed, usually 1.5% agarose or more, which slows down the rate of migration of very large DNA molecules. Most of the above comments on the separation of sample components by pulsed field methods also apply to RGE because essentially it is only the means by which different field orientations are applied that differs.

6.3.9 Rotating field electrophoresis (RFE)

RFE is really the mechanical analogue of PACE. The gel is stationary and instead of using sophisticated electronic switching to alter field strength and orientation in RFE moving curved electrodes are used (*Figure 7*). Since the electrodes can be much closer to the gel than in the CHEF/PACE apparatus, the RFE apparatus can be more compact and/or rather larger gels can be used. Square gels are employed, often up to 20 cm square, and the agarose can be quite dilute if required since there are no mechanical stresses on the gel. Although 0.4% gels can

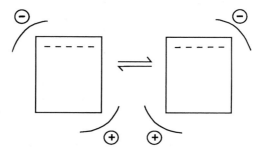

Figure 7. The electrode configuration for RFE.

be used they tend to exhibit reduced band sharpness, so 0.7% gels are preferable even over the 1–5 Mb DNA size range. As in the other designs of pulsed field apparatus, gels are run in a submarine format and buffer is recirculated through a cooling bath to enable runs to be performed at 10–15°C.

The electrodes are driven by electric stepping motors to any desired position so field orientation can be varied infinitely and even reversed and also, if required, altered during the run (sometimes referred to as 'angle ramping') as well as pulsed. Thus both pulse ramps and angle ramps can be applied in linear, logarithmic or any other mode, so the method is a truly very versatile one. Lanes of separated sample components run very straight down the gel facilitating inter-sample comparisons. Again many of the comments made above relating to the separation of sample components apply also to RFE, but due to its versatility it should be possible to optimize separations over very wide size ranges. Indeed, a principal disadvantage may be that since virtually all the separation parameters are variable, finding the best separation conditions may be quite time-consuming.

6.3.10 Summary of pulsed field methods

The earliest pulsed field techniques for separating large DNA molecules (PFGE and OFAGE) suffered from non-uniform fields across the gel which led to both the speed and direction of movement of DNA zones varying from lane to lane and with distance from the origin. For accurate interpretation of the results of many experiments such variations cannot be tolerated and both the later approaches to PFGE and all the other methods described above are in essence designed to improve field uniformity to give straighter lanes and hence to give better quality of results and/or to speed up separation times. In terms of the DNA size range that can be studied all the methods are quite similar and can handle molecules up to at least 7 Mb. They also follow the general rule that the longer the pulse time (frequency with which the field direction is altered) the larger are the sizes of DNA molecules subjected to optimum separation. Pulse times of seconds or minutes are used for DNA up to about 1.5 Mb and hours for larger DNA. Generally, it is also true that the larger the DNA molecules to be separated, the lower the field strength should be, so that while 10–15 V cm^{-1} works well for DNA up to about 1 Mb, above this much lower voltages (such as 2 V cm^{-1}) are required. This, of course, extends the separation times enormously and a week or so may be needed to separate the largest molecules. It is possible to speed up runs to some extent by raising the temperature but above about 20°C resolution begins to suffer and patterns become quite fuzzy by 25°C, so it is usual to perform runs at 10–15°C.

All the pulsed field methods described in this chapter have been developed in a period of less than seven years and rapid advances have been made in both our knowledge of the underlying processes by which separations of large DNA molecules occur in gels and in designs of apparatus for achieving them. Currently, this is a very active area of research and further improvements in all aspects are virtually certain, so while techniques such as PACE and RFE are probably the most versatile and advantageous methods to employ at the time of writing they may well be rapidly superseded by new innovations. It is clear, however, that the drive to separate larger and larger DNA species is strong and of great importance to our understanding of the very fundamentals of life.

References

1. Maurer, H.R. (1971). *Disc Electrophoresis and Related Techniques of Polyacrylamide Gel Electrophoresis*, 2nd edn. Walter de Gruyter, Berlin, New York.
2. Richards, E.G. and Lecanidou, R. 1971). *Analytical Biochemistry*, **40**, 43.
3. Hjertén, S., Jerstedt, S., and Tiselius, A. (1965). *Analytical Biochemistry*, **11**, 219.
4. Deininger, P.L. (1983). *Analytical Biochemistry*, **135**, 247.
5. Serwer, P. (1983). *Electrophoresis*, **4**, 375.
6. Bell, L. and Beyers, B. (1983). *Analytical Biochemistry*, **130**, 527.
7. Stellwagen, N.C. (1987). *Advances in Electrophoresis*, **1**, 177.
8. Slater, G.W., Rousseau, J., Noolandi, J., Turmel, C., and Lalande, M. (1988). *Biopolymers*, **27**, 509.
9. Hervet, H. and Bean, C. P. (1987). *Biopolymers*, **26**, 727.
10. Lumpkin, O.J., Déjardin, P., and Zimm, B.H. (1985). *Biopolymers*, **24**, 1573.
11. Serwer, P. (1989). *Electrophoresis*, **10**, 327.
12. Slater, G.W. and Noolandi, J. (1989). *Electrophoresis*, **10**, 413.
13. Viovy, J.L. (1989). *Electrophoresis*, **10**, 429.
14. Stellwagen, N.C. and Stellwagen, J. (1989) *Electrophoresis*, **10**, 332.
15. Willis, C.E., Willis, D.G., and Holmquist, G.P. (1988). *Applied and Theoretical Electrophoresis*, **1**, 11.
16. Hjertén, S. (1962). *Archives of Biochemistry and Biophysics Suppl.* **1**, 147.
17. Fawcett, J.S. and Morris, C.J.O.R. (1966). *Separation Studies*, **1**, 9.
18. Gelfi, C. and Righetti, P.G. (1981). *Electrophoresis*, **2**, 213.
19. Campbell, W.P., Wrigley, C.W., and Margolis, J. (1983). *Analytical Biochemistry*, **129**, 31.
20. Righetti, P.G., Brost, B.C.W., and Snyder, R.S. (1981). *Journal of Biochemical and Biophysical Methods*, **4**, 347.
21. Gelfi, C. and Righetti, P.G. (1981). *Electrophoresis*, **2**, 220.
22. Kingsbury, N. and Masters, C.J. (1970). *Analytical Biochemistry*, **36**, 144.
23. Andrews, A.T. (1986). *Electrophoresis; Theory, Techniques and Biochemical and Clinical Applications*, 2nd edn. Oxford University Press, Oxford.
24. Davis, B.J. (1964). *Annals of the New York Academy of Sciences*, **121**, 404.
25. Margolis, J. and Kenrick, K. G. (1968). *Analytical Biochemistry*, **25**, 347.
26. Caton, J.E. and Goldstein, G. (1971). *Analytical Biochemistry*, **42**, 14.
27. Jeppesen, P.G.N. (1974). *Analytical Biochemistry*, **58**, 195.
28. Uriel, J. and Berges, J. (1966) *Comptes Rendus Hebdomadaires des Séances de l'Academie des Sciences, Paris*, **262**, 164.
29. Uriel, J. and Berges, J. (1974). In *Electrophoresis and Isoelectric Focusing in Polyacrylamide Gel* (ed. R.C. Allen and H.R. Maurer), pp. 235–45. Walter de Gruyter, Berlin.
30. Peacock, A.C. and Dingman, C.W. (1968). *Biochemistry*, **7**, 668.
31. Serwer, P. (1983). *Electrophoresis*, **4**, 375.

32. Serwer, P. (1981). *Analytical Biochemistry*, **112**, 351.
33. Sugden, B., de Troy, B., Roberts, R.J., and Sambrook, J. (1975). *Analytical Biochemistry*, **68**, 36.
34. Loening, U.E. (1967). *Biochemical Journal*, **102**, 251.
35. Dahlberg, A.E., Dingman, C.W., and Peacock, A.C. (1969). *Journal of Molecular Biology*, **41**, 139.
36. Willoughby, E.W. and Lambert, A. (1983). *Analytical Biochemistry*, **130**, 353.
37. Goldman, D. and Merril, C.R. (1982). *Electrophoresis*, **3**, 24.
38. Merril, C.R., Dunau, M.L., and Goldman, D. (1981). *Analytical Biochemistry*, **110**, 201.
39. Prunell, A., Kopecka, H., Strauss, F., and Bernardi, G. (1977). *Journal of Molecular Biology*, **110**, 17.
40. Sutherland, J.C., Lin, B., Monteleone, D.C., Mugavero, J., Sutherland, B.M., and Trunk, A. (1987). *Analytical Biochemistry*, **163**, 446.
41. Chen, C.Z. and Sutherland, J.C. (1989). *Electrophoresis*, **10**, 318.
42. Schwartz, D.C., Saffran, W., Welsh, J., Haas, R., Goldenberg, M., and Cantor, C.R. (1982). *Cold Spring Harbor Symposia on Quantitative Biology*, **47**, 189.
43. Orbach, M.J., Vollrath, D., Davis, R.W., and Yanofsky, C. (1988). *Molecular and Cell Biology*, **8**, 1469.
44. Schwartz, D.C. and Cantor, C.R. (1984). *Cell*, **37**, 67.
45. Carle, G.F. and Olsen, M.V. (1984). *Nucleic Acids Research*, **12**, 5647.
46. Kenwick, S., Patterson, M., Speer, A., Fischbeck, K., and Davies, R. (1987) *Cell*, **48**, 351.
47. Smith, C.L., Econome, J.G., Schutt, A., Klco, S., and Cantor, C.R. (1987). *Science*, **236**, 1448.
48. Gardiner, K. and Patterson, D. (1989). *Electrophoresis*, **10**, 296.
49. Stellwagen, N.C. (1985). *Biopolymers*, **24**, 2243.
50. Slater, G.W. and Noolandi, J. (1986). *Biopolymers*, **25**, 431.
51. Serwer, P. (1988). *Applied and Theoretical Electrophoresis*, **1**, 19.
52. Birren, B.W., Hood, L., and Lai, E. (1989). *Electrophoresis*, **10**, 302.
53. Crater, G.D., Gregg, M.C., and Holzwarth, G. (1989). *Electrophoresis*, **10**, 310.
54. Carle, G.F., Frank, M., and Olsen, M.W. (1986). *Science*, **232**, 65.
55. Sutherland, J.C., Monteleone, D.C., Mugavero, J.H., and Trunk, J. (1987). *Analytical Biochemistry*, **162**, 511.
56. Craig, J., Fowler, S., Skinner, J.D., Burgoyne, L.A., and McInnes, J.L. (1988). *Applied and Theoretical Electrophoresis*, **1**, 23.
57. Gejman, P.V., Sitaram, N., Hsieh, W.T., Gelernter, J., and Gershon, E.S. (1988). *Applied and Theoretical Electrophoresis*, **1**, 29.
58. McPeek, F.D., Coyle-Morris, J.F., and Gemmill, R.M. (1986). *Analytical Biochemistry*, **156**, 274.
59. Carle, G.F. and Olson, M.V. (1985). *Proceedings of the National Academy of Sciences of the USA*, **82**, 3756.
60. Hightower, R.C. and Santi, D.V. (1989). *Electrophoresis*, **10**, 283.
61. Gardiner, K., Laas, W., and Patterson, D. (1986). *Somatic, Cell and Molecular Genetics*, **12**, 185.
62. Chu, G., Vollrath, D., and Davis, R.W. (1986). *Science*, **234**, 1582.
63. Chu, G. (1989). *Electrophoresis*, **10**, 290.
64. Clark, S.M., Lai, E., Birren, B.W., and Hood, L. (1988). *Science*, **241**, 1203.

65. Birren, B.W., Lai, E., Hood, L., and Simon, M.I. (1989). *Analytical Biochemistry*, **177**, 282.
66. Southern, E.M., Anand, R., Brown, W.R.A., and Fletcher, D.S. (1987). *Nucleic Acids Research*, **15**, 5925.
67. Vollrath, D. and Davis, R.W. (1987). *Nucleic Acids Research*, **15**, 7865.
68. Birren, B.W., Lai, E., Clark, S.M., Hood, L., and Simon, M.I. (1988). *Nucleic Acids Research*, **16**, 7563.
69. Anand, R. (1986). *Trends in Genetics*, **2**, 278.
70. Serwer, P. (1987). *Electrophoresis*, **8**, 301.
71. Sutherland, J.C., Emrick, A.B., and Trunk, J. (1989). *Electrophoresis*, **10**, 315.

6

Recovery of DNA from electrophoresis gels

PAUL TOWNER

1. Introduction

A task which any recombinant DNA worker will have to overcome is the recovery of DNA fragments from gel support media. Generally this will necessitate obtaining DNA fragments from agarose gels in a condition suitable for further manipulative procedures such as restriction analyses, labelling for use as probes and, most importantly, for ligations and transformations. These criteria are not easy to fulfil, and as a consequence a variety of techniques have been developed.

Native polyacrylamide gels can be used to separate DNA fragments of a few hundred base-pairs in much the same way as agarose, but no advantage in their use is conferred over high concentration agarose gels. Strand-separating or denaturing polyacrylamide gels contain 7 M urea and are used for the purification of chemically-synthesized oligonucleotides in mg quantities or for the retrieval of single-stranded nucleic acid of several hundred base-pairs for analytical purposes.

This chapter assumes the reader to have prior knowledge of the use of electrophoresis gels (Chapter 5), possibly having made some attempts at the recovery of DNA from them. The methods described here aim to draw all the successful recovery protocols together, none the less, the references to original articles and the information contained in descriptive texts (1–4) is well worth reading to compare and contrast the variety of procedures.

2. Agarose gels

Electrophoresis-grade agarose suitable for the separation of DNA fragments is obtainable from many suppliers. Two standard types are available, normal gelling temperature agarose, which is used for the majority of experiments, and low-melting-point (LMP) agarose. The latter is purer than the normal type, and even though more expensive is well worth the extra cost when used for retrieving DNA fragments because it can be easily molten at 70°C without detriment to the

DNA sample. Agarose gels can be prepared with 0.3 to 2.5% (w/v) agarose in TBE or TAE buffers (Chapter 5, *Table 1*). Some workers suggest that DNA recovery from TAE gels yields better quality DNA.

There is no limit to the size of double-stranded DNA fragments which can be recovered from agarose gels. However, fragments below 250 bp are poorly resolved and the large DNA species (>50 kb) separated by pulsed field gel electrophoresis are bound to become sheared during the extraction. However, no problems of shearing arise during the recovery of lambda vector arms of 22 kb.

2.1 Recovery of DNA from agarose gels

Several methods for retrieval of DNA from agarose gels have been developed and are based either on physical separation or electroelution. They all work quite well and it is a matter of finding the procedure which works best in your hands. The main criteria for choosing a suitable recovery method are the quantity of DNA required and the purpose for which it will be used. If a 0.2 μg sample is required, this can be separated on a gel using a normal well, and would probably result in a gel block of about 25 μl containing the DNA. This is a convenient size to use in the glass powder (Section 2.1.1.*iv*) or freeze-squeeze (Section 2.1.1.*ii*) procedure. On the other hand 50 μg DNA, which is a considerable amount used for the preparation of a DNA stock, would require electrophoresis on a 10-cm-wide lane and result in 2 to 3 ml of gel slice. This quantity of DNA would preferably be separated on a LMP gel and extracted using the melting/DE52 procedure (Section 2.1.1.*iii*).

The best strategy to adopt is to experiment with DNA fragments similar in size to the sample of interest, tagged with a ^{32}P-label (see Volume II, Chapter 4) for ease of detection using a scintillation GM tube. This will then enable the fate of the DNA sample to be followed semi-quantitatively. Electrophoresis of the DNA on a gel containing 0.01% (w/v) ethidium bromide, as described in Chapter 5, Section 4.4, allows visualization of the DNA bands when placed on a transilluminator. Gel slices containing the DNA fragments can then be conveniently excised with a single-edge razor blade, being careful not to damage the transilluminator surface. The block should then be trimmed on all six sides to remove portions of agarose devoid of DNA and the agarose containing the DNA placed into a suitable tube.

The fragment of DNA can then be separated from the gel matrix and the fate of the radiolabel traced. A good recovery of radioactive material should indicate a good recovery of DNA. However, this material might be totally unsuitable for cloning because it is of poor quality and probably contaminated with sulphonated polysaccharides from the agarose. It is generally necessary to further purify the DNA sample by adsorption chromatography, using reverse phase or using DE52-Sepharose followed by size exclusion for desalting (Section 2.2). Some of the available methods combine these steps and result in good recoveries of clonable DNA. It should be stressed that if small amounts of DNA are being recovered any contact with a glass surface is a potential route for loss of material.

All glassware, universal bottles, corex tubes, and glass wool must therefore be siliconized to reduce potential losses (see *Appendix 4*).

2.1.1 Physical separation methods

i. Phenol extraction

In this procedure the gel slice is broken up using a spatula and shaken in buffer to encourage solubilization of DNA. This mixture is extracted using phenol, so the DNA remains in the upper aqueous layer which is removed and treated with ethanol to precipitate the DNA. Although this is a straightforward procedure, which is more successful with TAE-based gels, the yield is poor. If LMP agarose is used the gel slice can first be melted and diluted with buffer prior to phenol extraction, which results in a much better yield.

Protocol 1. Phenol extraction of DNA from agarose gels

1. Electrophorese the DNA sample in LMP agarose preferably in TAE buffer (Chapter 5, *Table 1*).

2. Isolate the agarose block containing DNA and, depending on volume, place in either a 1.6 ml microfuge tube or a universal bottle.

3. Incubate at 70°C for 20 min to completely melt the gel.

4. Add 3 vol. of TAE buffer containing 0.1 M NaCl and incubate at 37°C for 5 min.

5. Add 4 vol. of phenol (equilibrated with 50 mM Tris–HCl pH 8.0, see Chapter 3, *Protocol 1A*) also at 37°C and vortex occasionally over 5 min to keep the mixture homogeneous.

6. Centrifuge at room temperature, 5 min at 15 000 g for microfuge tubes and 10 min at 2000 g in a swing-out rotor of a bench centrifuge for universal bottles.

7. Remove the upper aqueous layer containing the DNA without disturbing the interface of phenol–agarose. Transfer to fresh microfuge or corex tubes.

8. Add an equal volume of chloroform–isoamyl alcohol (24:1) to the aqueous layer and mix vigorously. Centrifuge as in step 6.

9. Remove the upper aqueous layer and add 0.1 vol. of 3 M sodium acetate pH 5.6 and 2.5 vol. of cold ethanol.

10. Incubate on ice for 30 min and centrifuge, microfuge tubes at 15 000 g, corex tubes at 12 000 g for 30 min at 4°C.

11. Discard the supernatant and carefully rinse the DNA residing in the tube with 70% (v/v) ethanol (it is worthwhile recentrifuging if the tube is even slightly jolted since there is a risk of discarding the DNA sample).

12. Dry the tube *in vacuo*.

Protocol 1. *Continued*

13. Dissolve the DNA precipitate in TE pH 7.4 (see *Appendix 4*) in the proportion of 0.25 vol. of TE per original volume of agarose block.

14. Assess the recovery by running a sample on a fresh gel; if it is poor try re-extracting the phenol layer at step 7. If you were encouraged by the amount of material visible in the tube at step 12 it is likely that agarose is present in the sample, this makes it difficult to redissolve the DNA and makes it necessary to purify the sample further.

---

ii. Freeze-squeeze

This method is best used with normal gelling temperature agarose and is especially reliable with small fragments of DNA (typically 200 to 800 bp). The procedure is most conveniently done by placing the gel slice on to a bed of glass wool in a small tube through which a hole is pierced (5). The tube is snap-frozen in liquid N_2 and placed in a larger tube then centrifuged. As the slice thaws the aqueous phase is exuded and flows into the larger tube while the agarose is retained on the glass wool. Yields can be almost 100% (6) but the material is unsuitable for cloning purposes. However, by passing the solution through a NENSORB 20 column (see Section 2.2.1) the DNA is then of exceptional quality.

Protocol 2. Isolation of DNA from agarose by 'freeze-squeeze'

1. Isolate the block of agarose containing the DNA of interest and place it into a bijou bottle with ten times its volume of 0.3 M sodium acetate pH 7.0, 1 mM EDTA. The block should not exceed a volume of 150 μl; if it does then divide it into several portions.

2. Gently agitate the tube for 30 min to effect equilibration of the buffer into the gel.

3. Transfer the agarose block containing the DNA from the bijou to a 0.6 ml microfuge tube which has been punctured at its base with a 19-G needle and plugged with a small quantity of glass wool.

4. Wrap the tube in aluminium foil and immerse in liquid N_2 for 30 sec.

5. Remove the foil from the tube quickly and place the small microfuge tube into a 1.6-ml microfuge tube and centrifuge for 15 min, 15 000 g at room temperature.

6. Remove the smaller tube and add 2.5 vol. of ethanol to the eluted solution containing the DNA, mix, incubate at $-20°C$ for 30 min, and then centrifuge at 4°C, 15 000 g for 30 min.

7. Carefully decant off the supernatant, add 1 ml of 70% (v/v) ethanol to the tube, gently invert to ensure the tube contents have mixed, centrifuge at 15 000 g for 30 min at 4°C, and pour away the supernatant.

Protocol 2. *Continued*

8. Dry the DNA sample under vacuum, and dissolve in TE pH 7.4 (*Appendix 4*) at the ratio of 20 μl for every 100 μl of original gel, assuming it contained up to 2 μg of DNA. Estimate the quantity of material recovered by running a sample in an agarose gel.

9. The material is unlikely to be of clonable quality, but its purity can be substantially improved by DE52 chromatography followed by a G25 spun column (see *Protocols 8* and *9*), in which case NaCl should be substituted for sodium acetate in step 1.

This obviates the need for ethanol precipitation because the eluted material (step 6) can be applied directly to the DE52. Alternatively, follow the procedure to step 8, and chromatograph the sample using a NENSORB 20 column (see Section 2.2.1).

---

iii. Gel melting

This procedure is applicable only to LMP agarose. The only disadvantage is that gels are fairly fragile and must be prepared in a cold room. It is a wise precaution to be generous with the agarose solution and not to dip combs too far into the gel tray when casting, so that the wells have less chance of splitting. The identified gel slice is melted at 70°C and diluted with several volumes of low salt buffer, cooled to 45°C and passed through a DE52 column. The DNA is eluted with a high salt buffer and then desalted through a G25 column. In my experience this is the most reliable method for the recovery of DNA from 0.5 to 10 kb in high yields and is routinely employed in the preparation of phosphatased vectors (see Chapter 7, Section 3.3).

Protocol 3. Recovery of DNA from low gelling temperature agarose

1. Remove the band of interest from the gel, place in a universal bottle and incubate for 20 min at 70°C in a water bath. The gel should be completely molten with no trace of particulate material.

2. Dilute the molten agarose with 25 volumes of wash buffer (0.2 M NaCl in TE pH 7.0, see *Appendix 4*) which has also been equilibrated at 70°C, and then leave to incubate at that temperature for a further 10 min.

3. Transfer the sample to a 45°C water bath and incubate for 5 min for temperature equilibration.

4. Load the solution at 2–4 ml per min on to an ion-exchange column which has been equilibrated in wash buffer at 45°C. The column can be prepared using any DE52-based support, such as DE52-cellulose (Whatman) or DEAE-Sephacryl or Sepharose (Pharmacia) and is preferably made in a commercially-available plastic column approximately 100 mm long and

Protocol 3. *Continued*

10 mm diameter fitted with a sintered disc. However, pipette tips and syringe barrels will suffice. A bed volume of 10 mm diameter and 25 mm high is sufficient to retrieve 20 μg of DNA. The column must be immersed up to its neck in the water bath, otherwise the agarose may cool slightly as it flows on to the DE52 and block the column. The eluant from the column is conducted to a suitable receptacle by fitting a flexible tube to the column tip which is fed through the water bath on to the bench. It is convenient if a modified organ bath, of the type used by pharmacologists, is available because the procedure can be quite messy. DNA is readily adsorbed to DE52-based supports in the presence of 0.2 M NaCl whereas substances within the agarose wash through.

5. Wash the column with 25 column volumes of wash buffer at 45°C; this step is important because it removes all traces of agarose.

6. Elute the DNA sample with 2.5 ml of 2 M NaCl in TE pH 7.0.

7. Pass the DNA through a PD10 column (Pharmacia) to desalt. Collect in 3.5 ml.

8. Precipitate the DNA in a 30-ml corex tube by adding 0.35 ml 3 M sodium acetate pH 7.0, 9 ml ethanol, and incubate at −20°C for 30 min. Finally, centrifuge at 4°C, 12 000 g for 30 min.

9. Discard the supernatant and carefully rinse the DNA in the tube with ice-cold 70% (v/v) ethanol and dry the tubes under vacuum.

10. Take up the DNA in the tube in 200 μl TE pH 7.4 and estimate the recovery by running a sample on an agarose gel.

11. This material is of clonable quality. The PD10 column can be avoided by adding ethanol directly to the eluant from the DE52 column. If this is done, the DNA present at step 9 should be desalted on a G25 spun column (see *Protocol 9*).

---

iv. Adsorption to glass powder

Tiny glass particles have an enormous surface area available for binding DNA and can be used with DNA-containing gel slices which have been dissolved using NaI solution (7). The glass powder is then washed several times with a buffer containing 50% (v/v) ethanol to remove impurities, and the DNA eluted with an aqueous buffer. This method works well with small quantities of gel and produces DNA of clonable quality. The crucial part of this protocol is that a glass powder with ideal characteristics is required. Suitable material is commercially-available in a complete kit known as Gene-Clean (BIO 101 Inc.). It is suggested that improved results are obtained using TAE-based gels.

Protocol 4. Recovery of DNA from agarose by the glass powder method

1. Excise the band from the gel and place in a 1.6 ml microfuge tube. Add 3 vol. of a saturated solution of NaI and incubate at 50°C for 3–5 min to dissolve the agarose completely.

2. Add 5 μl of glass powder suspension, vortex and leave for 5 min to enable the DNA to adsorb to its surface.

3. Centrifuge the tube by pulsing for 5 sec at full speed in a microfuge to pellet the glass particles to which the DNA has adsorbed. It is important not to over-centrifuge, otherwise a clump is formed which is difficult to resuspend.

4. Remove the supernatant and discard.

5. Add 0.2 ml of wash buffer to the glass powder and resuspend the pellet by vortexing, then centrifuge as in step 3 and discard the supernatant. Repeat this washing cycle twice. Wash buffer is prepared by mixing equal volumes of absolute ethanol and 20 mM Tris–HCl pH 7.2, 0.2 M NaCl, 2 mM EDTA; this mixture stores indefinitely at −20°C. Impurities in the DNA which had bound to the glass powder are washed away, but the DNA firmly adheres due to the presence of ethanol.

6. At the end of the final wash cycle ensure that all liquid is removed from the pellet by drying the inside of the tube with tissue, but do not disturb the pellet.

7. Add 20 μl of TE pH 7.4 (*Appendix 4*) to the pellet, then resuspend and incubate at 50°C for 5 min to elute the DNA from the glass powder. Pulse-centrifuge as in step 3 and remove the TE, containing the DNA. Add a further 20 μl TE and repeat the extraction procedure.

8. If sufficient DNA was used, run a sample on a gel. If this shows a satisfactory yield then the sample will probably be of clonable quality. If only traces of DNA are detectable then try re-extracting the glass powder. It is also possible that the DNA is still in the NaI solution (discarded in step 4) because of insufficient adsorption to the glass powder surface. This protocol is devised for 1 to 5 μg DNA; if considerably less material is being handled it is expedient to use less glass powder suspension; conversely, increased amounts of DNA necessitate more.

2.1.2 Electroelution methods

i. In situ *elution*

DNA can be isolated from whole gels by cutting a slot in the gel just ahead of the DNA band. The tank buffer is adjusted so that it is level with the surface of the gel. The slot is filled with fresh buffer and electrophoresis resumed. If UV-transparent apparatus is available the elution can be monitored, otherwise the solution containing the DNA is periodically removed and replaced with fresh buffer. It is important to capture the DNA before it has time to migrate through the buffer and re-enter the gel. This can be prevented by placing dialysis membrane on the anode side of the wall, but can be irksome because the DNA then needs to be electroeluted from the membrane.

Protocol 5. *In situ* electroelution of DNA from agarose gels

1. Electrophorese the DNA sample and observe the position of the fragment of interest within the gel by viewing through a UV-transparent gel plate.

2. Use a scalpel blade to excise a region of gel which will give a rectangular slot adjacent to the band of interest. This can be at the boundary of any of the four sides of the fragment but is usually done against the longest dimension so that the electroelution is quicker.

3. Place the gel back into the electrophoresis tank and remove the buffer until the level is just below the top of the gel. Place a glass plate on the gel to anchor it down but leave sufficient space to access the slot and observe migration of DNA if UV-transparent apparatus is available.

4. Replace any buffer in the slot with a fresh sample.

5. Electrophorese the DNA fragment toward the anode and the slot. By observing, or through a knowledge of the migration rate, remove the buffer containing DNA from the slot and replace with fresh, aiming for five replacements to obtain all the DNA sample.

6. The recovered DNA will not be very pure. The best course to follow is to adjust the buffer in the sample and pass it through a DE52-resin before further use (see *Protocol 8*).

---

ii. Electroelution on to DE52-paper

An incision is made in the gel just ahead of the band of interest and a sliver of ion-exchange paper put in place. Electrophoresis is resumed until all of the ethidium-staining material has transferred from the gel and is adsorbed on to the paper, and is conveniently observed with UV-transparent apparatus. The paper is washed with isopropanol to remove the ethidium bromide and the DNA eluted with a high salt buffer.

Protocol 6. Electroelution of DNA on to DE52-paper

1. Prepare and run the agarose gel in a UV-transparent mould; if a gel tank with a UV-transparent base is available, then all the better.

2. Identify the band(s) of interest and make an incision in the agarose on the anode side of the band using a scalpel blade. Open the slit in the gel and slide in a piece of pre-wetted DE52-membrane (Schleicher & Schuell) or DE81-paper (Whatman). This should be wider than the DNA band by 2 mm on either side.

Protocol 6. *Continued*

3. Resume electrophoresis and monitor DNA migration by observing with the transilluminator every few minutes. This is critical if several bands are closely spaced and unique fragments are to be isolated. The fluorescent band will disappear from the gel and be adsorbed on to the DE52-paper.

4. Remove the paper, which will have a fluorescent patch on it of intensity in proportion to the amount of DNA transferred, and rinse it in distilled water.

5. Place the DE52-paper to which the DNA is bound (ensure that traces of agarose are not transferred) into a 1.6 ml microfuge tube. Add 1 ml of isopropanol and invert several times to encourage dissolution of the ethidium bromide. Remove and discard the supernatant. Repeat this extraction twice.

6. Add 1 ml wash buffer (0.2 M NaCl in TE pH 7.4, *Appendix 4*) and incubate for 10 min at 37°C. Remove and discard the supernatant, then repeat once. This step will remove contaminants, but not DNA, from the paper.

7. Transfer the DE52-paper to a fresh tube and add sufficient elution buffer (2 M NaCl in TE pH 7.4) to cover the area of paper where the DNA is bound.

8. Incubate at 45°C for 60 min. Remove the supernatant containing the DNA and desalt it through a G25 spun column (see *Protocol 9*). Assess recovery, and if it is poor elute the DNA which is probably still bound to the paper with a larger volume of elution buffer and incubate at a higher temperature (e.g. 65°C). The presence of DNA still bound to the paper can be assessed by re-staining with ethidium bromide.

---

iii. Electroelution into a dialysis tube

The DNA contained within gel slices can be electroeluted by placing the slices in a dialysis tube in an electric field. The DNA elutes from the gel and adheres to the dialysis tube wall, from which it can be freed by reversing the current for a short time. The buffer containing the DNA is removed and either extracted with phenol–chloroform and precipitated using ethanol or further purified through an ion-exchange column.

Protocol 7. Electroelution of DNA into a dialysis tube

1. Transfer a gel slice containing the DNA fragment into a dialysis tube which has been rinsed in distilled water. The tubing is prepared by boiling for 10 min in 50 mM sodium carbonate, 1 mM EDTA (pH 8.0) and stored at 4°C.

2. Place 50 vol. of electrophoresis buffer per volume of gel slice into the tube and clamp the ends.

3. Place the tube in the electrophoresis tank such that the long dimension of the agarose block is parallel with the dialysis tube walls and to the electrodes of the tank. Electrophorese for 30 min using the conditions under which the

Protocol 7. *Continued*

original separation was performed. If UV-transparent apparatus is available the electroelution can be monitored every few min. The fluorescing band of DNA will be seen to transfer from the gel; however, it will not be apparent that the DNA is now firmly bound to the dialysis tube wall.

4. Without disturbing the position of the tube in the tank, reverse the current to enable the DNA to unlatch from the dialysis tube and enter the buffer solution; 60 sec is sufficient, otherwise the DNA may bind to the opposite wall of the tube.

5. Remove the buffer containing the DNA from the dialysis tube and rinse the tube with 1 ml TE pH 7.4 (*Appendix 4*), then combine the two solutions.

6. The DNA in solution must be further purified by DE52 chromatography (*Protocol 8*) to be of any quality.

7. With large quantities of DNA (>20 μg) recovery can be quite good, whereas samples of 0.5 μg tend to be lost during the manipulations.

2.2 Purification of DNA

2.2.1 NENSORB 20 chromatography

This type of column (marketed by DuPont) has been successfully used for some time, however, many other similar reverse phase mini-columns are on the market and perform equally as well. Any nucleic acid sample can be loaded on to the column as long as it is not too badly contaminated with detergent, phenol or chloroform, which decrease binding. After binding, the nucleic acid remains firmly attached to the resin and can be washed exhaustively. The advantage of NENSORB 20 is that the nucleic acid can be eluted in a small volume of 50% (v/v) methanol or other alcohol and dried down in a siliconized microfuge tube resulting in a completely salt-free sample.

2.2.2 DE52 chromatography

DE52-resins are available on a variety of support media and are marketed as DE52-cellulose (Whatman) and DEAE-Sepharose or Sephacel (Pharmacia). They are all equally as good for home-made mini-columns. Commercial units are supplied by BRL, who produce NACS columns, and Hybaid who supply Qiagen Tip 20 cartridges.

Protocol 8. Purification of DNA by DE52 chromatography

1. Prepare a slurry of DEAE-based resin according to the manufacturer's instructions in TE pH 7.4 containing 0.2 M NaCl (*Appendix 4*).

2. Prepare a column of the resin in either a small syringe barrel or a pipette tip after plugging with glass wool. The procedure described here is based on small

Protocol 8. *Continued*

Gilson pipette tips ('yellow tips') which can be filled to give a bed volume of 100 μl resin, sufficient to load up to 5 μg DNA.

3. Dissolve the DNA sample in TE pH 7.4 containing 0.2 M NaCl, or adjust it to 0.2 M NaCl if already in solution. A sample of 5 μg DNA should be in a volume of at least 200 μl.

4. Pipette the sample on to the column and discard the eluant.

5. Wash the resin with ten to twenty column volumes of TE pH 7.4, 0.2 M NaCl.

6. Elute the DNA sample with 80 μl TE pH 7.4, 2 M NaCl; repeat with a further 80 μl.

7. Desalt the DNA through a G25 spun column (see *Protocol 9*).

8. If the column bed is a much larger volume (as in a syringe barrel) elute with 2.5 ml buffer and desalt through a PD10 column.

2.2.3 Removal of low-molecular-weight contaminants from DNA

The use of DE52 for DNA purification results in samples which are quite pure but which contain 2 M NaCl. Other procedures, such as ethanol precipitation, will on occasion leave traces of salt contaminating the sample. This becomes a problem if the DNA is then dissolved in such a small volume of buffer that the salt is now of a significant concentration. In *Protocols 9* and *10* procedures are described which will effectively remove contaminating solutes from 80-μl and 25-μl samples respectively. Larger volumes (2.5 ml) require desalting using a PD10 (Pharmacia) column.

Protocol 9. Desalting DNA with a spun G25-size exclusion column

1. Prepare a slurry of G25-resin by swirling 5 g into 100 ml TE pH 7.4 (*Appendix 4*) and leaving to swell for several h.

2. Plug a 1-ml syringe barrel with glass wool and pipette the agitated G25 into it. Add more material until the column bed has completely filled the barrel.

3. Place the column into a plastic or glass test-tube and centrifuge in a swing-out rotor of a bench centrifuge at 2000 g for 4 min.

4. Place the column into a fresh tube and centrifuge as above to check that no further liquid passes from the column.

5. Cut away the cap from a 1.6 ml microfuge tube and slide it into the bottom of a test-tube of 125 × 13 mm internal diameter. Place the spun-column into this tube so that its tip feeds into the microfuge tube.

6. Pipette 50–80 μl of DNA to be desalted on to the column bed and centrifuge as in step 3.

Protocol 9. *Continued*

7. The sample of DNA in the microfuge tube will be reduced by a few microlitre. A bed height of 80 mm will completely desalt 80 μl of 5 M NaCl.

Protocol 10. Using Sepharose-CL6B to desalt DNA

1. Puncture a 0.6 ml microfuge tube with a 23-gauge needle and place into it sterile glass beads (200–400 μ) to give a bed height of 3 mm.

2. Carefully pipette 200 μl of Sepharose-CL6B on to the bead layer, place into a 1.6 ml microfuge tube and centrifuge horizontally for 2 min. This can be done at 1000 g with the tube combination perched on the top of a 10 mm diameter plastic tube in a swing-out rotor of a bench centrifuge. The centrifugation will result in a bed volume of 100 μl if the gel matrix was suspended in a twofold volume of buffer.

3. Replace the large microfuge tube. Carefully layer 20 μl of DNA sample on to the Sepharose surface and spin as above.

4. The ratio of sample to bed volume should not exceed 5:1 to ensure complete desalting.

2.3 Tips and strategies for recovery of DNA from agarose

Each of the procedures described has its proponents, but no overwhelmingly successful method has yet appeared which is quick and totally reliable. I favour the strategy of obtaining a high recovery of DNA regardless of its purity. The material can then be purified using reverse phase or ion-exchange chromatography. Several devices and product kits are on the market and can be used with success, nevertheless buying and using a kit will not necessarily be overwhelmingly better than relying on home-made units, which are considerably cheaper. For example, Schleicher & Schuell produce the 'Biotrap', which is a sophisticated electroelution device but its successful application is only a slight improvement upon using dialysis tubes in an electrophoresis tank. Similarly, the glass powder extraction procedure marketed as 'Gene-Clean' (BIO 101 Inc.) is convenient and reliable, but if suitable glass powder is available it is simple enough to follow the procedure in *Protocol 4*.

3. Acrylamide gels

Electrophoresis gels based on polyacrylamide can be used under native or denaturing conditions (Chapter 5, Section 4.1). Denaturing gels of 4–25% (w/v) acrylamide in TBE containing 7 M urea are used to separate and recover single-stranded DNA molecules. High percentage, 16–25% acrylamide, is used for the purification of synthetic oligonucleotides of 10 to 120 bases, whereas low concentrations, 4–10% can be used for the recovery of single-stranded nucleic acids several hundred bases long. Less frequently used are non-denaturing polyacrylamide gels of 2–10% which separate double-stranded DNA of 50–1000 bp. Polyacrylamide gels are electrophoresed vertically between two glass plates. Preparative gels are 200 × 200 mm with a gel thickness of 2.0 mm. A gel lane of 150 mm can then accommodate all of the material obtained from a 0.2 μmol deoxyribo-oligonucleotide synthesis. After electrophoresis the bands of interest can be visualized by UV-shadowing, directly on the backing plate if it is of the green-tinted fluorescent variety. If not, a fluorescent TLC plate placed under the polyacrylamide and irradiated with UV will show the band. The region of gel containing the full length oligonucleotide can then be sliced out accurately with a scalpel.

Analytical gels are 0.35 mm thick and are used to separate ng to μg amounts of nucleic acids which are detected by virtue of the presence of a radiolabel. Nucleic acids in such gels can be recovered by slicing out regions of polyacrylamide after aligning the autoradiograph with the gel and marking the position of the bands on the gel backing plate.

3.1 Recovery of DNA from polyacrylamide gels

To obtain unique, full-length oligonucleotides from a chemical synthesis it is often worthwhile electrophoresing the mixture in a polyacrylamide gel, from which the DNA can be recovered with 30 to 50% yield. This is usually good enough for most procedures and represents mg quantities. If oligonucleotides are required for cloning purposes the amounts needed are so small that any contaminants present are diluted infinitely. The method described in *Protocol 11* is also suitable for the recovery of much longer strands of nucleic acid and is routinely used to isolate RNA molecules of several hundred bases. Similarly, this method is suitable for the recovery of double-stranded DNA from non-denaturing polyacrylamide, but does not give very high yields.

Protocol 11. Recovery of DNA from polyacrylamide gels

1. Slice the band of interest from the gel and place it in a bijou or universal bottle if the material is from a 2-mm-thick gel. If the gel was 0.34 mm thick and up to 30 mm wide, the slice will fit into a 1.6-ml microfuge tube.

2. Oligonucleotide DNA from 2-mm-thick gels is extracted by adding sufficient 0.3 M sodium acetate pH 6.8, 1 mM EDTA to cover the gel in the bottle, which is left to agitate on a table shaker for 12–18 h at ambient temperature.

3. Polynucleotide samples from 0.35 mm gels of lower acrylamide concentration are extracted by adding buffer as in step 2 to just cover the slice and storing at 4°C for 3–6 h.

Protocol 11. *Continued*

4. Remove the buffer from the gel, being careful not to transfer any polyacrylamide particles, and transfer to a corex or microfuge tube depending on the volume. Gel samples in microfuge tubes should then be re-extracted with fresh buffer.

5. Add 3 vol. of cold ethanol to oligonucleotide samples, and 2.5 vol. to polynucleotides, incubate at $-20°C$ for 30 min then centrifuge at 0°C for 30 min, 15 000 g for microfuge tubes, 12 000 g for corex tubes.

6. Discard the supernatant and rinse the tube contents carefully with 70% ethanol.

7. Discard the supernatant and dry the tube *in vacuo*.

8. Take up the DNA sample in TE pH 7.4 (*Appendix 4*).

9. Assess the yield by taking an absorbance spectrum (for oligonucleotides; see Chapter 3, *Protocol 2*) or by assessing radioactivity recovered (for labelled polynucleotides).

Extraction of nucleic acid from polyacrylamide is best done on whole gel slices. If the slices are fragmented the yield of DNA does not improve, but problems ensue because particles of polyacrylamide contaminate the DNA when it is being precipitated.

If it is imperative that all the material in the gel slice is recovered then electroelution of the slice will remove all the nucleic acid, as opposed to the two-thirds obtained by buffer diffusion. Rather than doing this in a dialysis bag, where a great deal of material will be lost, it is preferable to cast the polyacrylamide gel slice into a LMP agarose gel and then electrophorese the DNA into the agarose where it can be recovered following the methods in the previous sections.

Acknowledgements

I would like to express my sincere thanks to all my colleagues at Bath who have tried and tested many of the techniques that I have described.

References

1. Perbal, B. (1988). *A Practical Guide To Molecular Cloning*, 2nd edn. Wiley Interscience, New York.
2. Sealey, P.G. and Southern, E.M. (1982). In *Gel Electrophoresis of Nucleic Acids* (ed. D. Rickwood and B.D. Hames), pp. 53–5. IRL Press, Oxford.
3. Ogden, R.C. and Adams, D.A. (1987). In *Methods in Enzymology*, Vol. 152 (ed. S.L. Berger and A.R. Kimmel), pp. 84–7. Academic Press, London.

4. Sambrook, J., Fritsch, E.F., and Maniatis, T. (1989). *Molecular Cloning, A Laboratory Manual*, 2nd edn. Cold Spring Harbor Laboratory, Cold Spring Harbor, New York.
5. Tautz, D. and Renz, M. (1983). *Analytical Biochemistry*, **132**, 14.
6. Koenen, M. (1989). *Trends in Genetics*, **5**, 137.
7. Vogelstein, B. and Gillespie, D. (1979). *Proceedings of the National Academy of Sciences of the USA*, **76**, 615.

7

Construction of recombinant DNA molecules

FRANK GANNON and RICHARD POWELL

1. Introduction

Although the central role of DNA in all living forms was well accepted by the 1950s it was very difficult to study it directly at that time. The information which was obtained about DNA came mostly from the interpretation of classical genetic experiments. The preponderance of these were performed using microorganisms whose relative simplicity and ease of manipulation made them the organisms of choice. The difficulty of obtaining a detailed understanding of a bacterial virus or phage such as lambda in which there are a mere 50 000 nucleotides (approximately) made the wish to understand the structure and function of higher organisms in which there are almost 10^5 times more nucleotides seem unattainable. Then in the mid-1970s a technological revolution occurred which has allowed this great chasm of accessible information to be bridged. As in all revolutions, developments in many unrelated areas came together in a way which magnified the individual components into a powerful new ensemble.

Research from many unrelated and frequently obscure areas of biochemistry, microbiology, chemistry, virology, and immunology provided the elements which we now recognize as recombinant DNA technology or genetic engineering. The essential result of this collage of techniques is that it is now possible to purify and amplify DNA fragments from any biological material. Direct analysis of DNA sequences is now a daily event in laboratories in all biological disciplines world-wide. The rapid growth of biotechnology as an industrial sector has been fuelled by the ability to express the proteins encoded by the isolated DNA fragments in host cells that grow readily in bioreactors. Our understanding of the detailed interactions which combine to allow the controlled expression of genes has contributed to areas as diverse as medicine and agriculture and molecular biologists are recognized to have valuable perspectives to bring to bear on many previously incalcitrant problems.

After a decade of practice the early steps in a genetic engineering experiment have been reduced to a simple equation:

vector + foreign DNA + host cell = gene library.

It is frequently a surprise to those entering the field that the experimental procedures involved, which have been analysed to an extent that they no longer warrant mention, are far from trouble-free. It should be reiterated that the skills and care of a biochemist are required to purify DNA free from all biological and organic factors which can interfere with subsequent steps, that the enzymes used are delicate tools which demand correct handling and that the biological hosts which serve as recipients for the foreign DNA require the knowledge and experience of a trained microbiologist. When all of these considerations are met then the construction of recombinant DNA molecules can and does work efficiently.

The demand for such molecules has given rise to a very large and supportive service industry. Every component for a recombinant DNA experiment can be purchased. Kits are available in which the skill required to obtain the molecule of interest is reduced to the ability to follow some very simple pipetting steps. Those who might find that too demanding can purchase recombinant DNA libraries ready made. However, most laboratories, even if they purchase ready made DNA clones, need to propagate, alter, sequence or generate new constructs from the DNA fragments. As such the A-B-C of digesting DNA, ligating it and using it in transformations has to be learned and appreciated by all. By becoming proficient in these skills the useful results which come from complex methods become readily attainable and reproducible.

2. Restriction enzyme digestions

In all aspects of genetic engineering, the strategy employed to reach the goal of the project comes from a sequence of major choices. The starting DNA is probably dictated by the sequence which is sought but the manner in which it is digested, the vector which is used and the host in which it is inserted are all inter-related choices. As the essential result from a genetic engineering experiment is the purification of a fragment of DNA which is more simple to analyse than that available in the animal, plant or microbial cell, the first step in processing purified DNA (see Chapter 3) is the generation of smaller and more manageable fragments. Almost without exception, this is now achieved by use of specific microbial nucleases called restriction enzymes.

The substrate nucleotide sequence at which each of several hundred restriction enzymes digest DNA is known (see *Appendix 5*). An examination of these will show that they are characterized by being palindromic. The manner in which these enzymes digest the DNA [leaving a 5′ protruding sequence, a 3′ protruding sequence, or blunt (flush) ends] is also known. Because of the palindromic nature of the restriction enzyme sites, fragments generated have complementary or 'sticky' ends. Ultimately these termini will be linked (ligated) to vector DNA (see Section 4.1) with homologous or compatible termini. Appropriate complementary termini can also be generated by two different enzymes. For example, digestion by *Bam*HI and by *Sau*3A both yield the complementary sequence $^{5'}$GATC$^{3'}$. Methods for routine restriction enzyme digestion by one or more restriction enzymes and the subsequent step of denaturing the restriction enzyme are shown in *Protocols 1–3*.

Protocol 1. Restriction enzyme digestion

1. Make up a 100 μl reaction mixture[a] by adding the reagents in order into a sterile 1.5 ml microfuge tube:

- 10 μl 10 × restriction enzyme buffer[b]
- x μl DNA
- y μl water

The value of x will depend on the concentration of your DNA solution.[c] The value of y should bring the volume up to 99 μl.

2. Add 1 μl of the restriction enzyme[d] and mix gently with the pipette tip.

3. Incubate in a water bath at the appropriate temperature for the desired length of time. The temperature and time depends on the enzyme and will be stated in the product guide provided by the supplier.

4. Store the reaction on ice while analysing the digestion by gel electrophoresis.

[a] When used for analytical purposes only, the reaction can be carried out in a volume of 20 μl.
[b] Use the buffer provided by the suppliers of the restriction enzyme, or make a buffer according to their instructions.
[c] Although concentrations of DNA as low as 10 ng can be visualized on an agarose gel, it is preferable to use 100–200 ng per digestion to facilitate easy detection of small DNA fragments.
[d] You should add about 1 unit of enzyme per microgram of DNA. If necessary dilute the enzyme stock with the storage buffer described in the supplier's product guide. It is acceptable to add more than 1 unit per microgram of DNA, but large excess amounts should be avoided.

Protocol 2. Sequential digestion by two restriction enzymes

1. Set up a reaction mixture as described in *Protocol 1*. This reaction mixture should be appropriate for the restriction enzyme requiring the lower salt concentration of the two enzymes to be used. If the reaction buffers of both enzymes are similar, the DNA may simply be incubated in the presence of both enzymes.

2. Incubate at the correct temperature for the desired length of time and place the reaction on ice for the time required for gel electrophoretic analysis.

3. Electrophorese a small sample of the reaction (Chapter 5, Section 4.4) and

Protocol 2. *Continued*

check for total digestion of the DNA. If the digestion is not total, re-incubate the reaction mixture for a further period of time. Alternatively, add a further aliquot of enzyme and re-incubate.

4. After electrophoretic analysis showing completion of digestion by the first restriction enzyme, change the salt concentration to that appropriate for the second restriction enzyme (see ref. 1) by adding a small aliquot (1–5 μl) of 5 M NaCl to the reaction mixture. Add the second enzyme and allow this reaction to progress before monitoring for correct double digestion by gel electrophoresis.

5. This method is adequate for many double digestions. However, in some cases the reaction conditions of the two enzymes may not be compatible to simple alteration of the salt concentration. Also the first enzyme may loose its DNA specificity and begin to cleave the DNA in an unexpected manner while under the reaction conditions for the second enzyme. In these cases the first enzyme may be completely inactivated after reaction by incubation at 65°C for 10 min before continuing with the second reaction. However, many enzymes are not heat-labile (see ref. 1) and therefore the DNA must be purified from the reaction mixture by phenol extraction and ethanol precipitation (*Protocol 3*) before continuing with the second restriction enzyme.

6. Complete double digestion of DNA by two enzymes recognizing restriction sites adjacent or very close to one another (as in a polylinker multicloning site) may not be distinguishable by simple gel electrophoresis from DNA cleaved by only one of the enzymes: both fragments may appear to be of similar size. A simple ligation reaction (*Protocol 9*) after DNA purification (*Protocol 3*) may be required to ensure correct double digestion. If the product of the double digestion is a DNA fragment with non-complementary termini, it cannot ligate to itself forming circular molecules as can a DNA fragment cleaved by one restriction enzyme. These differences are easily analysed by gel electrophoresis.

Protocol 3. Purification of DNA by phenol extraction and ethanol precipitation

1. Add an equal volume of phenol, prepared as described in Chapter 3, *Protocol 1A*, to the reaction mixture and vortex gently.

2. Separate the aqueous phase which contains the DNA from the organic phase by centrifugation in the microfuge, at 5000 r.p.m. for 5 min or at 10 000 r.p.m. for 1 min.

3. Remove the aqueous phase with care into a fresh microfuge tube and add an equal amount of chloroform–isoamyl alcohol (24:1). Repeat step 2.

Protocol 3. *Continued*

4. In order to precipitate the DNA, add a one-tenth volume of 3 M sodium acetate pH 5.5 to the aqueous phase and then 2 vol. of ethanol. Incubate at $-20°C$ overnight or for much shorter periods at $-80°C$ (20–30 min).

5. Pellet the precipitated DNA by centrifugation in the microfuge at 10 000 r.p.m. for 5–15 min. Remove the ethanol with care and dry the pellet in a dessicator or 50°C oven for 5 min. An extra wash with 70% (v/v) ethanol may be included to remove excess salt from the pellet. The dried DNA may be resuspended in sterile TE pH 8.0 (see *Appendix 4*) or water, and stored at 4°C for further manipulation.

6. This procedure denatures and removes contaminating protein from a DNA sample. A second useful method is drop dialysis, which can remove salt, SDS, and even enzyme inhibitors. As such, it can be used with the majority of the methods in this chapter involving DNA purifications before or after enzymatic reactions:

    (a) Gently place a drop dialysis filter (Millipore: cat. no. VSWP 02500), floating shiny-side up, on 10–20 ml dialysis buffer (TE pH 8.0 or water) in a Petri dish.

    (b) Pipette the DNA sample (10–100 μl) on to the filter.

    (c) Allow to dialyse for 1–2 h before removal for further analysis.

The target site for a restriction enzyme may be 4, 6 or more nucleotides in length (1). This fact dictates the size of the average DNA fragment which results from complete digestion. The probability of a restriction enzyme site occurring for an enzyme with a 4-nucleotide target site is 1 in 256 (there is a 1 in 4 chance of a given nucleotide occurring at any site, so the chances of a correct 4-nucleotide sequence occuring therefore is $1/4 \times 1/4 \times 1/4 \times 1/4 = 1/256$). This means that the average size of the DNA fragments which result from a complete digestion of DNA by a restriction enzyme with a 4-nucleotide target site is 256 bp.

Fragments of that size are too small for most purposes. They are significantly smaller than the average prokaryotic gene (if one assumes that an average protein contains approximately 500 amino acids which are encoded by 1500 nucleotides) and even more so for eukaryotic genes, most of which contain intronic sequences. Expressed in another manner, a library with fragments of that size would require approximately 10^7 constituents to include a total eukaryotic genome and this would complicate the act of screening the library for the sequence of interest.

Finally, although plasmid vectors do not have a lower limit for the size of the fragment integrated, lambda and cosmid vectors do. For all of those reasons, total digestions by restriction enzymes which recognize 4-nucleotide sites are not used for the generation of genomic libraries. Two alternatives can be considered: either the use of restriction enzymes which digest the DNA less frequently (at 6- or

8-nucleotide recognition sites) or the use of incomplete or partial digestions with a frequent-cutting enzyme. The former method has the double disadvantage that although the average size might be acceptable, fragments at the extremes of the spread will be either too small or too large and the location of the restriction enzyme site might be inopportune if it occurred within the gene of interest. For these reasons, partial digestions of the DNA which one wishes to use in the generation of the library are the usual if not universal choice.

Because of the frequency of occurrence of 4-nucleotide target sites the fragments incorporated into the vectors as a result of a partial digestion will closely approximate a random collection which includes the total target genome. Controlled partial digestions can be obtained by varying the amount of enzyme, the time of incubation or the conditions of the reaction from suboptimal to optimal. In all cases it is important to recall that the enzyme used is a sensitive reagent. When in very dilute solution it may be unstable and hence one-tenth of a unit of enzyme may not have one-tenth of the activity of 1 unit. Similarly, the activity of the enzyme may not be maintained without diminution over a long period of time. For these reasons it is normal to carry out a scaled-down preliminary experiment to establish the optimum conditions. This should respect as far as is possible the final full-scale experiment as regards units of enzyme per microgram of DNA and the concentration of enzyme relative to the volume of the reaction.

The enzyme used will depend on the requirements of the experiment. The restriction enzymes used for partial digestions are chosen because of the compatibility of the termini which they generate within the integration sites of the vectors. For this reason *Sau*3A (recognition sequence $^{5'}$-GATC-$^{3'}$), for example, is frequently selected for this step as the fragments which result from the digestion can be linked to *Bam*HI-digested ($^{5'}$-GGATCC-$^{3'}$) vectors.

The procedure for a partial digestion of the target DNA is shown in *Protocol 4*.

Protocol 4. Establishment of correct partial digestion conditions

1. Prepare a 100 μl reaction mixture (*Protocol 1*) optimal for the specific restriction enzyme and containing 10 μg of genomic DNA.

2. Dispense 20 μl of the mixture into microfuge tube 1 and 10 μl into the remaining tubes 2–9. Place all tubes on ice.

3. Add 1 μl (= 1 unit) of restriction enzyme to tube 1, mix gently and place on ice.

4. With a fresh tip, pipette 10 μl from tube 1 into tube 2, mix gently and replace on ice. Perform similar serial dilutions throughout the assay tubes, so diluting the enzyme by 50% between each tube, keeping the enzyme–DNA mixture on ice.

5. Incubate at 37°C for 1 h before placing the tubes on ice.

6. Add EDTA pH 8.0 to 20 mM to inactivate the enzyme.

Protocol 4. *Continued*

7. Electrophorese each sample alongside appropriate DNA size markers (Chapter 5, Section 4.4). Visualize the DNA with a UV source and analyse the DNA digestion pattern in each gel track, as compared to the DNA size markers, to determine which enzyme concentration produces the maximum amount of DNA fragments of the desired size.

Note: The addition of spermidine can be used to enhance restriction enzyme digestion of impure DNA. Add 0.1 M spermidine to a final conc of 4 mM and incubate the DNA mixture at 37°C for a few minutes (spermidine may precipitate long DNA). Then add the restriction enzyme after ensuring no DNA precipitation has occurred.

Although *Protocol 4* is simple and requires no unusual skills, the reality is that one often finds that the digestion is not satisfactory. In such circumstances a number of control experiments are recommended:

(a) Check that the enzyme and the buffer used are functioning correctly by using plasmid DNA which is known to be correctly digested by a different restriction enzyme.

(b) Distinguish between a problem with the target DNA and with the solution in which the DNA is present. This is done by adding plasmid DNA which can be digested to the target DNA and carrying out the restriction enzyme reaction. If the 'clean' DNA is digested it shows that the target DNA is the source of the problem. If it is not digested, then the first priority is to remove the phenol, polysaccharide, salt or unknown inhibitor of the restriction enzyme. This can be done by a variety of methods including drop dialysis (see *Protocol 3*), chromatography (see Chapter 6, Section 2.2) or dialysis. Simply repeating an ethanol precipitation step may not be fruitful if this had been included in the preparation of the target DNA because many inhibiting compounds are precipitated by ethanol.

(c) If the target DNA seems to be the cause of the problem then some of the steps described in the previous chapters should be repeated.

When the restriction enzyme digestion appears to be correct, the full-scale experiment can be performed. Because of the difficulty of correctly sampling small volumes from viscous solutions of undigested DNA, it is seldom that full-scale reactions require the same amount of enzyme or time. However, it is not possible to give an accurate indication of the precise conditions to use because this will ultimately depend on factors such as the stability of the restriction enzyme and the cleanliness of the DNA preparation. It will also depend on the subsequent uses intended for the DNA. For example, a cosmid library will require DNA which is on average very long (> 40 kb) and any over-digestion is a disadvantage.

After the digestion of the target DNA, an aliquot is used to analyse its size by electrophoresis on an agarose gel. If it is unsatisfactory, further enzyme may be added. If persistent difficulties are encountered the addition of 4 mM spermidine to the sample may prove beneficial.

The analysis of the restriction enzyme digestions is performed in agarose gels with DNA fragments of known sizes as size markers. The methods involved have been described in Chapter 5, Section 4.4. With partial digestions, a spread of DNA sizes is obtained. It is important to remember that 1 mole of a fragment of size 20 000 bp will give a signal on visualization with ethidium bromide which will be ten times as intense as that of a fragment of size 2000 bp. Ultimately it will be the number of molecules of DNA which is present in a ligation which will dictate the success of the experiment. It is advisable therefore to eliminate DNA fragments which are too small prior to further manipulations.

Some vector systems make this step seem superfluous (e.g. lambda systems which are non-viable with small DNA fragments, see Chapter 9, Section 2.2). However, the prior fractionation of DNA, usually by sucrose gradients (*Protocol 5*) after phenol extraction, has the twin benefits of size fractionation and elimination of traces of organic solvents and is recommended.

Protocol 5. Large-scale preparation and purification of partially-digested DNA

1. Selecting the optimal conditions established for correct partial digestion (*Protocol 4*), prepare a reaction mixture containing 50–100 μg of genomic DNA. Scale up the reaction mixture volume so that the reaction time, temperature and DNA concentration are identical to those established.

2. Add half the amount of restriction enzyme to that established for correct partial digestion. Few scale-up experiments produce identical results to pilot studies and the margin of error with this method favours under-digestion of the DNA by the enzyme and not useless over-digestion.

3. After incubation for the correct time period, place the reaction on ice and analyse a small sample by gel electrophoresis. If the sample is under-digested, replace at the correct temperature for a further 15 min before repeating the electrophoresis. Addition of more enzyme at this stage usually results in over-digestion.

4. Purify the digested DNA by phenol extraction and ethanol precipitation (*Protocol 3*).

5. Apply to a 10–40% (w/v) sucrose gradient and centrifuge at 26 000 r.p.m. for 24 h at 15°C in the Beckman SW40 rotor (see *Protocol 6*).

6. Collect fractions and analyse samples by gel electrophoresis to determine which fraction contains DNA of the correct size or within correct size limits for the cloning experiments.

7. Purify selected DNA by dialysis and concentrate to 0.1–0.5 μg μl^{-1} by ethanol precipitation (*Protocol 3*, steps 4–5).

After the sucrose gradient, fractions are collected and they are again analysed on agarose gels. Presumably because of differences in the sample buffers there is a tendency to overestimate the size of the DNA in these gels. Bearing in mind that small molecules tend to cause problems in molecular cloning experiments it is again advisable to have size markers appropriate for the fragment size sought and to chose fractions that appear to be slightly too large.

3. Preparation of vectors for molecular cloning experiments

The essential requirement of a vector is that it can replicate in the host organism. Phage lambda, cosmids, and plasmids fulfil this basic requirement and are widely used in cloning experiments (see Chapter 9). Plasmids are extrachromosomal elements which, when used as cloning vectors, typically carry resistance to an antibiotic and are present as multiple copies in the host organism. Plasmids are most frequently used to subclone large fragments or to generate small libraries such as would be required for microorganisms, viruses and some cDNA banks.

Genomic libraries for eukaryotic organisms are generally prepared with phage lambda as vector. Only the salient features of lambda when used as a vector will be noted here; for a more complete discussion see Chapters 2 and 9. Lambda can function in a lytic or lysogenic (integrated into the DNA of *E. coli*) mode. For molecular cloning, only the lytic functions are required as these ensure a high-copy number of the phage. Nature has conveniently arranged matters such that the lysogenic functions, which are unnecessary for lambda as a vector in the service of a molecular biologist, are grouped together in the middle of the phage. This means that digestion by appropriate restriction enzymes can allow separation of the 'arms' of lambda from the central 'stuffer' fragments. In the generation of genomic banks this fragment is replaced by the foreign DNA and all of the lytic functions required for replication and infection by the phage are retained. The initial transfer of the phage arms linked to foreign DNA into the host cell is effected by regenerating *in vivo* phage particles by the addition of lambda head and tail proteins. This procedure of 'packaging' DNA is described in Chapter 8.

The length of the DNA which can be enclosed in the phage head cannot be greater than approximately 47 000 bp. Similarily, phage which contain less than approximately 32 000 bp are also unstable. This places constraints on the size of insert which can be carried when phage lambda is the vector. The packaging of lambda DNA *in vitro* depends on the initial binding of capsid proteins to the *cos* sites of the phage. These must be presented in a linear array with the distance between the *cos* sites respecting the size limitations outlined above. This is achieved by the ligation of the arms of lambda and the target DNA at a high concentration to form concatamers (see Section 4.1). The packaged DNA is transferred to *E. coli* with very high efficiency. This fact allied to the ease of storage of lambda genomic libraries and the large number of lambda plaques which can be analysed on a single agar plate makes lambda genomic libraries particularly popular.

3.1 Phage lambda

When lambda is used as a cloning vector the details of the strategy used depends on the particular vector. With λgtWES for example, digestion by *Eco*RI followed directly by a size separation or with a second digestion with an enzyme (*Xba*I) which has a site only in the stuffer fragment prior to size separation yields arms of lambda which will accept foreign DNA.

Protocol 6. Preparation and purification of λgtWES arms

1. Carry out a total *Eco*RI digestion of 50–100 μg λgtWES DNA (*Protocol 1*).
2. Phenol extract and ethanol precipitate the DNA (*Protocol 3*).
3. Add MgCl$_2$ to 0.01 M to the resuspended DNA and incubate at 42°C for 1 h. This promotes annealing of the complementary *cos* sites thereby joining the λ arms.
4. Apply the DNA on to a 10–40% (w/v) sucrose gradient. These gradients are readily achieved by freezing and thawing twice a 25% sucrose solution in 1 M NaCl, 20 mM Tris–HCl pH 8.0, 5 mM EDTA in a swing-out ultracentrifuge tube or by using gradient-forming devices.
5. Centrifuge at 26 000 r.p.m. for 24 h at 15°C in the Beckman SW40 rotor.
6. Collect 0.5–1.0 ml fractions carefully and analyse a sample from each fraction by gel electrophoresis to determine which contain purified lambda arms with no contaminating lambda stuffer fragments.
7. Dialyse the appropriate fractions against sterile water overnight to remove the sucrose.
8. Concentrate the DNA by ethanol precipitation (*Protocol 3*, steps 4–5).
9. An alternative favoured by some workers is to substitute NaCl gradients for the sucrose gradient:
   (a) Prepare a 20% (w/v) NaCl solution in 50 mM Tris–HCl pH 8.0, 1 mM EDTA.
   (b) Freeze and thaw once at -70°C (-20°C is insufficient to form a correct gradient).
   (c) Apply the sample and centrifuge at 39 000 r.p.m. for 3 h at 15°C in the Beckman SW40 rotor.
   (d) Purification of selected fractions from the gradient is by dialysis against sterile water overnight at 4°C.

The λEMBL series of vectors (Chapter 9, Section 2.2.2.*ii*) are designed for even more simple use. The internal fragment of these vectors is flanked by polylinkers.

By digestion with two restriction enzymes the arms retain termini which can base-pair with the digested target DNA whereas the internal fragment has termini which cannot ligate to the arms. The small oligonucleotide which separates the arms and the linker is readily removed by isopropanol precipitation:

Protocol 7. Preparation and purification of λEMBL3 arms

1. Carry out a total *Bam*HI digestion of 50–100 μg of λEMBL3 DNA (*Protocol 1*).

2. Phenol extract and ethanol precipitate the DNA (*Protocol 3*).

3. Resuspend the *Bam*HI-digested DNA and prepare an *Eco*RI reaction solution (*Protocol 1*).

4. Phenol extract the *Bam*HI-*Eco*RI-digested DNA.

5. To selectively remove the polylinker, precipitate the double-digested DNA with a one-tenth volume of 3 M sodium acetate pH 5.5 and 1 vol. of isopropanol.

6. Incubate for 15 min at −20°C and centrifuge at 10 000 r.p.m. in the microfuge for 10 min to pellet the DNA.

7. Resuspend the DNA at 0.2–0.5 μg μl^{-1} in sterile TE pH 7.4 (*Appendix 4*) or water.

8. Store at 4°C or −20°C for prolonged storage.

3.2 Cosmids

Cosmid libraries (Chapter 9, Section 2.3) can be seen as an extension of libraries prepared using lambda as the vector. By inclusion of a small segment of lambda into a plasmid, the *in vitro* packaging system can be used with resultant high-efficiency transformation of *E. coli*. The size requirements for packaging remain unaltered and, as the plasmid element can be less than 4 kb in length, the size of DNA that can be transferred may be over 40 kb long. Once transferred to *E. coli*, cosmids (lacking as they do, all lambda functions) act as large plasmids. As they typically carry antibiotic resistance genes, colonies which contain them grow on selective agar plates. They have the advantage relative to phages of containing a large segment of contiguous genomic DNA in each selected clone but have the disadvantages of greater difficulty of storage and fewer units per plate which makes their screening moderately more cumbersome.

When cosmid vectors are used the methods involved are essentially the same as for plasmid vectors (see Section 3.3). Frequently the cosmids are digested with two restriction enzymes which generate non-homologous termini (see *Protocol 2*). If these vectors ligate to form a dimer, it is not large enough to be packaged by the lambda proteins and is therefore unable to be transferred to *E. coli*. For this reason the target DNA is frequently phosphatase-treated when cosmid libraries are generated (see *Protocol 8*). This ensures that only DNA which is contiguous on the genome is inserted into the vector. Given that cosmid vectors can incorporate fragments up to 40 000 bp in length and, as noted above, the visual estimate on an agarose gel of the number of molecules of small fragments can give a misleadingly low impression, particular care must be taken to eliminate DNA fragments that are smaller than those desired. Salt gradients are preferred by some when separating large DNA fragments (see *Protocol 6*) but any method in conjunction with phosphatase treatment of the DNA can be used to effectively diminish the number of cosmid clones with inappropriate inserts.

Protocol 8. Phosphatase treatment of DNA

A. *Bacterial alkaline phosphatase (BAP)*

1. Prepare a reaction mixture containing appropriate amounts of $10 \times$ reaction buffer ($10 \times$ BAP buffer $= 500$ mM Tris–HCl pH 8.0, 10 mM $ZnCl_2$) and DNA in a 100 μl volume.

2. Add 1 unit of BAP and incubate at 60°C for 30 min. This high temperature supresses any residual exonucleases in the enzyme preparation and is recommended when using BAP.

3. Stop the reaction by adding SDS to 0.1% (w/v) and proteinase K to 100 μg μl^{-1}. Incubate at 37°C for 30 min.

4. Purify the phosphatase-treated DNA by phenol extraction and ethanol precipitation (*Protocol 3*) and resuspend the DNA in sterile TE pH 8.0 (*Appendix 4*) or water.

5. Because of the stability of the enzyme and the consequences of its persistance in an experiment further purifications of the DNA before ethanol precipitation are often included. These include extra phenol extractions or column chromatography (Chapter 6, Section 2.2).

B. *Calf intestinal phosphatase (CIP)*

1. Set up a reaction mixture as in step A1 ($10 \times$ CIP buffer $= 200$ mM Tris–HCl pH 8.0, 10 mM $MgCl_2$, 10 mM $ZnCl_2$).

2. Add 1 unit of CIP and incubate at 37°C for 30 min.

3. Stop the reaction by heating the mixture at 75°C for 10 min, (CIP is heat-labile, BAP is not) before purifying the phosphatase-treated DNA by phenol extraction and ethanol precipitation (*Protocol 3*).

3.3 Plasmids

There is a very wide choice of plasmids which can be used for genetic engineering experiments (see Chapter 9). When used as vectors they are digested at one locus either by a single restriction enzyme or by two at a multi-cloning site to achieve insertion of target DNA in a defined orientation. The digestions are carried out as in *Protocols 1* and *2*. When digestions are verified as complete and correct by agarose gel electrophoresis the DNA is phenol-extracted and ethanol-precipitated (*Protocol 3*). For many experiments it is very advisable to treat the digested vector DNA with phosphatase (*Protocol 8*). By removing the 5′-phosphates from the plasmids, they can no longer be circularized by DNA ligase (see Section 4.1). This reduces the background of colonies which do not contain recombinant molecules in an experiment. It should be noted, however, that undigested supercoiled plasmids can transform *E. coli* with an efficiency approximately 50-fold greater than linearized DNA. As remnants of undigested DNA are frequently undetected by routine agarose gel electrophoresis their contribution to background problems in cloning experiments should not be underestimated. If they persist after alkaline phosphatase treatment, more extensive digestion by the restriction enzyme or, ultimately, purification of the linear form of the plasmid by centrifugation through a sucrose gradient is recommended.

After alkaline phosphatase treatment it is imperative to remove all traces of the enzyme as it will impede further steps in the generation of recombinant molecules.

4. Construction of recombinant molecules

The previous sections have described how to prepare appropriately digested target and vector DNA. The final step of linking these together in a recombinant molecule prior to transfer to *E. coli* can be achieved in different ways that are outlined below.

4.1 Ligation

The purification of the enzyme DNA ligase was important in developing the original concept of recombinant DNA. This enzyme links fragments of DNA to each other in a covalent manner (*Protocol 9*). The essential requirement for this ligation is that the DNA fragments present a 5′-phosphate group in close proximity to a 3′-hydroxyl group. When protruding complementary termini are present on the target and vector molecules they provide an obvious docking mechanism to bring the 5′- and 3′-ends of molecules together. However, ligation can also occur when there is no obvious mechanism for prior adhesion of fragments. Blunt-end ligation is not as efficient as ligation of complementary cohesive termini but it occurs readily when a higher concentration of DNA ligase is provided. It is ofter the method of choice for joining DNA molecules containing non-complementary protruding termini after these termini have been modified (*Protocol 10*).

Protocol 9. DNA ligation reaction

1. Prepare a ligation mixture. The $10 \times$ ligation buffer contains 0.66 M Tris–HCl pH 7.6, 50 mM $MgCl_2$, 50 mM dithiothreitol, 10 mM ATP. The volume of the ligation reaction and the DNA concentration usually depend on the type of ligation experiment. Use a 10 μl vol. with DNA at > 100 ng μl^{-1} for concatamer ligation products, or a 10 μl volume with DNA at < 10 ng μl^{-1} for circular ligation products.

2. Add T4 DNA ligase. For a cohesive-end ligation add 0.25 units of enzyme per μg of DNA, and for a blunt-end ligation add 2.5 units μg^{-1}.

3. Incubate the reaction mixture at 15°C for 1–16 h. Simple cohesive-end ligations are usually complete in 1–2 h.

4. Analyse for correct and complete ligation by gel electrophoresis with unligated material as a marker, or by transformation if the resulting DNA contains vector sequences.

Protocol 10. Preparation of blunt-ended DNA fragments

A. *Klenow polymerase–5′-overhang fill-in reactions*

1. Prepare a reaction mixture containing DNA and Klenow reaction buffer ($10 \times$ Klenow buffer $= 500$ mM Tris–HCl pH 7.5, 100 mM $MgCl_2$, 10 mM dithiothreitol) and 20 μM each dNTP.

2. Add 1 unit Klenow polymerase and incubate at 15–37°C for 15–30 min.

3. Stop the reaction by heating at 75°C for 10 min.

4. The progress of the reaction can be monitored by a simple ligation experiment followed by gel electrophoresis. DNA that originally had complementary termini should not now be ligated by low amounts of T4 ligase (*Protocol 9*). Instead, $10 \times$ the amount of T4 ligase should be required to achieve blunt-end ligation.

B. *T4 DNA polymerase–3′-overhang exonuclease reactions*

1. Prepare a reaction mixture containing DNA and T4 DNA polymerase reaction buffer ($10 \times$ T4 DNA pol buffer $= 500$ mM Tris–HCl pH 8.8, 50 mM $MgCl_2$, 50 mM dithiothreitol) and 20 μM each dNTP.

2. Add 1 unit of T4 DNA polymerase and incubate at 15°C for 15–30 min.

3. Stop the reaction by heating at 75°C for 10 min.

4. Analyse the reaction by ligation of a sample.

The standard ligation reaction outlined in *Protocol 9* is clearly a simple procedure. Experience shows that it frequently does not work efficiently. Some simple precautions are suggested. Principal among these is the usefulness of showing that the individual components can ligate to themselves. This control allows the researchers to:

(a) establish which fragment is the source of the problem; and

(b) show, by use of known materials, that the ligation enzyme and buffer are capable of carrying out the ligation.

If the problem is with one of the DNA fragments then either the restriction enzyme used in the digestion contained some exonuclease which makes cohesive ligation inefficient, or the fragment is in a solution which contains either organic solvents or other contaminants such as particles from agarose gels if they had been used in the purification. A mixing experiment using DNA fragments that are known to be capable of ligation will distinguish between these two possibilities.

The result of a ligation depends on the relative concentrations of the DNA molecules in the solution. Two types of product from ligation can be envisaged: linear molecules, which are multimers (concatamers) of the component fragments, or circular molecules. When cloning with lambda or cosmid vectors, the former are required. When using plasmid vectors the latter are preferred. If the concentration of DNA termini in a solution is very high then concatamers are more likely to be formed because of the statistical probability of two molecules being joined rather than one molecule circularizing and linking to itself. If solutions of DNA fragments are dilute the converse is true. The length of the DNA fragment also influences the outcome of the ligation step. Mathematical treatments of these factors have been presented (2). Practical consequences of these considerations are that the total DNA concentration in a plasmid cloning experiment should be maintained at less than 1 μg ml^{-1} and in a lambda cloning experiment at over 100 μg ml^{-1}.

4.2 Linkers and adapters

Although blunt-end ligation can work effectively it is significantly more simple to ligate large DNA molecules which have cohesive termini. Such termini may be added artificially to DNA fragments by the addition of small oligonucleotides, the sequence of which includes a restriction site. These oligonucleotides are of two types: linkers and adapters.

Linkers are complementary oligomers which form small double-strand DNA fragments which include a restriction enzyme site. They are ligated to blunt-end DNA by DNA ligase. Because of the high concentration of these small molecules present in the reaction, the ligation is very efficient when compared to blunt-end ligation of large molecules. The terminal restriction enzyme site is generated by digestion with the appropriate enzyme. Obviously, sites for the enzyme may be present in the target DNA fragment. As a result, any such sites are protected by prior methylation (see Volume II, Chapter 3, Section 4). This requirement limits the use of linkers to those for which modification enzymes (methylases) are available which correspond to the restriction enzyme target site.

Adapters are similar to linkers in that they are small oligonucleotides which are blunt-end ligated to the target DNA. They are not perfectly double-stranded, however, and are synthesized to present cohesive termini without any digestion. In this way the target DNA is ready for ligation to the complementary restriction enzyme termini of the vector.

Protocol 11. Addition of linkers/adapters to blunt-end DNA molecules

1. Prepare a ligation mixture containing methylated DNA and linkers in a 50 μl volume (*Protocol 9*). Keep a similar ratio of linkers to DNA, for example 1 μg DNA:1 μg linkers (this ensures a large molar excess of linkers).

2. Add 5 units T4 DNA ligase and incubate for 16 h at 15°C.

3. Stop the reaction by heating at 70°C for 10 min.

4. Increase the volume of the ligation mixture to 100 μl by adding 10 μl of the relevant 10 × restriction enzyme buffer and 40 μl water.

5. Add 100 units of restriction enzyme and incubate at 37°C for 12 h.

6. Heat at 70°C for 10 min to inactivate the enzyme.

7. Separate the large DNA with linkers at its termini from digested linkers by size fractionation on a Bio-Gel A-50m column (*Protocol 12*).

8. Purify and concentrate the large DNA by ethanol precipitation (*Protocol 3*, steps 4 and 5).

Adapters are becoming more favoured as a method of adding complementary termini to blunt-end DNA molecules. The procedure is similar to that of linkers. However, no restriction enzyme digestion is required thereby also removing the need to methylate the target DNA. The adapters containing single-strand complementary ends are ligated on to the DNA (step 1), and phosphorylated with T4 polynucleotide kinase directly in the ligation mixture. Excess adapters can be removed by gel filtration (*Protocol 12*).

Protocol 12. Removal of excess linkers/adapters: Bio-Gel A-50m column

1. Build the column in a sterile 5 ml or 10 ml plastic pipette (disposable). Plug the bottom of the pipette with sterile glass wool.

2. Pre-swell the Bio-Gel A-50m in column buffer (10 mM Tris–HCl pH 7.5, 100 mM NaCl, 1 mM EDTA) for 60 min at room temperature. Pour the resin into the pipette giving a bed volume of 5 ml. Wash the column with 50 ml column buffer. This removes a ligase inhibitor from the resin that may interfere with further cloning reactions.

3. Load the sample on to the column, wash with more column buffer and collect 20–30 100 μl fractions.

Protocol 12. *Continued*

4. Thorough washing of the column with more buffer allows re-use.

5. Analysis of the fractions can be carried out by various methods. Simple electrophoresis of a sample from each fraction may show the fractions containing the linkered/adaptered DNA. Alternatively trace amounts of radiolabel may be incorporated on the linkers (see Volume II, Chapter 4) and the fractions analysed using a scintillation counter. This shows two peaks of radioactivity: an earlier peak containing the linkered DNA and a final large peak containing excess labelled linkers/adapters. The column may also be equilibrated before use using known DNA size markers. Analysis by gel electrophoresis should determine which fraction contains DNA of a particular size.

4.3 Tailing

The addition of long homopolymers of deoxynucleotides to the ends of molecules was particularly popular in the initial years of cDNA cloning prior to the ready availability of linkers and adapters. The enzyme terminal deoxytransferase is an unusual polymerizing enzyme in that it does not require a template. If a deoxynucleotide triphosphate is provided the terminal deoxytransferase will add it to any available 3'-hydroxy site which serves as a primer. The kinetics of addition varies for each deoxynucleotide. Because of inhibiting secondary structures that result from polymers of deoxyguanosine, its elongation is self-limiting to approximately 30 bases. The optimum primer is a protruding 3'-sequence but blunt-end fragments can be effective in the presence of the cacodylate–cobalt buffer which allows transient single-strands to occur at the ends. A popular combination is the use of d(G) tails on the vector and highly efficient d(C)-tailing of the target DNA.

Protocol 13. Homopolymer tailing and annealing reactions

1. Prepare a tailing mixture containing DNA, tailing buffer (10 × tailing buffer = 1 M sodium cacodylate pH 7.0, 10 mM $CoCl_2$, 1 mM dithiothreitol[a]) and the nucleotide to be added on (20 μM dNTP) in 50 μl volume.

2. Add 10 units of terminal deoxynucleotidyl transferase and incubate at 37°C for 10 min. Under these conditions the enzyme will add 20 nucleotides to the 3'-end of the DNA in 10 min if the reaction includes 4 pmol of DNA. Trial experiments may be carried out before a cloning experiment with trace amounts of radioactive nucleotides to determine the exact extension of nucleotides on to the 3'-end.

3. Stop the reaction by heating at 75°C for 10 min. Purify the DNA by phenol extraction and ethanol precipitation (*Protocol 3*). It is also useful to dialyse

Protocol 13. *Continued*

the DNA against sterile water for 1–2 h to remove components from the tailing mixture that may inhibit bacterial transformation (*Protocol 3*, step 6).

4. Prepare an annealing mixture containing vector and target DNAs with complementary homopolymer tails in 100 μl of 50 mM Tris–HCl pH 8.0, 1 mM EDTA, 100 mM NaCl. The DNA concentration and ratio of vector: target depends on the actual cloning experiment (*Protocol 9*). Heat the sample at 70°C for 30 min in a water bath. Turn off the heat and allow the sample to cool in the water bath to room temperature (7–12 h).

5. The sample is now ready for transformation.

a Extreme care is required making up the 10 × tailing buffer. Add the chemicals in the order stated to prevent precipitation in the solution.

---

5. Conclusion

The generation of recombinant molecules requires a combination of methods that are frequently deceptively complex. The quality of the materials used at all stages, whether purchased or generated, is of the utmost importance. Although it is true that 'only one correct clone is needed', inefficiency at any of the steps described is usually severely punished. Attention to detail, which is the essence of good science, is well rewarded with libraries capable of providing answers to almost all questions in molecular biology.

Acknowledgements

The comments and recommendations of our colleagues are gratefully acknowledged. RP is supported by E.C. grant BAP-0125-IRL.

References

1. Brown, T.A. (1991). *Molecular Biology Labfax*. BIOS, Oxford.
2. Dugaiczyk, A., Boyer, H.W., and Goodman, H.M. (1975). *Journal of Molecular Biology*, **96**, 174.

8

Generation and identification of recombinant clones

T. A. BROWN

The central step in a gene cloning experiment is the introduction of recombinant DNA molecules synthesized *in vitro* into host *Escherichia coli* cells and the subsequent identification of recombinant clones. In outline the procedure is very straightforward:

(a) Bacteria and DNA are mixed together and some of the cells take up DNA molecules.

(b) The cells are plated on to a selective medium (or a series of media) that enable recombinant colonies (i.e. those that contain recombinant DNA molecules) to be identified.

The methodology is made more complex by the need to pre-treat either the DNA or the bacteria to enable uptake to occur, and by the variety of different selection strategies that are available, the appropriate one depending on the type of vector being used.

In this chapter I will first describe the means for introducing recombinant DNA molecules into bacteria, and then provide details for recombinant selection with the most important vectors. These protocols will deal solely with the use of *E. coli* as the host organism, *E. coli* being the host that is used for all basic gene cloning experiments.

1. Introduction of DNA into *E. coli* cells

There are two different approaches to the introduction of recombinant DNA molecules into *E. coli*:

(a) Uptake of naked DNA, which is called 'transformation' if the DNA is plasmid-based, or 'transfection' if it derives from a phage chromosome.

(b) '*In vitro* packaging', which involves construction of infective phage particles that contain the recombinant DNA molecules.

1.1 Transformation

To simplify things I will use the term 'transformation' to cover both transformation itself and transfection. The protocols are applicable to both procedures.

In nature, transformation is probably not a major process by which bacteria obtain genetic information (1). This is reflected by the fact that in the laboratory only a few bacteria (notably *Bacillus* and *Streptococcus* species) can be transformed with ease. Most species, including *E. coli*, take up only limited amounts of DNA under normal circumstances and have to undergo a physical and/or chemical pre-treatment before they can be transformed efficiently. Cells that have undergone such a treatment are said to be 'competent'.

1.1.1 Preparation of competent cells

Methods for preparing competent *E. coli* cells derive from the work of Mandel and Higa (2) who developed a simple treatment based on soaking the cells in cold $CaCl_2$. Why this treatment is effective is not known. Their original experiments were with λ DNA but the procedure was quickly shown to be applicable also to plasmid (3) and chromosomal (4) DNA. *Protocol 1* is a simple and easy method based on this original procedure.

Protocol 1. Preparation of competent *E. coli* cells

1. Inoculate 5 ml of LB medium (*Appendix 4*) with a colony of *E. coli*. Incubate overnight at 37°C with shaking.
2. Remove 300 μl of the overnight culture and inoculate 30 ml of fresh LB medium. Incubate at 37°C with shaking until the optical density (OD) at 550 nm reaches 0.4 to 0.5. This should take about 2 h.
3. Centrifuge the cells at 6000 *g* for 10 min at 4°C in a pre-cooled rotor. It is important that from this point onwards the cells are not allowed to warm up.
4. Discard the medium and resuspend the cells in 15 ml ice-cold, sterile 50 mM $CaCl_2$. Leave on ice for 15 min with occasional shaking.
5. Centrifuge the cells as in step 3 and discard the $CaCl_2$.
6. Resuspend in 3 ml ice-cold, sterile 50 mM $CaCl_2$. The cells are now competent.

The growth stage that the cells have reached is critical. At less than OD 0.4–0.5 there will be insufficient cells for transformation. At higher cell densities the procedure will not result in a good yield of competent cells.

The procedure can be scaled up to as much as 1 litre of starting cells. The initial pellet should be resuspended in a one-half volume of $CaCl_2$ and the competent cells taken up in a one-tenth volume of $CaCl_2$.

This procedure is very reliable and works well with most strains of *E. coli*. The cells can be used immediately or stored at $-80°C$ after the addition of 15% (v/v) glycerol. Stored cells retain their competence for several months as long as they are not allowed to warm up. They should be stored in 300 μl aliquots so that stocks do not have to be repeatedly thawed.

Cells prepared by the method shown in *Protocol 1* should yield more than 10^7 transformants per microgram of supercoiled plasmid DNA, which is satisfactory for most standard cloning procedures. For some applications though the amount of available DNA is so low that higher transformation efficiencies are required. A combination of highly-transformable strain (e.g. DH1, see *Appendix 3*) and more sophisticated technique can yield competent cells that provide a 50-fold improvement in transformation efficiency (5).

Protocol 2. Preparation of highly-transformable *E. coli*

- *In this procedure it is important to use pure, quality reagents and the cleanest glassware.*

1. Make a streak plate (Chapter 2, *Protocol 2*) on SOB agar (*Appendix 4*) with *E. coli* DH1. Ideally, the inoculum should come from a frozen stock.

2. Incubate the plate overnight at 37°C and then transfer five small colonies (about 2 mm diameter) into 1 ml SOB + 20 mM $MgSO_4$.

3. Gently vortex to break up the colonies, then inoculate into 100 ml SOB + 20 mM $MgSO_4$ in a 1 litre flask.

4. Incubate at 37°C with shaking until the culture reaches 10^8 cells per millilitre. This will correspond to about 2 h incubation and a final OD at 550 nm of 0.4 to 0.5.

5. Split the culture into 50-ml portions and transfer to pre-cooled centrifuge tubes. Cool on ice for 10 min, then centrifuge at 6000 g for 10 min at 4°C.

6. Discard the medium, including the last traces which can be removed by careful aspiration.

7. Gently resuspend the cells from each tube in 20 ml of ice-cold, sterile transformation buffer. The composition depends on whether the cells are to be used immediately or stored:
   - If the cells are to be used immediately the buffer contains 10 mM Mes (pH 6.3), 45 mM $MnCl_2$, 10 mM $CaCl_2$, 100 mM KCl, 3 mM hexamminecobalt chloride.
   - If the cells are to be stored at $-80°C$ the buffer contains 10 mM potassium acetate (pH 7.5), 45 mM $MnCl_2$, 10 mM $CaCl_2$, 100 mM KCl, 3 mM hexamminecobalt chloride, 10% (v/v) glycerol.

8. Stand the cells on ice for 10 min, then centrifuge as in step 5.

9. Remove the buffer, including the final traces, and carefully resuspend each pellet in 4 ml of transformation buffer.

Protocol 2. *Continued*

10. If the cells are to be used be immediately:

    (a) Add 140 μl of filter-sterilized DD to each suspension. DD = 1.53 g dithiothreitol, 9 ml dimethylsulphoxide (DMSO), 100 μl 1 M potassium acetate (pH 7.5), 900 μl water.

    (b) Stand on ice for 15 min, add a further 140 μl of DD, and stand on ice for a further 15 min. The cells are now ready to use.

11. If the cells are to be stored:

    (a) Add 140 μl DMSO to each suspension.

    (b) Stand on ice for 15 min, then add a further 140 μl DMSO.

    (c) Quickly dispense 50 μl aliquots into chilled microfuge tubes and freeze in liquid nitrogen. Store at $-80°C$.

As an alternative to preparing your own competent cells you can purchase them from any one of several suppliers. The quality is generally good, and a range of different *E. coli* strains is available. However, for reasons best known to the suppliers, the cost of commercial competent cells is prohibitively high and only the most overly-endowed labs should consider their purchase as routine. The exception to this is if you are working with one of the recently-developed host strains for cloning mammalian and plant DNA (e.g. *E. coli* SURE, see *Appendix 3*), which carry a number of mutations to prevent rearrangement of cloned DNA. A side-effect of these mutations is that the bacteria are difficult to culture and are somewhat refractory to standard procedures for rendering them competent.

1.1.2 DNA uptake by competent cells

Competent cells are induced to take up DNA by a short heat-shock. As with the $CaCl_2$ treatment, the biological basis to the technique is not understood. It is, however, relatively efficient.

Protocol 3. Uptake of DNA by competent cells

1. Add a suitable amount of DNA (see below) to 300 μl of competent cells (from *Protocol 1*) or 50 μl of competent cells (from *Protocol 2*).

2. Leave on ice for 30 min.

3. Transfer to a 42°C water bath for 2 min.

4. Return to ice. Move fairly rapidly to the next stage of the procedure (*Protocol 4* or Section 2).

The important question is how much DNA to add. In terms of volume it is unwise to add the DNA in a volume greater than 5% that of the competent cells (e.g. no more than 15 μl of DNA per 300 μl of competent cells). This is partly because the $CaCl_2$ concentration must be maintained at 50 mM or above (you could add extra to compensate) and partly because components of the ligation mixture may interfere with DNA uptake.

To a certain extent the amount of DNA to be added depends on how much is available and the type of vector being used. However, the number of transformants does not usually increase at DNA amounts above about 2 ng DNA for the volumes of competent cells used in *Protocol 3*. It is better to use several aliquots of competent cells, and subsequently a larger number of agar plates for recombinant selection, than to add excessive amounts of DNA into one tube.

1.1.3 Expression of antibiotic resistance genes carried by plasmid vectors

Several plasmid cloning vectors carry genes for antibiotic resistance and are plated on to antibiotic media for transformant and recombinant selection (Section 2.1). If cells containing these vectors are plated out immediately after DNA uptake the yield of colonies will be low as there will not have been time for expression of the antibiotic resistance genes before the cells actually encounter the inhibitor. To circumvent this problem the cells are usually incubated in broth for a short period before being plated out.

Protocol 4. Plasmid expression

1. Add 0.5 ml LB medium (*Appendix 4*), pre-warmed to 37°C, to each tube of cells immediately after the heat-shock (*Protocol 3*, step 3).

2. Incubate in a 37°C water bath for 1 h.

3. Return to ice.

1.1.4 Transformation and transfection: trouble-shooting

It is not possible to assay transformation and transfection until the cells are plated out and the transformation efficiency can be determined. However, it is worth considering the problems that can arise during preparation of competent cells and DNA uptake before going any further.

A total lack of transformants can only be explained by a trivial error such as no DNA, wrong strain, or suchlike. In practice, contamination of the host strain is usually to blame, so follow the guide-lines in Chapter 2, Section 6.2 to track down and eliminate the cause(s).

A more common problem is low transformation efficiency. This can be due to any of a number of factors:

i. The cells were grown to the wrong stage

The most important aspect of preparation of competent cells is to make sure the starting culture is neither too young nor too old. Do not be impressed by molecular biologists who judge the OD by eye: measure it properly!

ii. The cells were allowed to warm up

Cells that return to room temperature rapidly lose their competence. It is possible to prepare competent cells with a non-refrigerated bench-top centrifuge but transformation efficiencies will be low. Once out of the centrifuge the cells should be returned to ice and solutions added to the cells should be pre-cooled on ice.

iii. The DNA was added in too high a volume

Keep to the ratio of DNA to competent cells described above. If your DNA has to be in a large volume then adjust to 50 mM $CaCl_2$ by adding the appropriate volume from a 1 M stock.

iv. The heat-shock was ineffective

Two minutes at 42°C is sufficient for cells in standard 1.5-ml microfuge tubes. If you are using different tubes then the time will have to be altered. In particular, a 50-ml centrifuge tube will necessitate a longer period at 42°C to provide the cells with the same heat exposure.

Identifying the reason for low transformation efficiency can be difficult and it should be borne in mind that in general terms high efficiencies are not always needed. So long as you obtain sufficient recombinants you need not worry about the efficiency of your transformation system. This is not an excuse for poor technique but it is not time-effective to spend weeks optimizing a procedure if it is already providing you with the desired result.

1.2 *In vitro* packaging

Although DNA uptake by competent cells provides enough transformants for many applications, there are situations where the desired number of recombinants is so high that a more efficient system is required. This is particularly true for the construction of genomic and cDNA libraries in λ phage vectors (Volume II, Chapters 2 and 3), where 1000 to 100 000 recombinants are needed from limited amounts of starting material.

Transformation of competent cells is certainly inefficient when compared to the infectivity of wild-type λ phage. This fact has prompted research into ways of constructing λ phage particles *in vitro*, so that recombinant λ DNA molecules can be packaged into infective phage particles in the test-tube. Packaging requires a number of different proteins coded by the λ genome, but these can be prepared at high concentration from cells infected with defective λ phage strains. Two different systems are in use. The first makes use of a pair of defective strains (*E. coli* BHB2688 and BHB2690; *Appendix 3*), each carrying a mutation in one of the components of the phage protein coat (6). Infected cells synthesize and accumulate all the other components, but cannot assemble mature phage particles. However, a mixture of lysates from the two strains will contain all the required proteins and can be used for *in vitro* packaging. The second system makes use of *E. coli* SMR10 (*Appendix 3*), which has defective *cos* sites (7). This strain synthesizes all λ proteins but cannot assemble phage particles *in vivo* as it does not recognize its own DNA as a substrate for packaging. However, a lysate will package recombinant λ DNA molecules that carry suitable *cos* sites.

Procedures for lysate preparation from *E. coli* BHB2688 and BHB2690 and their use in *in vitro* packaging are given in Volume II, Chapter 2, *Protocols 8* and *9*. In our experience the two-strain system is more reliable than the one-strain procedure, but this may simply reflect the fact that the latter technique is relatively new. However, the preparation of effective packaging extracts is difficult and time-consuming whichever system is used, and it is advisable to purchase ready-made extracts from a commercial supplier. In general, the efficiency of commercial extracts is lower than those made in the laboratory (assuming you have mastered the technique), but still satisfactory. The use of a commercial packaging extract is described in Volume II, Chapter 3, *Protocol 11*.

2. Plating out and recombinant selection

After transformation of competent *E. coli*, or infection with packaged λ genomes, the proportion of cells that have actually taken up DNA will be relatively low. In *Protocol 3* DNA is added to about 10^8 cells but even under the best conditions only 10^7 transformants are expected. In most experiments a proportion of these will not be recombinants but will contain self-ligated vector molecules. Even if the vector was phosphatased to prevent self-ligation (Chapter 7, *Protocol 8*) a few molecules will have escaped the treatment and re-circularized without insert DNA.

The medium chosen for plating the cells should therefore be designed with two criteria in mind:

- non-transformed cells should not be able to grow at all; and
- recombinant colonies should be distinguishable from non-recombinant transformants.

The precise strategy depends on the genetic markers carried by the vector. As there are a large number of different vectors (Chapter 9) it might be expected that an equally large number of different selection strategies must be learnt. Fortunately, this is not the case as most vectors are constructed along the same lines and only a few types of selection are used. The most important ones are:

(a) For plasmid vectors
- antibiotic resistance
- inactivation of β-galactosidase activity

(b) For M13 vectors
- inactivation of β-galactosidase activity

(c) For λ vectors
- selection on the basis of genome size
- selection through inability to infect a P2 lysogenic strain of *E. coli*
- inactivation of the *c*I gene
- inactivation of β-galactosidase activity

2.1 Selection of plasmid vectors carrying antibiotic resistance genes

The first cloning vector to gain widespread use was pBR322 (Chapter 9, *Figure 1*). pBR322 carries two genes, one that codes for a β-lactamase that provides resistance to ampicillin, and one (actually a set of genes) that codes for tetracycline resistance. When pBR322 is used as a cloning vector the insert DNA is placed within one of these two genes, which one depending on the restriction site used. If, for instance, the *Bam*HI site is used then the tetracycline resistance gene is inactivated. The results of DNA uptake can therefore be determined by assessing the ampicillin and tetracycline responses of the individual cells:

- untransformed cells are amp^s tet^s;
- recombinant cells are amp^r tet^s;
- cells transformed with self-ligated pBR322 are amp^r tet^r.

The selection strategy therefore involves plating cells on to ampicillin agar, which screens out untransformed cells, and then transferring to tetracycline agar to distinguish the recombinants.

Protocol 5. Recombinant selection with pBR322

1. Prepare and dry 4 LB-ampicillin plates, 1 LB-tetracycline plate, and 1 LB plate (*Appendix 4*). Also prepare an overnight culture of the host *E. coli* strain in 5 ml LB.

2. After expressing the cells (*Protocol 4*), make a short dilution series in microfuge tubes:
   - Tube 1   200 μl cells
   - Tube 2   20 μl cells + 180 μl LB
   - Tube 3   2 μl cells + 198 μl LB

3. Spread the contents of each tube on to an LB-ampicillin plate. For the spread plate technique see Chapter 2, *Protocol 3*.

4. Controls:
   (a) Spread 200 μl of transformed cells on to an LB plate.
   (b) Spread 200 μl of the overnight culture (untransformed cells) on to an LB-ampicillin plate.

Protocol 5. *Continued*

5. Incubate all the plates overnight at 37°C.

6. Check the results. The LB plate should show confluent growth, indicating that the cells are viable. The untransformed cells should not grow on the LB-ampicillin plate. The transformed cells plated on to LB-ampicillin should produce discrete colonies, derived from individual amp^r cells.

7. Choose the LB-ampicillin plate that shows the largest number of separated colonies and make a replica-plate on to LB-tetracycline. Alternatively, use toothpicks to transfer small amounts of numbered colonies from the LB-ampicillin plate on to LB-tetracycline.

8. Incubate both plates at 37°C overnight.

9. Colonies that grow on the LB-tetracycline plate are amp^r tet^r and so contain self-ligated pBR322 molecules. Colonies that do not grow are recombinants – amp^r tet^s. These can be recovered from the LB-ampicillin plate, and inoculated into broth for further study.

2.1.1 Trouble-shooting

i. No transformed colonies are obtained

If no colonies appear on the LB-ampicillin plate then check the LB control. If this shows confluent growth then there is no problem with the cells and one of the problems affecting preparation of competent cells and DNA uptake is the cause. Refer to Section 1.1.4. The only other possibility is too much ampicillin in the medium.

ii. Confluent growth on the LB-ampicillin plates

This is due to inactivation of the ampicillin when the plates are poured. Ampicillin is very sensitive to heat and can be substantially degraded if it is added to the molten agar before this has cooled down sufficiently. Make sure the instructions in *Appendix 4* are followed.

iii. All the colonies from the LB-ampicillin plate grow on
    LB-tetracycline

No recombinant molecules have been constructed. Check you have correctly followed the protocols in *Chapter 7*.

iv. Colonies on the LB-ampicillin plate are surrounded by haloes of
    smaller colonies

The β-lactamase enzyme produced by amp^r bacteria is extra-cellular and diffusible, so will spread into the agar around resistant colonies. The local ampicillin concentration in the agar can therefore be depleted allowing nearby non-transformed cells (which are not killed by ampicillin, merely prevented from dividing) to grow and produce small colonies. The problem is particularly prevalent with high copy number vectors, presumably because these direct the synthesis of relatively large amounts of β-lactamase. An effective solution can be difficult to find. Try increasing the ampicillin concentration to 60 or 70 μg ml^{-1} and make sure no inactivation is occurring during medium preparation. Do not incubate the plates for too long.

2.2 Lac selection of plasmids

Many plasmid cloning vectors employ a system called Lac selection, which centres around the plasmid-borne gene *lacZ'*, coding for the first 146 amino acids of β-galactosidase. This is the enzyme responsible for converting lactose to glucose plus galactose in the normal *E. coli* bacterium. The segment coded by *lacZ'* is not by itself sufficient to catalyse the conversion, but it can complement a host encoded fragment to produce an active enzyme. Enzyme activity can be assayed with a chromogenic substrate, such as 5-bromo-4-chloro-3-indolyl-β-D-galactoside (X-gal), which is colourless but is converted to an intense blue product as a result of β-galactosidase activity. The assay is very sensitive and unambiguous.

A typical vector of this type is pUC18 (Chapter 9, *Figure 3*). pUC18 also carries an ampicillin resistance gene, so transformants are plated on to ampicillin agar, on which all cells containing a vector molecule are able to grow to produce colonies. The cloned DNA is inserted into a restriction site within *lacZ'*, which means that:

- recombinants are amp^r *lacZ*$^-$; and
- non-recombinants are amp^r *lacZ*$^+$.

The two types of colony can be distinguished by including X-gal in the agar, as recombinants will be white and non-recombinants blue. Unlike pBR322, this system therefore allows recombinants to be identified during the first plating-out.

Protocol 6. Recombinant selection with pUC18

1. Prepare and dry 4 LB-ampicillin plates and 1 LB plate (*Appendix 4*). Also prepare an overnight culture of the host *E. coli* strain in 5 ml LB.

2. After expressing the cells (*Protocol 4*), make a short dilution series in microfuge tubes:
   - Tube 1   200 μl cells
   - Tube 2   20 μl cells + 180 μl LB
   - Tube 3   2 μl cells + 198 μl LB

3. Prepare 200 μl of 2% (w/v) X-gal in dimethylformamide.[a] Add 50 μl to each tube, along with 10 μl 100 mM isopropyl-β-D-thiogalactopyranoside (IPTG).[b]

Protocol 6. *Continued*

4. Immediately spread the contents of each tube on to an LB-ampicillin plate. See Chapter 2, *Protocol 3* for the spread plate technique.

5. Controls:[c]

   (a) Spread 200 µl of transformed cells on to an LB plate.

   (b) Spread 200 µl of the overnight culture (untransformed cells) on to an LB-ampicillin plate.

6. Incubate all the plates overnight at 37°C.

7. The controls should show confluent growth of the transformed cells on LB, and no growth of the untransformed cells on LB-ampicillin. The transformed cells should produce discrete colonies on LB-ampicillin, some of these colonies being blue (non-recombinants) and some white (recombinants).

[a] It is best to prepare fresh X-gal for each experiment as solutions (and the solid) are light-sensitive. Stocks can, however, be stored with care at −20°C in light-tight tubes. Follow the relevant safety precautions when handling dimethylformamide.

[b] IPTG is a non-metabolizable inducer of the *lac* operon and is therefore needed to switch on expression of *lacZ'*. Stocks can be stored at −20°C with no problem.

[c] If you have used a phosphatased vector (Chapter 7, *Protocol 8*), so expect no non-recombinants, then you will need an additional control to check that the colour reaction is working. Transform an aliquot of cells with 1 ng unrestricted pUC18 vector and treat in the same way as one of the test dilutions.

2.2.1 Trouble-shooting

Recombinant selection with a *lacZ'* plasmid is subject to the same problems as pBR322 (Section 2.1.1). The X-gal system also presents its own difficulties.

i. All the colonies are white

If the colour reaction does not work then probably the X-gal has been degraded. Use a fresh stock and make sure the microfuge tubes are clean. Alternatively, the IPTG may be at fault but this is less likely unless the stock is more than three months old or has been left to stand at room temperature.

ii. The colour reaction is faint

Some vectors do not produce a dense blue coloration. To enhance the reaction leave the plates at 4°C for 4–6 h after the colonies have grown. If the colour change is still ambiguous then suspect partial inactivation of the X-gal (see above). Some companies market related compounds [Indigal (United States Biochemical Corp.), Bluo-gal (Gibco-BRL)] that may give a more intense colour than X-gal.

iii. The colonies are not uniformly coloured

The periphery is more densely coloured than the centre

Usually this is a non-recombinant.

The colony is white but there is a faint blue region in the centre

Typically a recombinant.

2.2.2 Positive negatives and negative positives

A general problem with Lac selection is that it is not entirely trustworthy. The *lacZ'* gene is very accommodating and can function even though quite large pieces of DNA (up to 100 bp) are inserted into it, so long as the reading frame is maintained. In fact this is one reason for the popularity of *lacZ'* vectors, as synthetic polylinkers carrying restriction sites for cloning purposes can be inserted into the gene without inactivating it (Chapter 9).

Unfortunately it also means that some recombinants will still produce active β-galactosidase and so will appear blue on X-gal plates. If you are cloning small fragments of DNA, do not assume that blue colonies are uninteresting and use a second method (e.g. colony hybridization: Volume II, Chapter 5, Section 3.10) to check them.

A second problem is that occasionally a white colony will not contain inserted DNA. Sequence examination of the vector usually reveals a small deletion in the *lacZ'* gene, probably as a result of excision of the inserted DNA at some stage in colony growth. This is a general problem that can arise with any cloning vector, especially if the host strain is not chosen with care (*Appendix 3*), or if the inserted DNA can form stem-loops that will enhance recombination and rearrangement events. However, with *lacZ'* vectors the problem can be worse because the presence of the polylinker stimulates stem-loop formation under some circumstances.

2.3 Recombinant selection with M13 vectors

All M13 vectors carry the *lacZ'* gene (Chapter 9, Section 3.1) and so recombinants are selected on X-gal plates in a manner similar to that described for pUC18. The plating-out procedure is different as the transformed cells will give rise not to colonies but to plaques, produced by M13 infection at foci on a lawn of bacteria. Each plaque represents a different transformant and its colour—blue or clear—indicates whether it is a recombinant or non-recombinant.

Usually, 1 ng of supercoiled M13 vector will give rise to about 10 000 blue plaques. After construction of recombinant molecules about 1000 blue and clear plaques are expected per ng DNA, although this assumes that all the original vector molecules were restricted. As the transfection efficiency is so much greater with unrestricted vector a small proportion of unrestricted molecules will cause the resulting plates to be swamped with blue plaques. This makes it difficult to predict how much DNA should be used in the transfection experiment. A single plate should ideally have about 500 plaques in order to achieve good separation. Start by transforming with 1 ng of recombinant molecules but be prepared to increase or decrease this figure in subsequent experiments if necessary.

Protocol 7. Recombinant selection with M13mp18

- M13 vectors do not carry antibiotic resistance genes so plasmid expression is not applicable. This protocol therefore follows directly from *Protocol 3*.
- You will need an exponential culture of host *E. coli* cells.[a]

1. Prepare and dry two YT plates. Melt 2×3 ml YTS agar in the microwave and place in a 50°C water bath. See *Appendix 4* for recipes.

2. To each tube of transformed cells add 50 μl 2% (w/v) X-gal in dimethylformamide, 10 μl 100 mM IPTG (see *Protocol 6*) and 200 μl of exponential host *E. coli* cells.

3. Pour the contents of each tube into 3 ml molten YTS agar and pour on to a YT plate. See Chapter 2, *Protocol 10* for the pour plate technique.

4. Allow the top agar to harden (about 5 min at room temperature) and then incubate the plates at 37°C overnight. Plaques will in fact be fully developed after about 10 h.

[a] If the competent cells are being made immediately before transformation then take 50 μl of the exponential culture from *Protocol 1*, step 2 and inoculate into 5 ml LB. Incubate at 37°C whilst you complete *Protocols 1* and *3*.

2.3.1 Trouble-shooting

The general problems discussed in Section 2.1.1, as well as the specific problems with Lac selection (Section 2.2.1), apply to M13 vectors. Other additional problems that may arise include the following:

i. The colour reaction is faint

M13 plaques do not give such an intense blue coloration as colonies obtained with a vector such as pUC18. It should nonetheless be possible to distinguish blue and clear plaques unambiguously. As well as the possibilities described in Section 2.2.1, the type of agar medium used influences the apparent colour of the plaques. A rich medium such as DYT (*Appendix 4*) is ideal for growth of plaques but its relative opaqueness makes discrimination of plaque colour more difficult. *Protocol 7* uses YT agar, which is less opaque than DYT. LB-agar is a suitable alternative or as a last resort a solid base of pure agarose can be used.

ii. The plates are smeary

Whorls and other artistic patterns are diagnostic of moist plates. Make sure they are dried thoroughly before use. Less dramatic smearing is due to the plates being moved before the top agar has hardened.

iii. *There are lumps on the plate*
This is a trivial problem that is due to the top agar not being completely molten. It will not solidify when held at 50°C but neither will it melt if it is not already completely molten. Even small lumps will ruin the resulting plate.

iv. *The plaques are not uniformly distributed*
A common problem is that plaques appear only on one half of the plate. This is generally ascribed to the plates being on an uneven surface when the top agar is poured on, as can happen if the plates are stacked as they are poured.

2.4 Recombinant selection with λ vectors

Lambda vectors are generally used for specialized purposes such as genomic and cDNA cloning. Descriptions of these procedures appear elsewhere in this book (Volume II, Chapter 2: genomic cloning; Volume II, Chapter 3: cDNA cloning) and include protocols for recombinant selection with the appropriate λ vectors. In this Section I will present just a summary of the strategies that can be employed. Additional details are given in Chapter 9, Section 2.2, where specific vectors utilizing these selection strategies are described.

2.4.1 Selection on the basis of size
As with most viruses the λ phage particle has a specific requirement for DNA molecules of a particular size. Only linear DNA molecules between 37 and 52 kb, and carrying *cos* sites at the termini, will be packaged into λ phage heads. This fact has been utilized in the construction of λ vectors, many of which produce molecules less than 37 kb in length if ligation occurs without DNA insertion. Only molecules that contain an insert will be packaged, so only recombinant phage will be produced.

Although size selection might appear to be foolproof it is rarely sufficient on its own for complete exclusion of non-recombinant molecules. This is because non-recombinant vector molecules must themselves be propagated in order to produce vector DNA, and so are constructed with a dispensable segment (the 'stuffer' fragment) between the two arms of the vector itself. The stuffer fragment is excised by restriction and the inserted DNA ligated between the arms in place of it. The problem that arises is that stuffer fragments may be present in the ligation so can themselves re-insert into the vector, producing non-recombinants that are appropriate sizes for packaging. This can be avoided by using ethanol precipitation to remove the stuffer fragment, or by digestion with additional restriction enzymes to produce sub-fragments of the stuffer that have termini incompatible with the vector arms (Chapter 7, Section 3.1). Complete removal is not usually possible however and it is normal for λ vectors to use a second type of selection to back up the size system.

2.4.2 Selection using the Spi phenotype

Vectors such as λEMBL3 and λEMBL4 (Chapter 9, Section 2.2.2.*ii*, and Volume 2, Chapter 2) employ a system called Spi selection. This is based on the fact that wild-type λ phage are unable to infect a bacterium which already carries a P2 prophage—they are *S*ensitive to *P2 I*nterference). The Spi phenotype is due to the activity of the *red* and *gam* genes on the λ genome, and if these genes are deleted the phage becomes Spi$^-$ and able to infect a P2 lysogen of a suitable host strain. Vectors that use the Spi system carry the *red* and *gam* genes on the stuffer fragment so that non-recombinants are Spi$^+$ and recombinants Spi$^-$.

2.4.4 Insertional inactivation of the λ*c*I gene

Not all λ vectors are replacement vectors with dispensable stuffer fragments. Some are analogous to plasmid and M13 vectors, carrying restriction sites that allow insertion of new DNA and inactivation of a selectable gene. The vector λgt10 (Chapter 9, Section 2.2.2.*i*), as well as a few others, carry cloning sites within the *c*I gene, which is part of the immunity region and codes for one of the regulatory proteins that control the λ infection cycle. *c*I$^+$ phage are highly efficient at forming lysogens in certain permissive host strains and as a result relatively few phage particles are produced. In contrast *c*I$^-$ phage are inefficient at lysogeny and larger amounts of phage particles are produced. The *c*I genotype can be assessed by eye as *c*I$^+$ plaques are turbid, because they contain large numbers of intact lysogenic bacteria, whereas *c*I$^-$ plaques are clear. Alternatively if a broth culture of a suitable host is infected with packaged phage, the bulk of the phage particles that are produced will be *c*I$^-$, as the *c*I$^+$ phage will form lysogens and be 'trapped' inside the cells.

2.4.3 Selection using *lacZ'*

Several λ vectors carry the *lacZ'* gene and recombinants are identified by the blue-clear colour reaction on X-gal agar, as described for M13 vectors (Section 2.3). In λgt11 (Chapter 9, Section 2.2.2.*i*), as well as a number of other λ vectors, the *lacZ'* gene carries restriction sites that allow insertional inactivation in the standard way. A few vectors carry the gene on the stuffer fragment, so that recombinants are *lacZ'*$^-$ due to complete loss of the gene.

3. Vectors combining features of both plasmids and phages

In recent years the development of novel vector types has blurred the distinction between plasmid- and bacteriophage-based systems. Two classes of vectors, cosmids and phagemids, combine features of both plasmids and phage chromosomes and have to be handled in special ways.

3.1 Cosmids

A cosmid (8) is essentially a plasmid that carries λ *cos* sites and is packageable into λ phage heads (Chapter 9, Section 2.3). Their advantage is that because they lack virtually all of the λ genome they can accommodate very large pieces of insert DNA, up to 45 kb or more, which is more than can be handled by an orthodox λ vector.

After *in vitro* packaging and infection of host *E. coli* cells a cosmid vector or recombinant molecule behaves like a plasmid, so a colony is produced, usually on an antibiotic plate, possibly utilizing Lac selection for recombinant identification. New 'phage' particles are not produced because the cosmid does not carry any of the standard λ genes.

The main application of cosmid vectors is in the construction of genomic libraries, where the large insert capacity is an advantage in minimizing the number of clones required for a complete library. Procedures for using a typical cosmid vector, pcos6, are given with other techniques for genomic libraries, in Volume II, Chapter 2.

3.2 Phagemids

A phagemid combines features of plasmid and M13 vectors (9). The main use of an M13 vector is in production of single-stranded (ss) DNA (Chapter 9, Section 3.1), which is required for several procedures, notably DNA sequencing (Volume II, Chapter 6). Unfortunately, the M13 genome cannot be modified to any great extent with impairing its basic genetic functions, so M13 vectors are not themselves very flexible. In addition they have quite small size capacities, and are inefficient at cloning fragments more than 2 kb in size.

Phagemids provide an alternative means of obtaining ss DNA (Chapter 9, Section 3.2). Phagemids carry two replication origins, one being a standard plasmid origin, and the other being derived from M13 or a related phage such as f1. The phage origin is the key component in the synthesis of ss DNA, though this also requires enzymes and coat proteins coded by phage genes, which the phagemid lacks.

A phagemid can be treated in exactly the same way as a standard plasmid vector, using the transformation and selection strategies described in Sections 1.1, 2.1, and 2.2. Alternatively cells containing a phagemid vector can be 'super-infected' with a 'helper phage', which itself contains a modified genome, but one that retains the genes for ss DNA production. The helper phage is therefore able to convert the phagemids into single-stranded DNA molecules, which are assembled into defective phage particles and secreted from the cell.

When using a phagemid for ss DNA production it is important to use both an appropriate host and the correct helper phage (*Appendix 3*). The technique varies slightly depending on the combination that is required. *Protocol 8* provides details for the phagemid pUC118, which is generally cloned in *E. coli* MV1184 and superinfected with the helper phage M13KO7.

Protocol 8. Cloning with pUC118

1. pUC118 is derived from pUC18 and so carries the *lacZ'* gene as well as the gene for ampicillin-resistance. Select recombinants as described in *Protocol 6*.

2. To produce ss DNA versions of the recombinant phagemids inoculate a single recombinant colony in 5 ml DYT (*Appendix 4*).

3. Add 10^8 plaque forming units of M13KO7.[a]

4. Incubate at 37°C for 1.5 h. Strong agitation is needed to get good yields of phage particles. We usually use test-tubes angled at 45 degrees on a platform shaker, so the cultures are vigorously aerated as well as being shaken.

5. Add 10 μl of kanamycin solution (35 mg ml^{-1}; *Appendix 4*). Continue the incubation overnight. M13KO7 carries a kanamycin-resistance gene so this treatment selects for superinfected cells.

6. Harvest extracellular 'phage' and prepare single-stranded DNA as described in Chapter 3, *Protocol 11*.

[a] M13KO7 phage are usually maintained as plaques on a lawn of *E. coli* MV1184 cells. Prepare a phage stock for superinfection by transferring a single isolated plaque into 2 ml of DYT + 70 μg ml^{-1} kanamycin (*Appendix 4*). Incubate with average agitation overnight at 37°C. Centrifuge the culture at 10 000 g for 5 min to pellet the bacteria and store the phage supernatant at 4°C. Measure the phage titre (Chapter 2, Section 5.3), which should be greater than 10^{11} per ml.

Phagemids are now very popular because of their combination of the best features of both plasmids and M13 vectors. Most of the new plasmid vectors becoming commercially-available are in fact phagemids (e.g. the pGEM series, see Chapter 9, Section 4.2.2.*ii*) and it is possible that in the future they will entirely supersede M13 cloning systems.

References

1. Smith, H.O., Danner, D.B., and Deich, R.A. (1981). *Annual Reviews in Biochemistry*, **50**, 41.
2. Mandel, M. and Higa, A. (1970). *Journal of Molecular Biology*, **53**, 159.
3. Cohen, S.N., Chang, A.C.Y., and Hsu, L. (1972). *Proceedings of the National Academy of Sciences of the USA*, **69**, 2110.
4. Oishi, M. and Cosloy, S.D. (1972). *Biochemical and Biophysical Research Communications*, **49**, 1568.
5. Hanahan, D. (1983). *Journal of Molecular Biology*, **166**, 557.
6. Scherer, G., Telford, J., Baldari, C., and Pirrotta, V. (1981). *Developmental Biology*, **86**, 438.
7. Rosenberg, S.M. (1987). *Methods in Enzymology*, Vol. 153 (ed. R. Wu and L. Grossman), pp. 95–103. Academic Press, New York.
8. Collins, J. and Hohn, B. (1978). *Proceedings of the National Academy of Sciences of the USA*, **75**, 4242.
9. Dente, L., Cesareni, G., and Cortese, R. (1983). *Nucleic Acids Research*, **11**, 1645.

Survey of cloning vectors for *Escherichia coli*

P. H. POUWELS

1. *Escherichia coli* as the host organism for recombinant DNA research

1.1 Introduction

One of the major goals of recombinant DNA research is to clone specific genes, determine their nucleotide sequence, and study the (over)-expression of the gene. In many research projects aiming at one or more of these goals the gene of interest has first to be isolated before investigations of its properties can begin. Most studies therefore start by making a gene library of the genome from which the gene is derived, in order to isolate the gene. After its isolation, the gene generally needs to be subcloned to eliminate irrelevant nucleotide sequences and to facilitate determination of the structure and genetic properties of the gene. In a next step, expression of the gene is studied, sometimes because one is interested in the product of the gene, and in other cases because the regulation of expression of the gene is the subject of study. Finally, it may be desirable to introduce specific alterations into the gene, to modify the properties of the gene product or to change the pattern of expression of the gene. In all these cases vectors are required which are specifically tailored to the desired objectives.

In this chapter I present a survey of vectors that can be used for such purposes as making gene libraries, expressing a gene and suchlike. I will start, in Section 1, by giving the reasons why *Escherichia coli* is the organism in which most recombinant DNA experiments are carried out, what types of vector exist and what their basic features are, and which are the criteria for choosing one vector above another. In Section 2, I will describe the general features of vectors that have been designed for making gene libraries and for sub-cloning of DNA fragments. For each category, individual vectors will be described in detail. In Section 3, I will then describe the general characteristics of vectors that were primarily designed to allow determination of the nucleotide sequence of a gene and modification of the structure by *in vitro* mutagenic treatment. This section is concluded by presenting a few examples of such vectors. In Section 4, the properties of expression vectors are discussed, and examples of several categories of expression vectors are given.

1.2 Advantages of *E. coli* as a host organism

There are several reasons why most people use *E. coli* as a host organism to make vector constructions, even if the vectors are meant to eventually be used in other organisms. Much knowledge has been gathered during the past decades with regard to the structure and properties of bacteriophage DNA and plasmids, DNA elements replicating extra-chromosomally in one or more copies per bacterium. In addition, the genetics and biochemistry of bacteriophage replication, plasmid replication and plasmid transfer are well understood, as well as that of control of gene expression.

Genetic studies with plasmids from Gram-negative bacteria have identified regions essential for replication of the plasmid, for transfer of plasmids between bacteria by means of conjugation, and for mobilization of plasmids. In the latter process the concerted action of the functions from more than one plasmid is required. Mobilization of a plasmid from one bacterium to the other requires the presence in the bacterium containing the mobilizable plasmid, of a second plasmid providing in *trans* the functions for transfer of the plasmid (1). In addition, functions (e.g. *par*; 2) have been identified that control the distribution of plasmids to daughter cells at the time of cell division. Other functions have been identified in regions of plasmid DNA that control site-specific recombination (e.g. *cer*; 3), a process implicated, through the interconversion of monomeric and multimeric forms of plasmid DNA, in plasmid stability, or to a region variably called *rom* or *rop* (repressor of primer; 4) which controls plasmid copy number.

The mechanisms that control gene expression in *E. coli* have been extensively studied and are, generally speaking, well understood. Gene expression is largely controlled at the level of initiation of transcription. For a large number of genes and operons the sites at which initiation of transcription takes place have been identified and their nucleotide sequences have been determined. In a number of cases the control of gene expression is also exerted at the level of transcription termination and/or anti-termination. In addition, regulation of gene expression takes place at the level of RNA processing and RNA degradation, a process intimately coupled to translation. Last but not least, gene expression is controlled at the level of initiation and elongation of translation. For a number of genes like the genes of the lactose (*lac*) operon, the tryptophan (*trp*) operon and the genes of the leftward (p_L) and rightward (p_R) operons of bacteriophage λ we have detailed knowledge of the nucleotide sequences of the genes pertaining to these control mechanisms. Expression from these promoters can be modulated by negative or positive control elements, like repressors and activators, and the structure and function of these elements is known in great detail (5). This knowledge is being exploited to try to achieve efficient expression of genes in *E. coli* and to fine-tune the levels of expression.

Other advantages of using *E. coli* as a host organism are that a great variety of well-characterized laboratory strains with useful features are available and that such strains can, in general, be cultivated in well-defined media. Of the many *E. coli* strains that exist, several are frequently used for cloning and expression of genes, sometimes because such strains have proven to yield high transformation frequencies, sometimes because they contain genetic characters that are essential for the use of a particular vector, and in other instances because expression and/or stability of a gene product is superior in one strain to that in others. A list of *E. coli* strains most frequently used, including a description of their genetic markers, is given in *Table 1*. See also *Appendix 3*.

Table 1. Genetic markers of *E. coli* strains

Strain	Genetic markers	Ref.
General purpose cloning vectors		
C600	F⁻ *supE44 thi1 thr1 leuB6 lacY1 tonA21*	74
DH1	F⁻ *gyrA96 recA1 relA1 endA1 thi1 hsdR17 supE44*	75
HB101	F⁻ *proA2 recA13 ara14 lacY1 galK2 xyl5 mtl1 rpsL20 supE44 hsdS20* $r_B^- m_B^-$	76
JA221	F⁻ *hsdR lacY leuB6 trpE5 recA1*	77
RR1	F⁻ *proA2 ara14 lacY1 galK2 xyl5 mtl1 rpsL20 supE44 hsdS20* $r_B^- m_B^-$	10
Single-stranded DNA vectors		
JM83	*ara* Δ*(pro-lac) rpsL thi phi80dlacZ* ΔM15	15
JM109	*recA1 endA1 gyrA96 thi hsdR17 supE44 relA1* Δ*(lac-proAB)* [F′ *traD36 proAB⁺ lacIq lacZ*ΔM15]	36
JM110	*rpsL thr leu thi lacY galK galT ara tonA tsx dam dcm supE44* Δ*(lac-proAB)* [F′ *traD36 proAB⁺ lacIq lacZ*ΔM15]	36
Bacteriophage λ vectors		
BHB2688	N205 *recA* (λ *imm434 clts b2 red3 Eam4 Sam7*)/λ	79
BHB2690	N205 *recA* (λ *imm434 clts b2 red3 Dam15 Sam7*)/λ	78
Y1088	*supE supF metB trpR hsdR$_K^-$ M$_K^+$ tonA21 strA* ΔlacU169 *proC*::Tn5(pMC9)	30
Y1090	*supF* ΔlacU169 Δ*lon araD139 strA trpC22*::Tn10(pMC9)	30

Finally, it is also worth mentioning that recombinant DNA research with *E. coli* is in most countries classified in the low-risk category and is, in many cases, even exempt from regulations. This means that the great majority of experiments with *E. coli* can be carried out under standard conditions (GMT = General Microbiological Techniques, see Chapter 2, Section 1.3) and, thus, do not require the more expensive containment facilities needed for research classified as CI or CII.

1.3 *Escherichia coli* is not an ideal host

Although *E. coli* has many attractive features for cloning and expression of genes (see *Table 2*), not the least being the capacity to very efficiently (over)-express cloned genes, it is not an ideal or universal host organism. The fermentation properties of *E. coli* have been studied in much less detail compared to typical industrial organisms like yeast, *Aspergillus niger* and *Bacillus subtilis*. Another limitation of the use of *E. coli*, in particular for the industrial production of pharmaceuticals and food additives, is the presence of an endotoxin which may diminish general public acceptability. In addition, *E. coli* is unable to perform a number of specific functions. Amongst the weak points of *E. coli* is its incapability to remove introns from intron-containing RNA. As a consequence, *E. coli* cannot properly express genes from eukaryotic organisms, most of which harbour introns in their DNA. This problem can still be surmounted by replacing the genomic gene by a cDNA copy. But more serious and less easy to circumvent, is the incapability of *E. coli* to modify proteins. The organism is unable to carry out N- or O-linked glycosylations, to completely remove the N-terminal methionine residue from a number of proteins and to acetylate, phosphorylate, or palmitate proteins. Although for a number of proteins it has been reported that the absence of, for example, N-linked glycosyl residues does not effect the biological activity of the proteins, it remains that a protein made in this way is not identical to its natural counterpart and that the absence of modified residues may have an effect on the immunological properties of the product or on its stability. Because *E. coli* has an internal reducing environment, proteins that in their active conformation contain disulphydryl bridges cannot fold properly and will thus form biologically-inactive proteins. For proteins whereby such modifications are a prerequisite for their proper functioning, the use of *E. coli* as an expression organism may be ruled out.

Table 2. Strong versus weak points of *E. coli* host vector systems

Positive	Negative
• Genetics and biochemistry well understood	• Little knowledge on fermentation properties
• Plasmid stability	• No GRAS[a] organism
• Many types of vector	• No capacity for intron splicing
• High yield of protein products	• Secretion of proteins into the medium is limited
	• Proteins not glycosylated
	• No disulphydryl bridges formed
	• Insolubility of proteins
	• Instability of proteins

[a] GRAS = generally regarded as safe.

1.4 Types of vector

In 1973 Cohen *et al.* (6) described the first successful construction of a recombinant DNA vector, the plasmid pSC101, for use in *E. coli*. Since then thousands of vectors have been constructed. Most of those have served as intermediates in making other recombinant DNA vectors or genetically-modified organisms. Such vectors have, in general, been used only by the investigators who made the vector, and have been mentioned briefly in the literature or have not been described at all. A much smaller class, but still encompassing more than one thousand vectors, comprises vectors that have been used in more than one experimental situation and in most cases in different laboratories. These vectors have been described in greater detail. *Cloning Vectors* (7) contains a compilation of most relevant vectors of this class described in the literature up until 1985. In three subsequent supplements *Cloning Vectors* has been updated. It describes the structure, properties and the constituents of construction of more than 1000 vectors, the majority of which were designed for use in *E. coli*.

Principally, there exist, dependent on the mode of replication in the bacterial host, four types of vector: plasmid vectors, cosmid vectors, and single- and double-stranded DNA bacteriophage vectors. A fifth category combines properties of plasmids and single-stranded DNA bacteriophage vectors. The general properties of these types of vector will be detailed in subsequent sections.

1.5 Selection of vectors

How should one make the choice if there are that many to choose from? When should one take a plasmid vector and when a bacteriophage λ vector for the experiment? What are the advantages of bacteriophage λ vectors over those of cosmids, or vice versa? And if the choice has been made for plasmid vectors, which is the best one? Decisions should primarily be based on the type of investigations one wishes to perform once the recombinant vector has been constructed. Of secondary importance, and less crucial, is the ease with which the goals can be reached. A third consideration is whether or not detailed knowledge is already available about the structure of the vector and more in particular about its nucleotide sequence. If it is the purpose of the experiment to construct a vector that can be used to screen a gene library made in a plasmid vector, one should, preferably, use a single-stranded or double-stranded phage vector, rather than a plasmid vector. Otherwise one runs the risk that the results will be confused by the homology that might exist between the two types of vector used. Whether the construction of a phage vector is easier or more difficult than that of a plasmid vector, is of lesser importance. What counts is that the results of the screening experiments have to be reliable.

In this overview I present a selection of vectors rather than a comprehensive account of all existing vectors, with the intention of helping the reader in selecting a vector for a specific purpose from the plethora of vectors that exist. I have chosen those vectors that already have been used successfully in the author's laboratory or elsewhere, or show great potential because of their improved properties. For each individual vector I will indicate the purposes for which it has been designed and how it can best be used to achieve the goals desired. In addition, for a few vectors, I will illustrate how useful they are, by describing experiments in which these vectors were successfully used. Finally, I will summarize specific features of the vectors, including information about the position of cloning sites, of selection markers, and other genetic traits, of promoters, translation-initiation regions, and other important features. In the description of the vectors, the restriction enzyme sites that occur once will be mentioned, and a selection of target sites for the most frequently used enzymes will be indicated in the accompanying illustrations. Of the many enzymes that are used in recombinant DNA experiments, only enzymes recognizing a hexanucleotide or longer sequence will be mentioned. Additional information about restriction enzyme sites and relevant nucleotide sequence information will be presented in the figure legends.

In this survey I will treat vectors designed for use in *E. coli* only. In a number of cases derivatives have been made of vectors described in this chapter which can be shuttled between *E. coli* and the cells of higher eukaryotes or yeast cells. These types of vector will not be treated here.

2. Construction of gene libraries; subcloning of DNA fragments

A generally accepted strategy for the isolation of a gene is the construction of a gene bank in a plasmid vector, a bacteriophage λ vector, or a cosmid vector. If the genome from which the gene has to be isolated is relatively small (e.g. from a prokaryotic organism) plasmid vectors are most frequently used, because they are easier to handle than the other two types of vector. For the isolation of genes from more complex genomes, in particular from eukaryotic cells, bacteriophage λ vectors or cosmid vectors are to be preferred, because the size of DNA fragments that can be incorporated into plasmid vectors is much smaller than in the other two types of vector. Consequently, the number of individual clones needed to obtain a representative gene library will be much smaller for bacteriophage λ vectors and cosmid vectors than for plasmid vectors. In the following sections the general properties of these three classes of vectors will be presented and individual vectors will be described in detail.

2.1 Plasmid vectors

2.1.1 General properties

Most plasmid-based vectors currently used in recombinant DNA experiments are derivatives of the naturally occurring plasmid ColE1 or from the ColE1-

related plasmid pMB1. Like the parent plasmids themselves, these plasmids, and their derivatives, are multicopy plasmids present at 20–80 copies per chromosome per bacterium. Other plasmid vectors containing an origin of replication of p15A or pSC101 are low-copy plasmids present at a few copies per bacterium only. The size of DNA fragments that can be cloned in plasmid vectors is virtually unlimited, but the stability of very large recombinant plasmids may be diminished. In addition, plasmid copy number and transformation capacity of plasmids are also unfavourably affected by the insertion of very large DNA segments.

Replication of plasmid DNA derived from ColE1 or pMB1 is uni-directional and is dependent on the presence in the plasmid of a nucleotide sequence of just a few hundred base-pairs, which includes the origin of replication (*ori*). Plasmid replication does not require the presence of plasmid-encoded proteins. Advantage can be made of this property to increase the plasmid copy number, by cultivation of the bacteria under conditions such that protein synthesis is arrested; for example, by addition to the culture medium of chloramphenicol (see Chapter 3, Section 4.2). By this treatment the bacterial content of plasmid DNA can be increased to a level such that 50% or more of all DNA is plasmid DNA. Since under such conditions several thousand copies of the plasmid are present per bacterium, relatively large quantities of plasmid DNA can be obtained from small amounts of bacteria. Amplification, by means of the cessation of protein synthesis, of the copy number cannot be achieved for all types of plasmid. So-called 'stringent' plasmids cannot, in contrast to 'relaxed' plasmids, be replicated in the presence of chloramphenicol. This is due to the fact that stringent plasmids, like pSCl01, R6K, and R1*drd*19 are dependent on protein synthesis for replication.

The plasmid content of a bacterium can also be drastically increased by raising the temperature of the culture medium in the case of plasmids carrying a mutation rendering replication uncontrolled at the higher temperature (8). Plasmids containing such run-away replicons are quite convenient for isolating large amounts of plasmid DNA.

Plasmid replication is affected by plasmid-borne genes encoding two RNA molecules, RNA-I and RNA-II (9), that determine the frequency of initiation of DNA replication at *ori* and a third gene located immediately adjacent to the RNA-encoding genes, *rop*, which codes for a 63-amino-acids long protein that negatively controls plasmid replication. In a number of cases the *rop* gene has deliberately been removed in part or completely, to inactivate the Rop protein. This has resulted in an increased copy number of the plasmids, making it easier to isolate large quantities of the recombinant DNA plasmid. An added benefit from such treatment is that gene expression, as a result of an increased gene-dosage, is likely to be enhanced also.

Plasmids that are closely related usually cannot be stably maintained together within the same bacterium. For example, plasmids derived from ColE1 or pMB1 and their derivatives cannot co-exist within the same bacterium. Such plasmids are called incompatible. Plasmids that can be maintained within the same bacterium together with ColE1-related plasmids are called compatible and belong to different compatibility groups. Examples of such plasmids are p15A, F'*lac*, R6K, and R1*drd*19. In instances where one wishes to introduce into the same bacterium two genes or regulatory elements present on different plasmids, one of which has a ColE1-like origin of replication, one has to resort to a plasmid of another compatibility group, to stably introduce the second DNA element into the bacterium.

Most plasmids that are currently used, and essentially all plasmid vectors treated in this survey, are non-conjugative plasmids lacking genes for mobilization. This means that genetic material can be transferred to other bacteria only if the cell contains other plasmids providing for mobilization and conjugation functions. A second requirement for transfer of genetic material to occur, is the presence in the plasmid vector of elements called *bom*, for *basis of m*obilization. In some vectors this element has purposely been deleted, to limit further the adventitious spread of cloned sequences.

Because most transformation procedures are relatively inefficient (under optimal conditions only a few per cent of all bacteria are transformed) it is essential that one can discriminate transformed and non-transformed bacteria. This is in most cases accomplished by making use of dominant selection markers; for example, genes carrying an antibiotic resistance marker which, if present on the plasmid, confer to the bacterium resistance to that antibiotic. The most commonly used marker is that for ampicillin resistance (*amp*) originating from transposon Tn3. Other markers that are currently in use are the tetracycline (*tet*) gene from transposon Tn10, the kanamycin resistance (*kan*) gene from transposon Tn903 and the neomycin resistance (*neo*) gene from transposon Tn5. All these markers are similar in that direct selection for plasmids containing one of the markers is easy and sensitive. *E. coli* bacteria are sensitive to very low concentrations of ampicillin (a few μg ml^{-1}) but become resistant to very high levels (up to 1 mg ml^{-1}), if the *amp* gene is expressed. Particularly attractive are vectors containing two or more dominant selection markers. If a DNA fragment is inserted into one of these markers, selection can be based on the other marker and inactivation of the first marker is evidence for the successful cloning of the DNA fragment (see Chapter 8, Section 2.1).

A special class of vectors, so-called direct selection vectors, permit direct selection of transformed bacteria by selecting against the parent plasmid. With this type of vector, replica-plating to verify the presence of an insertion in an antibiotic-resistance gene is not necessary any more. In general, this type of vectors can only be used in combination with specific bacterial strains (see ref. 7 for a compilation of these vectors).

Systems which allow easy detection of a recombinant plasmid amongst a large population of parent plasmids are based on the capacity of enzymes like β-galactosidase to hydrolyse chromogenic substances such as X-gal (5-bromo-4-chloro-3-indolyl-β-D-galactopyranoside), which results in the formation of a dark blue, insoluble compound. Bacteria containing a functional β-galactosidase gene (*lacZ*) of *E. coli* form blue colonies on agar plates supplemented with X-gal. By making use of appropriate bacterial strains, discrimination is possible between bacteria transformed with the parent vector and those transformed with the recombinant vector, by visual screening of the colonies formed (see Chapter 8, Section 2.2). If a Lac⁻ host bacterium contains a plasmid with a functional *lacZ* gene blue colonies will be formed. When the *lacZ* gene is disrupted in the recombinant plasmid by introduction of an exogenous DNA fragment, a mixture of blue and colourless colonies are to be expected. The blue colonies represent bacteria harbouring the parent plasmid, while colourless colonies contain recombinant plasmids.

2.1.2 Specific vectors

The major use of the vectors described in this section is the cloning and amplification in *E. coli* of exogenous DNA fragments. In a number of cases, however, the vectors can also be used for more specialized purposes, like expression of foreign proteins and DNA sequence analysis.

i. pBR322

Plasmid pBR322 (4363 bp; *Figure 1*; 10) is the most widely-used cloning vector presently available. It encodes two selectable genetic markers, resistance to ampicillin (*amp*) and resistance to tetracycline (*tet*). Replication of pBR322, and all its derivatives, takes place in a 'relaxed' fashion, implying that the plasmid copy number can be increased from 20 to 1000 copies per chromosome per bacterium, by cultivating the bacteria in the presence of chloramphenicol.

Plasmid pBR322 and its derivatives lack the *cer* site, resulting in the formation of plasmid multimers (3). Because the copy number of plasmid multimers is reduced, they are more easily lost than monomers. The plasmid also lacks a partition function needed for proper segregation of plasmid vectors to daughter cells during cell division. So, in the absence of selection pressure, uneven segregation and consequent loss of plasmids may occur. This instability can be largely overcome by incorporation into the vector of the partition locus of plasmid pSC101 (11).

Due to the presence of the *bom* site, plasmid pBR322 can be mobilized by the presence of conjugative plasmids, like ColK, which deliver in *trans* the mobilization functions. Since the probability that recombinant pBR322 vectors are transferred to other bacteria is very low, the vector has been certified by the NIH Recombinant DNA Advisory Committee as an EK2 vector (12).

ii. pBR327

Plasmid pBR327 (3274 bp; *Figure 2*; 13) is a versatile cloning vector derived from pBR322 by deletion of a sequence of 1089 bp between nucleotide 1428 and nucleotide 2517. Due to this deletion, pBR327 makes no functional Rop protein and, consequently, has an increased copy number. Since the deletion has also

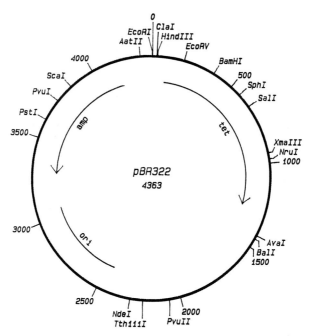

Figure 1. Structure of pBR322. The plasmid carries unique cloning sites (apart from those shown in the diagram) in the *tet* gene for *Bsp*MI (nucleotide 1054), *Eco*NI (nt 626), and *Nhe*I (nt 229), and in the *amp* gene for *Eco*31I (nt 3429). These sites can be used for insertion of DNA fragments and selection of recombinants by insertional inactivation. The vector contains unique cloning sites for *Bsm*I (nt 1353), *Bsp*MII (nt 1664), *Ssp*I (nt 4170), *Sty*I (nt 1369) and *Xca*I (nt 2248) elsewhere in the plasmid. The basic constituents of the vector are the origin of replication from pMB1, the *amp* gene from transposon Tn3 carried on plasmid pRSF2124 and the *tet* gene from pSC101. A detailed account of the structure, including the nucleotide sequence, and properties of pBR322 and its derivatives is given in Balbas *et al.* (9). The nucleotide sequence of pBR322 can be retrieved from the EMBL database.

removed the *bom* site, pBR327 is non-mobilizable, constituting a further improvement in safety, compared with pBR322. Plasmid pBR327 has been certified by the NIH Recombinant DNA Advisory Committee as an EK2 vector.

Except for the loss of some unique cloning sites, plasmid pBR327 has the same cloning properties and genetic markers as pBR322.

iii. pBR327par

To overcome segregational instability encountered with plasmids derived from pBR322, under non-selective conditions, a derivative was constructed, pBR327*par* (14), in which the *par* locus from pSC101 was incorporated into the *Ava*I site as a 0.4 kb *Bam*HI fragment. Except for its improved stability properties, pBR327*par* and pBR327 have the same characteristics.

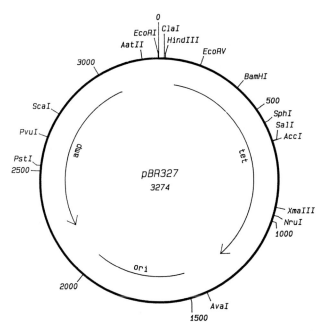

Figure 2. Structure of pBR327. The plasmid has unique cloning sites (except for those shown in the drawing) in the *tet* gene for *Bsp*MI (nucleotide 1054), *Dsa*I (nt 528), *Eco*NI (nt 626), *Nhe*I (nt 229) and *Nsp*HI (nt 566), in the *amp* gene for *Asp*700I (nt 2878), *Eco*31I (nt 2340) and *Ksp*632I (nt 3074). Plasmid pBR327 contains unique cloning sites for *Bsm*I (nt 1359), *Hgi*EII (nt 1972), *Ssp*I (nt 3083) and *Sty*I (nt 1369) elsewhere in the plasmid. Plasmid pBR327 was constructed from pBR322 by removal of an *Eco*RII fragment between the *tet* gene and the origin of replication, and religation, after rendering the cohesive ends blunt with nuclease S1. The nucleotide sequence of pBR327 has been published (13) and can be retrieved from the EMBL database.

Another vector, pAT153 (3658 bp; 79) with properties very similar to those of pBR327, was constructed by deletion from pBR322 of a 705 bp *Hae*II fragment (nt 1648–2353). Like pBR327, plasmid pAT153 lacks the *bom* site. It shows no detectable plasmid transfer in the presence of the plasmids ColK and R64*drd*11, which provide for mobilization and conjugation functions. The deletion has also removed the *rop* gene, causing an increase in copy number, compared with pBR322. Plasmid pAT153 has the same unique cloning sites as pBR327, except for the absence of unique *Dsa*I and *Ksp*632I sites, and for the presence of a unique *Bal*I (nt 1446) site. The nucleotide sequence of pAT153 has been deduced from that of pBR322. According to the UWGCG Vecbase database, there is conflicting information on the structure of pAT153.

iv. pUC series

Plasmids of the pUC series (15–17) have proven very useful for cloning of DNA fragments. Their most characteristic feature is the presence of part of the *E. coli lacZ* gene, the so-called *lacZ'* fragment, with multiple cloning sites (a 'polylinker') downstream of the translation-initiation codon ATG. The *lacZ'* fragment contains the promoter-operator region and 59 codons of the beginning of the *lacZ* gene. This part is sufficient to complement the M15 deletion of the *lacZ* gene. Insertion of DNA fragments into one of these sites disrupts, in general, the coding information, and leads to the loss of a functional *lacZ'* gene product. This loss can be directly visualized, by a change of the colour of the colonies from blue to colourless (18), and thus does not require replica-plating as in the case of insertional inactivation of an antibiotic-resistance marker (see Chapter 8, Section 2.2).

Plasmids of the pUC series can be used for DNA sequence analysis, by either the chain termination method of Sanger (19) or the Maxam–Gilbert procedure (20). In the former procedure, both DNA strands can be sequenced simultaneously by making use of two DNA primers that are complementary to regions of the vector flanking the polylinker region.

To further increase the versatility of the vector, restriction sites in the polylinker region have been made unique by *in vitro* mutagenesis of existing restriction sites elsewhere in the plasmid. A *Pst*I site and a *Hinc*II site in the *amp* gene, and an *Acc*I site near the *ori* region were removed, without loss of essential functions. The pUC vectors exist as pairs of two vectors, carrying the polylinker sequence with multiple cloning sites in opposite orientation. Derivatives have been made of the pUC vectors in which the *amp* gene has been replaced by the *kan* gene (21). DNA fragments can thus be easily subcloned from a vector with an *amp* gene as selection marker to a vector with a *kan* gene as selection marker, and vice versa.

Because the pUC vectors lack a functional *rop* gene, they have an increased copy number compared with pBR322. The exceptionally high copy number (500–700 copies per cell) of the pUC plasmids is, in part, also due to a fortuitous single base-pair mutation in the *ori* region encoding RNA-I. As a result of this mutation an RNA-I molecule is formed which is three nucleotides smaller than the equivalent transcript from pBR322. The lack of three nucleotides at the 5'-end of RNA-I probably has resulted in an increased copy number.

The structure of pUC18 (*Figure 3*) is given as an example of the plasmids of this series.

Experimental approaches
To appreciate the great potential pUC vectors have for cloning strategies, I will describe an experiment carried out by P. J. Punt (MBL-TNO). The experiment is part of a research project aiming at the identification and characterization of the signals that determine expression of the glyceraldehyde 3-phosphate dehydrogenase (*gpdA*) gene of *Aspergillus nidulans* (22).

The experiment shows that, due to a combination of properties, pUC vectors are attractive for cloning experiments, because they allow:

(a) selection of recombinant plasmids;

(b) determination of the nucleotide sequence of an inserted DNA fragment, without the need to sub-clone the fragment;

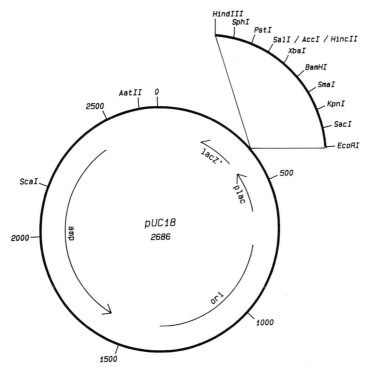

Figure 3. Structure of pUC18. Plasmid pUC19 is the same as pUC18, but has the polylinker region in the opposite orientation. These plasmids have other unique cloning sites outside the *lacZ'* gene for *Aat*II (nucleotide 2621), *Afl*III (nt 806), *Bbe*I (nt 239), *Dra*II (nt 2675), *Eco*78I (nt 237), *Nar*I (nt 236), *Nde*I (nt 184), *Pss*I (nt 2678), *Ssp*I (nt 2503), unique cloning sites within the polylinker region (except those shown in the drawing) for *Asp*718I (nt 408), *Ava*I (nt 412), *Ban*II (nt 406), *Bsp*MI (nt 442), and unique sites in the *amp* gene for *Asp*700I (nt 2298), *Cfr*10I (nt 1779), *Eco*31I (nt 1760), and *Gsu*I (nt 1769). The nucleotide sequence has been deduced from known sequences and is given in the EMBL database. The nucleotide sequence of the polylinker region of plasmids pUC18 and pUC19 is identical to that of M13mp18 and M13mp19 (see *Figure 13*).

(c) determination of the polyadenylation site and length of the poly(A)-tail of mRNAs (via cDNA cloning);

(d) determination of the presence of introns near the 3'-end of mRNA.

In order to characterize the *gpdA* gene, the nucleotide sequence of *gpdA* mRNA had to be determined and compared with the sequence of the *gpdA* gene. To this end, cDNA prepared by reverse transcription of *A. nidulans* mRNA with oligo(dT) as a primer was cloned into pUC18. Recombinant plasmids could be identified on the basis of their LacZ⁻ phenotype. Inserts containing *gpdA*-cDNA sequences were identified by hybridization with a *gpdA*-specific probe. In the experiment the polyadenylation site of *gpdA* mRNA was determined by carrying out a so-called 'T-track analysis' of the DNA fragments inserted into pUC18. In a typical experiment the relative position of T residues was determined for four plasmids with *gpdA*-specific inserts. As is shown in *Figure 4* the cDNA inserts start with a poly(T)-tail, followed by the same, *gpdA*-specific sequences. The gel pattern shows some heterogeneity with respect to the site at which the A residues are added. By comparison of this sequence with that of the *gpdA* gene, the polyadenylation site of *gpdA* mRNA could be unambiguously determined. In addition, the presence of an intron near the 3′-end of the gene could be established.

v. pACYC177

Plasmid pACYC177 (3941 bp; 23) is one of the first vectors that was constructed, but, due to its attractive properties, it is still a very popular vector. Its main value lies in the presence of an origin of replication from plasmid p15A, that is related to

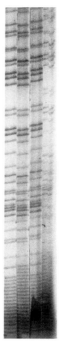

Figure 4. T-track analysis of *gpdA* cDNA cloned in pUC18. The relative position of T-residues in the inserts in pUC18 was determined with the double-stranded DNA sequencing method (T-reaction only) of Chen and Seeburg (80). DNA was synthesized by extension of a primer-oligonucleotide that hybridized to a region of pUC18 immediately adjacent to the insert. The DNA synthesized was analysed by electrophoresis on a polyacrylamide gel. The experiment shows T-tracks of inserts from four plasmids. Each T-track starts with 15–25 neighbouring T residues which correspond to the A residues of the poly(A)-tail. The site of polyadenylation varies by a few nucleotides in the different cDNA inserts. The *gpdA* mRNA thus shows some heterogeneity at its polyadenylation site. A comparison of the sequences shown here with that of the *gpdA* gene (not shown) indicates the presence of an intron near the 3′-end of the *gpdA* gene.

ColE1-like plasmids, but is compatible with them. This offers the possibility of introducing into the same bacterium, genes present on a pBR322 derivative and other genes inserted into pACYC177, and to study interaction of the plasmids or the products thereof. Plasmid pACYC177 (*Figure 5*) is small and carries two selectable markers, encoding resistance to ampicillin and kanamycin, and has numerous unique cloning sites, both inside and outside these genes.

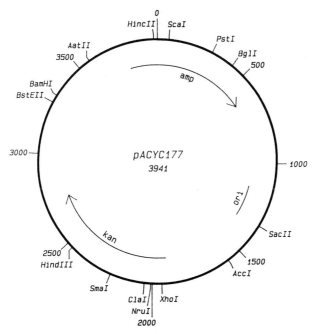

Figure 5. Structure of pACYC177. Apart from the sites shown in the drawing, unique sites are present for *Apa*LI (nucleotide 3811), *Ban*I (nt 589), *Eco*31I (nt 476), *Eco*57I (nt 3817), *Fsp*I (nt 319), *Gsu*I (nt 473) in the *amp* gene, for *Ban*II (nt 2006), *Eco*NI (nt 2263), *Pfl*MI (nt 2614) and *Sci*I (nt 1951) in the *kan* gene, and for *Alw*NI (nt 989), *Asu*II (nt 850), *Bbv*II (nt 3499), *Bst*EII (nt 3275), *Dra*II (nt 3502), *Dra*III (nt 1820), *Nhe*I (nt 1585), *Nsp*HI (nt 1014), *Pss*I (nt 3505), and *Xca*I (nt 1576) elsewhere in the vector. Plasmid pACYC177 comprises the *amp* gene from transposon Tn3 (nt 3024–762), the *kan* gene from transposon Tn903 (nt 1587–3023) and the *ori* region from p15A (nt 763–1586). The nucleotide sequence of the plasmid is known and can be retrieved from UWGCG vecbase.

vi. pGB2
After cloning of a DNA fragment into a vector, it is sometimes desirable to screen gene libraries for the presence of sequences related to the cloned gene. In such cases it is necessary that the cloning vector does not hybridize to the vector in which the library was constructed. To obviate this problem, the cloned DNA is normally excised from the vector and used as a probe. It is difficult, however, to completely remove traces of vector DNA, the presence of which might confuse the results of the hybridization experiments.

Plasmid pGB2 (3.8 kb; *Figure 6*; 24) was constructed to alleviate these problems. The vector shows no homology, as detected by DNA–DNA hybridization, to several widely-used plasmid vectors, like pBR322 and pUC and bacteriophage λ vectors. The plasmid contains the replication origin of pSC101, linked to a DNA fragment specifying resistance to spectinomycin (*spc*) and streptomycin (*str*). The plasmid has a low-copy number which cannot be amplified with chloramphenicol. A polylinker region that includes several unique restriction sites was inserted into the vector.

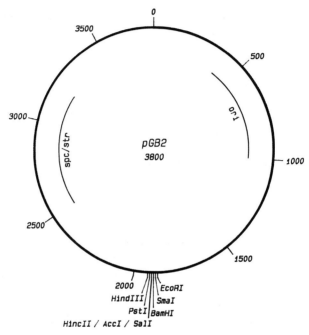

Figure 6. Structure of pGB2. Plasmid pGB2 comprises a 1.8 kb *Eco*RI-*Hind*III fragment carrying the *spc/str* gene of R100, a 2.0 kb *Hinc*II-*Rsa*I fragment from pSC101 containing the *ori* region, and an *Eco*RI-*Hind*III fragment of pUC8 (15) with the polylinker region. The *Eco*RI site at the junction of the R100 fragment and the pSC101 fragment has been deleted. The nucleotide sequences of the constituent parts of plasmid pGB2 have been determined. They can be retrieved from the EMBL database.

2.2 Bacteriophage vectors with double-stranded DNA

2.2.1 General properties

Viral vectors containing double-stranded DNA are all derived from the temperate bacteriophage λ. Because wild-type λ DNA contains several target sites for most commonly-used restriction enzymes, derivatives of λ have been made that do not suffer from this inconvenience. They either have a unique restriction enzyme site at which exogenous DNA fragments can be inserted (insertion vector), or they contain pairs of restriction sites flanking on both sides a region that can be replaced by exogenous DNA (replacement vector). Insertion vectors have proven to be very useful for the cloning of relatively small DNA fragments and screening of plaques for the presence of a given DNA sequence. Replacement vectors on the other hand are particularly useful for the construction of gene banks or libraries. For the latter purpose, the use of λ replacement vectors has certain advantages over that of plasmid vectors. These advantages are more pronounced if it concerns the preparation of a gene bank from a complex genome like that of an animal cell.

Bacteriophage λ has a double-stranded DNA genome of approximately 49 kb, more than 40% of which is dispensable and can be replaced by exogenous DNA. This means that DNA fragments up to 20–25 kb in size can be cloned into phage λ. In order to obtain a representative gene bank from a mammalian genome with 99% probability, a population of 10^6 recombinant phage with an average insert of 15 kb is required. Since the efficiency of the *in vitro* packaging system (25) compares favourably with that of most transformation procedures with plasmid DNA (see Chapter 8, Section 1.2), and the average size of the DNA fragments that can be stably incorporated into λ is far greater than that of plasmid vectors, it has become common practice to construct a gene bank with lambdoid vectors rather than with plasmid vectors. An added benefit of λ vectors is that the number of phage particles in a plaque can be as high as 10^6–10^7, facilitating screening, by means of nucleic acid hybridization or immunochemical staining techniques, of the phage libraries for the presence of a certain gene.

The strategy for the use of a replacement vector to prepare a gene library is illustrated in *Figure 7*. By treatment with an appropriate restriction enzyme, the vector DNA is split into three parts. One part, called the left arm, contains all the functions for morphogenesis of the virus particle, while a second fragment, called the right arm, harbours all functions for multiplication of viral DNA and expression of viral genes. The central fragment, called the stuffer fragment, contains no essential functions and is dispensable. In the commonly-used replacement vectors, nucleotide sequences coding for one or more restriction enzyme sites have been introduced at both sides of the stuffer fragment, so that this fragment can be simply uncoupled from the vector arms by treatment with a single restriction enzyme. Phage λ DNA contains 5'-protruding ends of 12 nucleotides which can anneal, to form a duplex molecule in which the left and right arm are joined via these cohesive ends (*cos*) (*Figure 7*). When an exogenous DNA fragment is ligated to the vector, a catenane of vector and exogenous DNA is generated. Viable phage can then be formed from these catenanes in an *in vitro* packaging system. In order to be packaged into viable phage, the DNA segment between the two *cos* sites should be larger than 40 kb but smaller than 52 kb. These are the lower and upper limits, respectively, for the size of a λ genome, in order to be encapsidated into a phage head. The vector fragment comprising left and right arms without stuffer is much smaller than 40 kb. Consequently, it will not be packaged into phage heads, unless an exogenous DNA fragment has been

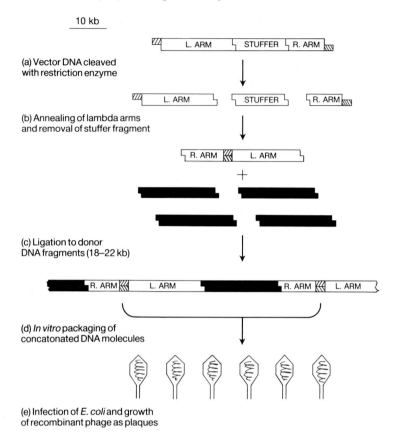

10 kb

(a) Vector DNA cleaved
with restriction enzyme

(b) Annealing of lambda arms
and removal of stuffer fragment

(c) Ligation to donor
DNA fragments (18–22 kb)

(d) *In vitro* packaging of
concatonated DNA molecules

(e) Infection of *E. coli* and growth
of recombinant phage as plaques

Figure 7. Schematic representation of cloning of DNA fragments into bacteriophage λ. [Reproduced from K. Kaiser and N. Murray (1985). In *DNA Cloning: A Practical Approach* (ed. D.M. Glover), Vol. 2, pp. 1–47. IRL Press, Oxford.]

ligated to it. Details of how to use a λ replacement vector are given in Volume II, Chapter 2.

For the introduction of small DNA fragments, ranging in size between 0 and 10 kb, insertion vectors are used. These vectors have been engineered in such a way that exogenous DNA can be incorporated into a unique restriction enzyme site present in the non-essential region of the genome, without loss of the capacity to form phage. Since phage λ can encapsidate not more than 105% of its normal amount of DNA, extra space has been made available by deletion of some non-essential regions, to accommodate a maximum amount of exogenous DNA. If the vector has retained its integration functions (*int*), recombinant phage remain fully integration proficient and are, therefore, useful to efficiently transduce *E. coli* strains (26).

An essential difference between insertion vectors and replacement vectors is that the two fragments, after cleavage of an insertion vector with a restriction enzyme, can form viable phage when no exogenous DNA has been incorporated, while a replacement vector from which the stuffer fragment has been removed, cannot. Thus, while there is a natural selection against parental replacement vectors, in the absence of the stuffer fragment special selection systems have to be introduced, to discriminate between a recombinant phage and a parental insertion vector. The selection for recombinant phage or against parental phage is, in general, based on insertional inactivation of a phage gene or a gene that was purposely introduced into the vector, leading to a change in phenotype (see Chapter 8, Section 2.4). If, for example, exogenous DNA is cloned into a site within the immunity region, as is the case with the 'immunity insertion vectors' (27), recombinants can be easily and efficiently selected. First, recombinant phage multiply faster than parent phage, because of their enlarged genome. They form larger plaques and rapidly outgrow the parent phage. More importantly, however, recombinant phage can be visually-screened, because inactivation of the immunity region results in clear plaques, while the parent phage shows the characteristic turbid plaque morphology, due to growth in the plaque area of bacteria that have undergone the lysogenic response to phage infection. Lysogenic bacteria produce the λ repressor (product of gene cI) that negatively controls the expression of other phage genes, thereby preventing the lytic cycle of the phage. In *E. coli* carrying the *high frequency lysogeny* mutation (Hfl$^-$), the λ repressor is over-produced to the extent that plaque formation is suppressed (28). However, cI^- phage forms plaques with normal efficiency on an Hfl$^-$ strain. Recombinant phage made from 'immunity insertion vectors' can thus readily be selected on such a strain.

Recombinant phage can also be selected on the basis of their ability to be multiplied in bacteria lysogenic for phage P2 (29). Such phage are said to have the Spi$^-$ phenotype, for lack of sensitivity to P2 interference (30). Wild-type phage λ cannot grow on such a host, but mutants from which the *red* and *gam* genes in the non-essential region of the genome have been removed or inactivated, and, in addition, have incorporated an octanucleotide sequence, called Chi, normally produce plaques on a P2-lysogen. If, in the stuffer fragment of a replacement vector carrying Chi, functional *red* and *gam* genes are present, it is not necessary to physically separate the stuffer fragment before ligating the vector arms to the exogenous DNA fragment. If the resulting phage are plated on a P2 lysogenic host, only recombinants that are Red$^-$, Gam$^-$, Chi$^+$ will produce plaques, while the parent phage will not, because it is Red$^+$, Gam$^+$.

2.2.2 Specific vectors

In the following paragraphs I will describe some examples of insertion vectors and replacement vectors. Bacteriophage λgt10 and λgt11 are widely used as insertion vectors for making gene libraries. Their properties, structure and modes of application have been described in detail elsewhere in the *Practical Approach* series (31). A summary of the essential features of λgt10 and λgt11 will be given here.

i. *λgt vectors*

Phage *λ*gt10 was specifically designed for the cloning of relatively short DNA fragments, in particular cDNAs (see Volume II, Chapter 3). The vector can accept DNA fragments with *Eco*RI cohesive ends of a size between 0 and 7.6 kb. Insertion of a DNA fragment into the unique *Eco*RI site inactivates the *λ* repressor gene generating a *cI*⁻ phage, which forms a clear plaque. The parent phage *λ*gt10 forms turbid plaques. Since *λ*gt10 does not require an insert fragment to yield a packageable DNA molecule, cDNA libraries cloned in *λ*gt10 initially consist predominantly of parent phage, if plated on wild-type *E. coli* host bacteria. Selection against the parent phage is possible during amplification of the library, by plating the phage on a Hfl⁻ strain, which permits plaque formation of *cI*⁻ phage only (see Chapter 8, Section 2.4).

The structure of *λ*gt10 is given in *Figure 8*.

Bacteriophage *λ*gt11 (*Figure 9*) is also a very useful vector for the construction of cDNA libraries. The vector was designed to allow insertion of DNA fragments up to 7.2 kb in length in a unique *Eco*RI site located near the C-terminal end of the *E. coli lacZ* gene. A DNA segment which is properly aligned with the reading frame of the *lacZ* gene, will be expressed as a fusion protein. A characteristic feature of the vector is its capability to express a cloned DNA fragment and distinguish the recombinant vector from the parent one on the basis of the product that is formed. The presence of a gene in the library can be easily detected by binding of the protein expressed from the cloned gene to antibodies raised against it. The expression takes place from the *lac* promoter. To maximize genetic

λgt10

Figure 8. Cloning of a DNA fragment into the unique cloning site (*Eco*RI) inactivates the immunity gene (i^{434}) resulting in a clear plaque. The letters *A* and *J* indicate structural genes involved in bacteriophage morphopoiesis; *b*527 refers to a deletion of part of the *λ* genome.

λgt11

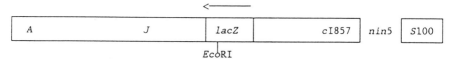

Figure 9. Cloning of a DNA fragment into the unique cloning site (*Eco*RI) allows for the formation of proteins fused to the C-terminal end of *β*-galactosidase. The symbols *A, J, cI*857, and *S*100 refer to *λ* genes and *nin*5 to a deletion of the *λ* genome.

stability and phage yield, recombinant phage are propagated on the special host strain, Y1088 (see *Table 1*). The *lac* deletion in this strain prevents host-phage recombination, *supF* suppresses the *S*100 amber mutation, and the plasmid pMC9 provides a high yield of *lac* repressor to inactivate the *lac* promoter. To minimize degradation of the fusion proteins formed, a protease minus strain (*lon*), Y1090 (*Table 1*), is used for expression studies.

Bacteriophage λgt18 (32) harbours a polylinker region with unique cloning sites for *Eco*RI, *Sac*I, *Sal*I, and *Xba*I, instead of the *Eco*RI site, making the vector somewhat more versatile for cloning cDNAs. To further improve the usefulness of the vector, a Chi sequence was inserted into the vector as well as a unique *Not*I site in the polylinker region, yielding λgt22 (33).

ii. λEMBL vectors

The λEMBL phage (34) constitute a family of replacement vectors with the generally useful properties of high capacity, polylinker cloning sites and genetic selection for recombinant phage. The phage λEMBL3 and λEMBL4 are particularly useful for the construction of gene banks by cloning partial *Sau*3AI digests of genomic DNA at the *Bam*HI sites of the vectors (see Volume II, Chapter 2). Alternatively, DNA fragments with cohesive ends matching those of *Bam*HI, *Eco*RI or *Sal*I can be inserted into the vector. The size of the fragments that can be incorporated is minimally 10 kb and maximally 20 kb. When the vector is treated with either *Bam*HI, *Eco*RI or *Sal*I, the central fragment can easily be removed and recombinant phage can be selected on P2-lysogenic host bacteria, due to their Spi⁻ phenotype. In λEMBL3 the *Sal*I sites in the polylinker region are nearest to the vector arms, whereas λEMBL4 has the linkers reversed and the *Eco*RI sites outermost.

The structure of λEMBL3 is given in *Figure 10*.

λEMBL3

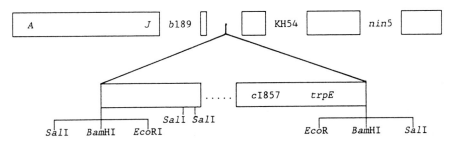

Figure 10. Phage λEMBL3 contains pairs of cloning sites (*Sal*I, *Bam*HI, *Eco*RI) flanking at both sides a region of the λ genome which can be replaced by foreign DNA. The symbols *A, J, c*I857 refer to λ genes, *trpE* encodes the *E. coli* enzyme anthranilate synthase and *b*189, KH54, and *nin*5 refer to deletions of the λ genome.

199

2.3 Cosmid vectors

2.3.1 General properties

Like bacteriophage replacement vectors, cosmid vectors are particularly useful for the construction of gene libraries. They have features of plasmid vectors, including the possibility to replicate at a high-copy number and to be amplified in the presence of chloramphenicol. Cosmids also carry a selectable genetic marker and have appropriate cloning sites. Since they also contain *cos* sequences, they share with λ vectors the advantage of *in vitro* packaging for efficiently recovering recombinant vectors. After ligation of the vector to exogenous DNA fragments, catenanes of vector and exogenous DNA fragments are formed. Recombinant cosmids can then be packaged with the *in vitro* packaging system into phage heads, just like phage vectors, if the DNA fragment between two *cos* sequences in the catenane has the right size (Chapter 8, Section 3.1). To obviate the need to prepare vector arms, a laborious and time-consuming procedure, cosmid vectors have been constructed with two *cos* sequences. With these vectors one simply has to insert the DNA fragments into a site located between the two *cos* sequences. The phage particles formed can be used to transduce *E. coli* bacteria, in which the recombinant cosmid can be propagated as a plasmid. Recombinant cosmids can either be handled as plasmid DNA and used to transform *E. coli* bacteria, or repackaged into phage heads and used for transduction experiments. It should be remembered, however, that transformation frequencies of large recombinant cosmids are smaller than those of plasmid vectors and, more importantly, that transformation does not select against recombinant cosmids which have suffered from deletions, as does the *in vitro* packaging system. As most cosmids are relatively small (approximately 5 kb), they can accommodate up to 45 kb of exogenous DNA, an amount significantly greater than can be included in λ replacement vectors. A greater amount of DNA present in a single cosmid not only reduces the number of cosmids required for a representative gene library, but also enhances the likelihood that a cloned gene is intact, if it is very large, as is often the case with eukaryotic genes.

Although much more DNA can be included in a cosmid vector compared with a λ replacement vector, a favourable feature of cosmid vectors, this latter type of vector also has some disadvantages. An important drawback of cosmid vectors is that the efficiency of the *in vitro* packaging system is 100- to 1000-fold lower for cosmid vectors than for phage vectors (10^5–10^6 versus 10^8). Also screening, by means of colony hybridization techniques, for the presence of a gene in a cosmid library, is much less sensitive compared to screening of a phage library. This is, in part, due to the much lower concentrations of recombinant cosmids in a colony as a result of the reduced copy number of these very large plasmids, compared with the concentration of phage in a plaque. Finally, preparation of a cosmid library requires the presence of much larger DNA (>100 kb) as starting material than is necessary for construction of a phage library.

2.3.2 Specific vectors

In the following paragraphs I shall describe three cosmid vectors differing in the type of origin of replication, number of *cos* sequences, selection markers and cloning sites.

i. pJB8

Plasmid pJB8 (5.4 kb; 35), which has been extensively used as a cosmid vector, is based on a pMB1 replicon and carries one *cos* sequence and harbours the *amp* gene as selection marker. The vector is probably most useful for the construction of gene banks by cloning of size-selected *Sau*3AI- or *Mbo*I-digested genomic DNA at the unique *Bam*HI site, packaging *in vitro* and selecting amp[r] transductants. Due to its small size, fragments up to 45 kb can be inserted into the vector and packaged into phage heads. Recombinant cosmid DNA can be amplified by incubating bacteria in the presence of chloramphenicol (Chapter 3, Section 4.2), to assist in screening of libraries by colony hybridization. The structure of pJB8 is given in *Figure 11*.

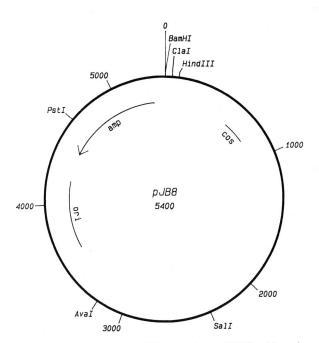

Figure 11. Structure of pJB8. Plasmid pJB8 comprises pAT153 with an insertion into the *Bam*HI site of a 1.78 kb *Bgl*II fragment carrying the *cos* sequences from bacteriophage λ. A synthetic DNA fragment with a *Bam*HI site was introduced into the *Eco*RI site, which resulted in duplication of the *Eco*RI site. The nucleotide sequences of the constituents of pJB8 are known and can be retrieved from the EMBL database.

ii. pcosEMBL vectors

The pcosEMBL (6.1 kb; 36) family of vectors was designed to simplify the isolation of specific recombinants from cosmid libraries and enhance the speed of isolating large (greater than 100 000 bp) regions of complex genomes in an ordered array of overlapping recombinant DNA clones ('chromosome walking'). An essential feature of these vectors is the possibility to apply genetic selection in obtaining specific clones, besides screening by means of colony hybridization. Gene banks can easily be made by cloning of size-selected *Sau*3AI or *Mbo*I fragments into the unique *Bam*HI site, causing inactivation of the *tet* gene, which can be taken as a measure of the quality of the library. The presence of two *cos* sequences eliminates the need to prepare vector arms using different restriction enzymes, because they can be simply obtained by treatment with *Pvu*II.

The pcosEMBL vectors are based on the R6K replicon and have a selection marker (*kan*) not present in the pUC vectors. The advantages of these features are twofold. No detectable homology exists between pcosEMBL and pUC vectors, so sequences cloned into pUC may be used as probes without the need for excision of vector sequences. The absence of homology between the vectors is also important in manipulation of clones by homologous recombination in *E. coli*, required for introduction of specific mutations in cloned sequences or for chromosome walking. Recombination between a cosmid clone and a pUC clone can only occur when the cloned DNAs in the two vectors have sequence homology. Such recombination generates a packageable cosmid DNA with the additional drug-resistance marker (*amp*) from the pUC plasmid. Selection for this plasmid isolates the desired cosmid from the gene bank.

The isolation of a specific sequence from the library runs through the following steps. First, the cosmid library is transferred to a Rec$^+$ host which carries the pUC plasmid. After a period, sufficient for recombination to occur between the cosmid vector and the selector plasmid, the recombinant cosmid vectors are repackaged and used to infect a RecA$^-$ host, with selection for ampicillin resistance. Subsequently, the original vector structure is restored, by a reversal of the recombination events. The reversion can be followed by screening for the absence of the *lac* operator sequence of the pUC vector. In a Lac$^+$ strain carrying a multicopy plasmid with the *lac* operator, the chromosomal *lac* genes are expressed, because the level of repressor molecules is insufficient for repression to occur. A Lac$^+$ strain (see *Table 1*) carrying the pUC-based cosmid thus yields blue colonies on X-gal plates, while cosmid clones which have lost the pUC sequences form colourless colonies.

The members of the pcosEMBL family differ with respect to the nature of the cloning sites, and the number of *cos* sequences (one or two). The structure of pcos2EMBL is given in *Figure 12*.

A third type of cosmid vector, the Lorist family of cosmid vectors, will be described in Section 4.2.3.

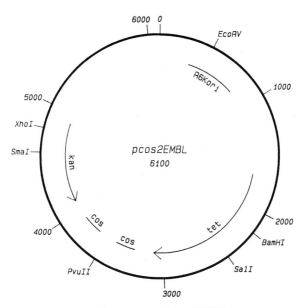

Figure 12. Structure of pcos2EMBL.

3. Vectors for DNA sequence analysis and *in vitro* mutagenesis experiments

3.1 Bacteriophage vectors with single-stranded DNA

In almost every research project in which recombinant DNA techniques are applied, DNA sequencing analysis of cloned DNA fragments has to be carried out. Sometimes this information is needed in order to characterize the cloned fragment, in other cases, sequence information is required for further genetic manipulation of the gene. As we have seen before (Section 2.1.2.*iv*) DNA sequence analysis can be carried out after subcloning of the fragment in plasmid vectors, such as pUC vectors. However, in most cases DNA sequence analysis is performed with vectors that have been specifically designed for this purpose. To date DNA sequence analysis can most conveniently be carried out with vectors carrying single-stranded bacteriophage DNA or plasmid vectors harbouring a replicon of single-stranded bacteriophage. In the next Sections such vectors will be treated in more detail.

3.1.1 General properties

The DNA of filamentous bacteriophage has a number of attractive properties which make it particularly suited as starting material for the construction of cloning vectors. The genomes of filamentous phages like M13, fl and fd consist of circular single-stranded DNA. Upon infection of sensitive, F$^+$ *E. coli* bacteria, viral DNA is converted into a double-stranded replicative form (RF), which is extensively replicated to yield a pool of double-stranded viral DNA. At the end of the replicative cycle, the formation of single-stranded viral DNA ensues which eventually results in the formation of progeny virus. Infection of *E. coli* by filamentous phage is not a lytic process. Bacteria continue to multiply for a number of generations, while extruding phage into the culture fluid. Since the concentrations of replicative intermediate and of progeny phage produced are very high (10^{11} to 10^{12} phage particles per millilitre), it is relatively easy to isolate large quantities of both forms of DNA (see Chapter 3, Sections 3.2 and 4.2). By insertion of a DNA segment into a non-essential region of the double-stranded viral DNA, the DNA segment can be amplified, after transfection of, in general, F^- host bacteria. Since the coat of filamentous phage can be adjusted to accommodate extra DNA, advantage can be taken of this property, to clone DNA segments and propagate them either as single- or double-stranded DNA. Although there is little constraint on the size of the fragments that can be incorporated in filamentous phage vectors, the upper limit in reality lies at about 3 kb of inserted DNA. Larger DNA inserts tend to be unstable (37).

Of the three filamentous phage mentioned above, M13 is used most frequently for the construction of single-stranded recombinant DNA phages. The second phage, fl, has proven extremely useful for the construction of plasmid vectors that can replicate both as double-stranded and as single-stranded DNA (Section 3.2). Vectors derived from the third type of phage, fd, are also used, but to a much lesser extent.

The cloning of DNA fragments into RF DNA was facilitated by introduction into the so-called intergenic region of the bacteriophage fl and fd of antibiotic-resistance genes (38), and for phage M13 of a fragment of the *E. coli lac* operon. The latter fragment (*lacZ'*) comprises the promoter, operator and the N-terminal region of the *lacZ* gene. The *lacZ'* fragment codes for an amino-terminal fragment of *β*-galactosidase, called *α*-protein. The *α*-protein can complement bacteria with a mutation in the N-terminal region of *β*-galactosidase. Bacteria containing an appropriate mutation in the *α*-region of *β*-galactosidase (e.g. ΔM15) can be complemented, producing blue plaques on indicator plates supplemented with X-gal, if repression of the *lac* operon is lifted by addition of the gratuitous inducer of the operon, isopropyl-*β*-D-thiogalactopyranoside (IPTG). The versatility of the M13 vectors was further enhanced by introduction into the structural region of *lacZ'* of a synthetic DNA fragment with an array of restriction enzyme cleavage sites. Introduction into one of these cloning sites of a DNA fragment will, in general, disrupt reading of the *α*-fragment, and will, consequently, change the phenotype of the resultant plaques from blue to colourless (Chapter 8, Section 2.3). Only in some cases were the reading frame of the *α*-fragment and inserted DNA fragment are correctly aligned, will a fusion protein be formed that retains enzymatic properties and, thus, gives rise to blue plaques.

The availability of single-stranded DNA vectors has eliminated the need of strand separation, a cumbersome and in many cases impossible task required for liquid DNA–RNA hybridizations, or the need to rely on the chemical DNA degradation method of Maxam and Gilbert (20) for determining the nucleotide sequences of DNA. Provided that DNA fragments can be cloned in M13 vectors in both orientations (occasionally only one orientation can be stably maintained in *E. coli*), single-stranded DNA of the sense and of the antisense orientation can be obtained. The applications of this strategy are numerous, such as DNA sequence analysis by the chain termination method of Sanger *et al.* (19), preparation of single-stranded DNA probes of either the insert or the vector, *in vitro* mutagenesis (16, 39), S1 nuclease mapping (40) and primer extension methods.

3.1.2 Specific vectors

M13mp18 and M13mp19 (7250 bp; *Figure 13*; 37) form a pair of vectors that are given as examples of the M13 family of single-stranded bacteriophage vectors. They are designed to facilitate DNA sequencing by allowing cloning of DNA fragments at a polylinker region, in both possible orientations. Because the polylinker region in the two vectors has an opposite orientation, DNA fragments can be cloned in one orientation in M13mp18 and in the other orientation in M13mp19, in the same restriction site. Alternatively, a DNA fragment cloned in one of the two vectors can be excised, using enzymes giving different cohesive ends, and introduced into the other vector. The resulting recombinant phage contain either the sense or the antisense strand of the cloned fragment. DNA fragments cloned into the M13mp vectors can readily be used as templates for DNA sequencing reactions or for the production of single-stranded DNA probes, using a standard oligonucleotide primer and the Klenow fragment of DNA polymerase I.

Other members of the M13 family have essentially the same structure and properties, but have different polylinker regions. Some members of the M13 family, such as M13mp10 and M13mp11 (16), carry amber mutations in essential genes. The presence of these mutations facilitates the selection of a recombinant phage obtained after *in vitro* mutagenesis of an inserted DNA fragment.

3.2 Plasmid vectors with the replication origin of single-stranded phage

3.2.1 General properties

A class of vectors that is of special interest and is rapidly increasing in popularity, shares the advantages of both plasmid vectors and of single-stranded DNA phage vectors in that they can be replicated both in a double-stranded and a single-stranded mode. Because they can be replicated as double-stranded plasmids, much larger DNA fragments can be stably incorporated than is the case with the filamentous phage vectors.

If plasmids, carrying the origin of replication of phage fl, are superinfected with

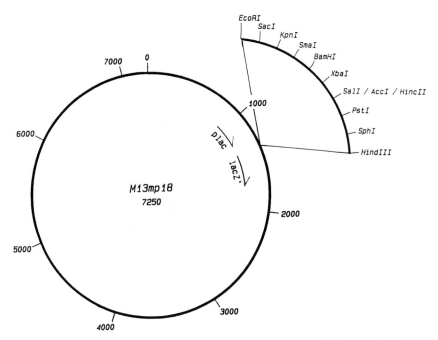

Figure 13. Structure of M13mp18. With the M13mp series the vector DNAs contain an 0.77 kb *Hae*II fragment carrying the *E. coli lac* promoter and the genetic information for the first 145 amino acids of β-galactosidase, cloned by blunt-end ligation at a *Bsu*I site in the intergenic region between genes II and IV of M13. The polylinker sequences containing multiple cloning sites are inserted between codons 5 and 6 of the *lacZ* sequence. Restriction sites present in the polylinker have been eliminated from elsewhere in the vector DNA. The polylinker retains the *lacZ* reading frame, and the vector gives allelic complementation of a LacZα⁻ host strain, yielding blue plaques on plates containing IPTG and X-gal. Recombinant phage containing inserts that destroy the reading frame or otherwise interfere with expression of the LacZα peptide are revealed as colourless plaques. The nucleotide sequence of the M13mp cloning vectors is known and can be retrieved from the EMBL database. The sequence of the polylinker region is given below. Numbers below the nucleotide sequence indicate the amino acids of the β-galactosidase.

M13mp18

	Thr	Met	Ile	Thr	Asn	Ser	Ser	Ser	Val	Pro	Gly	Asp	Pro
ATG	ACC	ATG	ATT	ACG	AAT	TCG	AGC	TCG	GTA	CCC	GGG	GAT	CCT

	1	2	3	4		*Eco*RI		*Sac*I		*Kpn*I		*Sma*I	*Bam*HI
	Leu	Glu	Ser	Thr	Trp	Arg	His	Ala	Ser	Leu	Ala	Leu	Ala
CTA	GAG	TCG	ACC	TGC	AGG	CAT	GCA	AGC	TTG	GCA	CTG	GCC

*Xba*I		*Sal*I		*Pst*I		*Sph*I		*Hin*dIII			7	8
		*Acc*I										
		*Hin*cII										

M13mp19

	Thr	Met	Ile	Thr	Pro	Ser	Leu	His	Ala	Cys	Arg	Ser	Thr
ATG	ACC	ATG	ATT	ACG	CCA	AGC	TTG	CAT	GCC	TGC	AGG	TCG	ACT

	1	2	3	4		*Hin*dIII		*Sph*I		*Pst*I		*Sal*I
												*Acc*I
												*Hin*cII

Leu	Glu	Asp	Pro	Arg	Val	Pro	Ser	Ser	Asn	Ser	Leu
CTA	GAG	GAT	CCC	CGG	GTA	CCG	AGC	TCG	AAT	TCA	CTG

*Xba*I		*Bam*HI		*Sma*I		*Kpn*I		*Sac*I		*Eco*RI	7

phage fl, the virions extruded into the medium contain either single-stranded fl DNA or single-stranded plasmid DNA (41). The amounts of single-stranded phage DNA and plasmid DNA produced are, in general, considerably lower than expected from the results obtained with filamentous phage vectors. It appears that plasmid replication interferes with phage multiplication, which in turn negatively affects the formation of single-stranded plasmid DNA. The relative amounts of phage DNA and plasmid DNA may also vary and depend both on the type of plasmid and helper phage used. To optimize the yield of single-stranded plasmid DNA, phage mutants have been isolated that show interference resistance. These mutants can increase the yield of single-stranded plasmid by tenfold and concurrently increase the level of phage by a similar amount.

In general, pairs of plasmid vectors have been constructed which differ in the orientation of the origin of replication of the filamentous phage. The orientation of the origin determines which of the two strands will be encapsidated in the viral particle. A '+' sign indicates that the polarity of the origin of plasmid and phage replication is the same. A '−' sign signifies that the two origins have the opposite polarity.

3.2.2 Specific vectors

In the following paragraphs examples are given of vectors that can replicate both as double-stranded and single-stranded DNA. Plasmids pEMBL18 and pEMBL19 (4.0 kb; *Figure 14*; 42) form a pair of plasmids that differ only in the orientation of the cloning sites. Both the sense and antisense strand of a DNA fragment can be obtained by cloning of the fragment in the two vectors in one of the unique cloning sites.

The series of pBGS (4.4 kb; 43) plasmids is more versatile than the pEMBL plasmids, in the sense that they not only allow the production of the sense strand and antisense strand by cloning of a DNA fragment in two orientations (cloning in pBGS18 and pBGS19), but also permit the synthesis of the two strands by cloning in one orientation in the + and − version of the vectors [pBGS18+ and pBGS18−; pBGS19+ (*Figure 15*) and pBGS19−].

Still another possibility is provided by plasmid pKUN19 (4.0 kb; *Figure 16*; 44) which comprises in one vector origins of replication of two single-stranded phage, fl and Ike. Because the polarity of the *ori* regions of these phage is different, the sense strand or anti-sense strand of a cloned DNA fragment will be produced depending on whether the bacteria are infected with phage fl or Ike.

All vectors have in common that they contain the *lacZ'* region and polylinker sequence from M13mp18 or M13mp19, which provides multiple cloning sites and recognition of recombinants by inactivation of the *lacZ'* function. The pEMBL vectors and pKUN19 carry the *amp* gene as selection marker, while the pBGS vectors harbour the *kan* gene for selection of transformants.

The nucleotide sequence of the constituent parts of the vectors is available and can be retrieved from the EMBL database. The nucleotide sequences of the

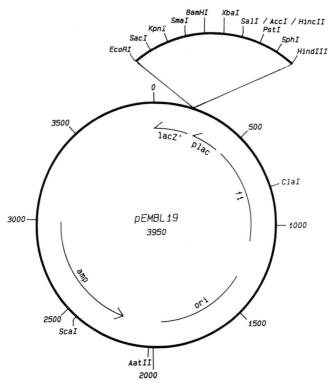

Figure 14. Structure of pEMBL19. Plasmids pEMBL18 and pEMBL19 comprise a 1.3 kb *Eco*RI fragment with the origin of replication of fI cloned, after filling-in of the cohesive termini, at the unique *Nar*I site of pUC18 and pUC19, respectively.

polylinker region of the vectors are identical to those of M13mp18 and M13mp19, and are given in the legend to *Figure 13*.

Improved versions of some of the vectors described in this Section have recently been developed. These vectors, which carry elements for *in vitro* transcription of cloned DNA, will be discussed in Section 4.2.2.

4. Expression vectors

Gene expression in *E. coli* is primarily determined at the level of transcription and translation, by signals located before the gene. For many genes, and in particular eukaryotic genes, it has been found that such sequences are not recognized by the transcription and translation machinery of *E. coli*. Expression of these genes, therefore, requires that the genuine control elements are replaced by analogous signals from *E. coli*. In this section, I will briefly outline what is known about control of gene expression and how this knowledge has been exploited in designing expression vectors. In addition, I will describe other

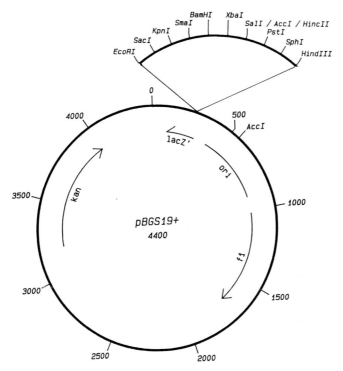

Figure 15. Structure of pBGS19+. Plasmids pBGS18+, pBGS18−, pBGS19+, and pBGS19− contain a 0.7 kb *Ssp*I fragment with the *ori* region from fl, inserted in two possible orientations, an 0.5 kb *Hae*II fragment from the pUC plasmids with the *lacZ'* gene and polylinker region, and a 3.2 kb *Eco*RI-*Pvu*II fragment with the origin of replication and *kan* gene.

factors that contribute to the yield of gene product formed and which have to be taken into account when selecting an expression vector.

Of all parameters that affect gene expression, the promoter is the most important one. Promoters have been classified as strong or weak, primarily on the basis of a comparison of the amounts of gene product made, and partly as a result of analyses of *in vitro* rates of RNA synthesis. Some constitutive promoters of bacteriophage T3, T5, and T7, and the promoter of the lipoprotein gene (*lpp*) of *E. coli* are considered to be strong promoters. The same is true for two highly controllable promoters, p_L and p_R, of the leftward and rightward operon, respectively, of bacteriophage λ, and for the regulatable promoter of the tryptophan (*trp*) operon of *E. coli*. In addition, two hybrid promoters have been constructed *in vitro* which can be classified as strong promoters. The first promoter, p_{3016}, is constitutive and comprises sequences from the *tet* promoter and from the *E. coli trpB* gene (7). The second one, p_{tac}, is a controllable promoter that is made out of sequences from the promoters of the *trp* and *lac* operons (45, 46). The *tac* promoter has proven very useful for the synthesis in *E. coli* of foreign proteins.

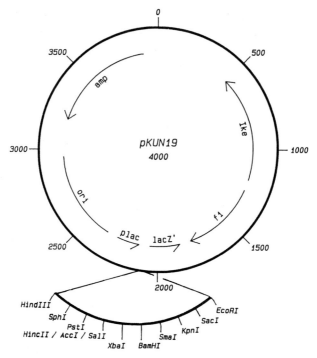

Figure 16. Structure of pKUN19. Plasmid pKUN19 comprises a 0.8 kb *Eco*RI-*Bam*HI fragment containing the Ike sequences and a 0.5 kb *Rsa*I fragment with the fl sequences, inserted into the *Nde*I and *Nar*I sites, respectively, of pUC19.

The second major determinant for gene expression is translation-initiation. Efficient translation requires the presence of a sequence (Shine–Dalgarno sequence: S–D) that is complementary to the 3'-end of 16S RNA. The S–D sequence has to be properly spaced from the translation-initiation codon ATG (5–13 nucleotides before the ATG) and should, preferably, contain A's and T's, rather than C's and G's (47). The efficiency of translation is also strongly affected by secondary structure around the site of initiation of translation. Although the experimental evidence is not conclusive, it strongly suggests that the absence of secondary structure in that area favours initiation of translation. A factor that may negatively affect translation of a cloned gene, is the presence, in front of the genuine start codon ATG, of a second S–D sequence and ATG codon. Translation, initiated at this site, will interfere with translation of the gene, unless a translation-stop codon is introduced, immediately upstream of the true start codon (48).

In order to get optimal expression of a cloned gene, the type of promoter and ribosome-binding site (RBS), and the presence of suitable cloning sites downstream from these control sequences, have to be taken into account, in making a choice in favour of an expression vector. An additional point of concern should be the presence of superfluous genetic information on the vector. *Escherichia coli* is capable of directing approximately 60–65% of its protein synthesizing capacity to a single protein. If more than one gene is cloned, the total amount of plasmid-encoded protein will not change, but the synthesis of a protein from a second and third gene proceeds at the expense of that of the first gene (49). It thus seems advisable to make use of expression vectors that contain no genes other than the selection marker.

A point equally requiring consideration, is that different types of protein product require different types of vector. For the synthesis of an enzymatically-active product, the protein has, in general, to be expressed in an unfused form. This implies that the gene should include a S–D sequence and ATG codon, or that the translation-initiation codon ATG of the gene should be juxtaposed to a S–D sequence downstream from the promoter in the vector. If, on the other hand, the aim of the experiment is to produce a protein, to raise antibodies against it, it is advisable to express the protein as a fusion with a protein that may serve as a carrier, thereby increasing the immune response. In that case, the gene, without translation-start signal, should be properly aligned with the protein coding sequence in the vector.

A third consideration regards the use of constitutive or inducible/repressible promoters. Vectors with either type of promoter are available, but, depending on the protein to be expressed, one is preferred to the other. The advantage of a constitutive promoter is the ease with which the experiment can be performed. A constitutive promoter requires no change of culturing conditions during the experiment, nor the utilization of perhaps expensive chemicals as inducers or repressors. The other side of the coin is that continuous expression of proteins that are toxic may be lethal to the cell. If the protein is unstable, continuous synthesis under control of a constitutive promoter is also not very effective. In most studies, therefore, promoters are used that can be effectively shut off and can be instantaneously switched on. The advantages of this procedure are evident. Bacteria can be cultivated to high cell density, prior to induction of the synthesis of the desired protein. During the relatively short period of synthesis that ensues, breakdown of the protein is minimized and any toxic effect of the protein on cell metabolism will no longer be seriously felt. It may be considered as a drawback that induction of protein synthesis sometimes requires a change of growth medium (e.g. removal of excess tryptophan, to induce the *trp* promoter) which is cumbersome, especially with large volumes of medium. Application of a heat-shock to induce the promoters p_L or p_R is also accompanied by some unwanted side-effects. First, with large volumes of culture medium, a rapid and uniform temperature-shift is difficult to achieve. Second, the high temperature may inhibit protein overproduction by destabilizing the desired protein, by enhancing proteolytic degradation, or by decreasing the solubility of the protein product.

In subsequent sections a variety of expression vectors will be described. The great majority of them are plasmid vectors, but a few bacteriophage λ and cosmid expression vectors will also be given. The vectors are divided into vectors designed for the (over)-expression of protein products and of vectors that can be used to synthesize *in vitro* large quantities of RNA. Because the vectors carry different promoters to drive the expression of cloned genes, a further subdivision is made according to their promoter. In other sections, expression vectors will be described that are specifically useful for studying transcription control elements, like promoters and terminators.

4.1 Plasmid vectors designed to effectively synthesize proteins *in vivo*

4.1.1 Vectors with promoter p_L from bacteriophage λ

In the following section I will describe two expression vectors with the promoter p_L of bacteriophage λ driving the expression of cloned genes. These two examples were chosen because they demonstrate how fused and unfused proteins can be made and how the environment around the start-site of translation can be modified in an attempt to increase gene expression.

Both vectors have in common that the promoter they carry can be fully repressed by cultivating the bacteria at 28–30°C, in the presence of the λ repressor, the product of gene *c*I. Mutants are available that synthesize a thermolabile repressor. By raising the temperature to 42°C, the promoter is fully induced, as a result of inactivation of the repressor. The two plasmids differ in the way repression is accomplished. In plasmids of the pJLA series (4.9 kb; 50), the gene encoding the thermolabile repressor, *c*I857, is present on the vector itself. The second plasmid, pPLcAT10 (2747 bp; 51), does not carry the *c*I857 gene. This plasmid should, therefore, be used in combination with a bacterial strain carrying the thermolabile repressor gene (M5219 or K12 HI), or in combination with a compatible plasmid carrying the *c*I857 allele of the λ *c*I gene (pcI857; 52).

The vector pPLcAT10 (*Figure 17*) was chosen because it allows for a transcription fusion of a gene with the promoter p_L, for a translational fusion of a coding sequence to the translation-start codon ATG or to codons immediately downstream from the ATG codon, and for adjustment of the distance between the S–D sequence and the initiator codon ATG. In plasmid pPLcAT10 an artificial RBS is used. Its composition is based on a comparison of a great number of published promoter sequences from which a consensus sequence was derived. Another characteristic feature of pPLcAT10 is the presence of a unique cloning site for *Bam*HI (^{5}'GGATCC^3') which partially overlaps the initiator codon ATG. Since treatment with *Bam*HI cleaves the DNA strand behind the G residue of the ATG codon, exogenous DNA fragments may be directly fused with the ATG codon, if they have the same 5'-protruding ends (^{5}'GATC^3'). Alternatively, they may be joined to the ATG codon, after removal of 5'- or 3'-protruding ends with single-strand specific nuclease, or, after filling-in of 5'-protruding ends with DNA polymerase. The presence of an *Xba*I site between the S–D sequence and the initiator codon ATG provides the possibility to lengthen or shorten this region and, in addition, alter its composition. This may have drastic effects on the level

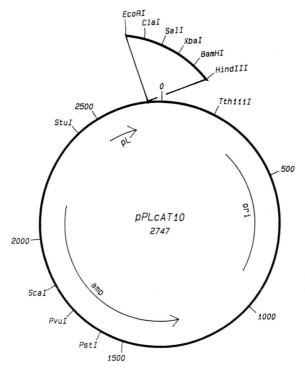

Figure 17. Structure of pPLcAT10. Plasmid pPLcAT10 is composed of a sequence from pBR322 (nucleotides 2068–4180) carrying the *ori* region and *amp* gene, a 0.4 kb fragment from the *kan* gene of transposon Tn903, a 247 bp DNA fragment with the promoter and operator from the leftward operon of bacteriophage λ, and a synthetic DNA fragment with an array of restriction enzyme sites. The nucleotide sequences of pPLcAT10 are known. The sequence of the region around the S–D region is given below.

pPLcAT10	GAATTCATCGATGTCGACCAAGGAGGTCTAGATG	GAT	CCG	GCC	AAG	CTT
	*Eco*RI *Cla*I *Sal*I S–D *Xba*I		*Bam*HI			*Hin*dlll

of expression of a gene. Finally, the presence of unique *Cla*I, *Eco*RI and *Sal*I sites 5′ to the S–D sequence renders the vector amenable to genetic engineering of the region upstream of the S–D sequence in the final version of the expression vector.

Characteristic features of the pJLA series of plasmids are the presence, in a tandem arrangement, of the promoters p_L and p_R, both repressible by the λ repressor. An essential difference, compared with pPLcAT10, is the presence in the pJLA plasmids, immediately downstream from the promoter, of a natural RBS that is known to be very effective. It originates from the *E. coli atp* operon. It has been shown for the *atp* RBS that nucleotides upstream and downstream from the S–D sequence play an important role in determining the efficacy of translation of cloned genes. The entire region spanning the S–D sequence and ATG codon was introduced into the vector. In plasmid pJLA502 (*Figure 18*)

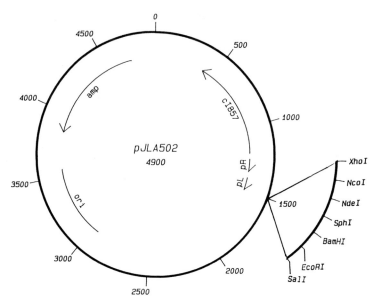

Figure 18. Structure of pJLA502. In plasmids pJLA503, pJLA504, and pJLA505 the ATG codon overlaps the *Nde*I, *Sph*I, and *Bam*HI site, respectively, of the polylinker region.

protein-coding sequences can be directly fused to the translation-initiation codon ATG of the *atpE* gene, because the ATG codon overlaps a unique *Nco*I cloning site, or to one of the unique cloning sites lying downstream from this site. To avoid possible negative effects on the yield of product of transcription that reads into the *ori* region, a transcriptional terminator from bacteriophage fd was introduced between the region with multiple cloning sites and the *ori* region.

4.1.2 Vectors with *trp* promoter

Plasmids carrying the *trp* promoter to drive expression of exogenous DNA are widely in use, because the *trp* promoter is strong and can be properly controlled. The *trp* promoter is repressed in the presence of excess tryptophan. This repression is not complete if a plasmid is used of which more than 50 copies are present in the bacterium. Because the number of repressor molecules is limited, the amount of repressor is in such case presumably insufficient to saturate the operators on all plasmids. As a consequence, a basal level of gene expression amounting to 1–3% of the level under fully induced conditions is observed (53). The *trp* promoter can be derepressed by either a shift to a medium without tryptophan, or the addition of a gratuitous inducer, 3-indolylacrylic acid. When a tryptophan-auxotrophic strain of *E. coli* is cultivated in a medium with a limited amount of tryptophan, induction is automatically achieved upon exhaustion of the supply of tryptophan in the medium.

The *trp* expression vectors have unique promoter-distal cloning sites downstream from either the S–D region of the first gene of the *trp* operon, *trpE*, the initiator codon ATG of *trpE*, or after the S–D sequence of the *trpL* gene. The *trpL* gene encodes a small regulatory protein (54) and is located between the promoter and the attenuator region, before the *trpE* gene. With these vectors fused and unfused proteins can be efficiently synthesized.

An example of the first type of vector is pCT-1 (4.9 kb; *Figure 19*; 55). It contains the *trp* promoter, operator, attenuator region and the S–D region of the *trpE* gene. Immediately behind this S–D sequence are unique cloning sites for *Cla*I and *Hind*III. Cloned genes can be expressed to form unfused proteins, if they contain an ATG codon which is positioned at an appropriate distance from the S–D sequence, by cloning of the fragment into either the *Cla*I or *Hind*III site. Unfused proteins can also be obtained if the gene includes the entire translational regulatory region.

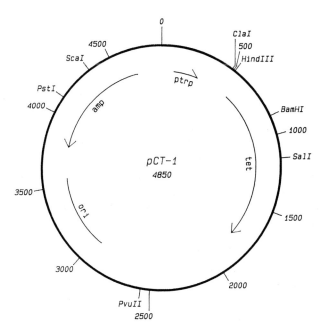

Figure 19. Structure of pCT-1. Plasmid pCT-1 was constructed by insertion of a 0.5 kb *Hin*fI fragment carrying the *trp* regulatory region into the *Eco*RI site of pBR322. To juxtapose the S–D sequence of the *trpE* gene to the *Hind*III site present in pBR322, nucleotides downstream from the S–D site were removed with nuclease Bal31, and a linker containing a *Cla*I site was introduced between the S–D sequence and the *Hind*III site. Plasmid pCT-1 confers on *E. coli* resistance to both ampicillin and tetracycline. The nucleotide sequences of the constituents of the vector are known and can be retrieved from the EMBL database. The sequence of the S–D region and the cloning sites is given below.

pCT-1	GAGAATCGATAAGCTT
	S–D *Cla*I *Hin*dIII

215

Plasmid pWT571 (3779 kb; *Figure 20*; 56) is particular useful for the synthesis of unfused proteins. It contains the *trp* promoter-operator region, harbours selection markers for ampicillin resistance and tetracycline resistance, and has unique promoter-distal cloning sites for *Cla*I, *Eco*RI, and *Hind*III. The *Eco*RI site partially overlaps the initiator codon ATG of the *trpL* gene. DNA fragments with a 5′-protruding end having the sequence $^{5'}$-AATT-$^{3'}$ can be directly attached to the ATG codon. Alternatively, DNA fragments can be coupled to the ATG codon after removal of the 5′-cohesive ends with a single-strand specific nuclease.

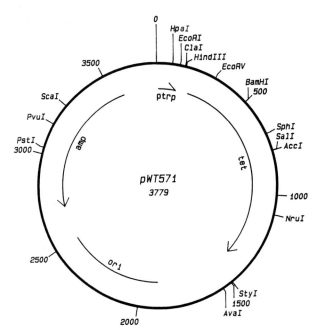

Figure 20. Structure of pWT571. Plasmid pWT571 comprises a 117 bp *Eco*RI fragment with the *trp* promoter and translation-initiation region of *trpL*, inserted into the *Eco*RI site of pAT153. To render the *Eco*RI site behind the S–D sequence unique, the second *Eco*RI site before the promoter was eliminated, by filling-in of the cohesive ends, followed by recircularization. The nucleotide sequences of the constituents of pWT571 have been determined and can be retrieved from UWGCG Vecbase. The sequences of the ribosome-binding site and adjacent cloning site are shown.

pWT571	AAGGGTAGCGACAATATG AAT TC
	S–D EcoRI

4.1.3 Vectors with other promoters

i. pUEX and pEX vectors

The pUEX series of vectors (6728 bp; 57) was designed to allow the synthesis of polypeptides for epitope mapping or raising antibodies against the polypeptide.

In the pUEX vectors (*Figure 21*) a polylinker sequence with unique cloning sites for *Bam*HI, *Eco*RI, *Pst*I, *Sal*I, and *Sma*I is present, near the 3′-end of the *lacZ* gene. This permits the formation of fusions of the polypeptide to the C-terminal end of β-galactosidase. Since the reading frame of the polylinker region in the three pUEX vectors is different, any DNA fragment can be aligned with the *lacZ* gene in one of the vectors.

Expression of the fusion proteins is controlled by the promoter p_R. The vectors harbour the allele of the thermolabile repressor *c*I857. When bacteria containing the pUEX vector are maintained at low temperatures (28–30°C), expression of β-galactosidase fusion protein will not take place. The promoter p_R can be induced by a shift of the temperature of the culture fluid to 42°C. To achieve efficient translation of the fusion proteins, a DNA sequence comprising the RBS up to the ninth codon of the *cro* gene of bacteriophage λ, was placed in front of the *lacZ* gene. The *cro*-gene and *lacZ*-gene sequences were joined through a

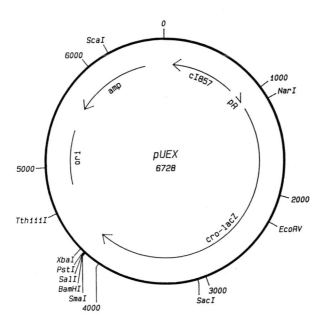

Figure 21. Structure of pUEX. The pUEX vectors confer resistance to ampicillin. The nucleotide sequence of the pUEX and pEX vectors can be obtained from the EMBL database. The sequences of the polylinker region of the pUEX vectors are given below. Those of the pEX vectors are the same.

pUEX1	GCC	CGG	GGA	TCC	GTC	GAC	CTG	CAG	CCA	AGC	TTG	CTG	ATT	GAT	TGA
		*Sma*I	*Bam*HI		*Sal*I		*Pst*I							

pUEX2	GAA	TTC	CCG	GGG	ATC	CGT	CGA	CCT	GCA	GCC	AAG	CTT	GCT	GAT	TGA
	*Eco*RI		*Sma*I		*Bam*HI		*Sal*I		*Pst*I					

pUEX3	GAA	TTA	ATT	CCC	GGG	GAT	CCG	TCG	ACC	TGC	AGC	AAG	CTT	GCT	GCT	TGA
			*Sma*I		*Bam*HI		*Sal*I		*Pst*I					

semi-synthetic oligonucleotide carrying sequences from the *lacI* gene. The *lacZ* sequence in the vector runs from codon 24 to codon 1005.

The fusion products that are made tend to be insoluble (58). This facilitates purification of the proteins, necessary for further immunological character- ization. Behind the polylinker region, translation-stop codons were placed in the three possible reading frames, followed by two transcriptional terminators, in tandem, from phage fd.

A set of vectors, pEX, has properties similar to those of the pUEX vectors. The pEX vectors lack the *cI857* gene and should therefore be maintained in bacteria carrying this gene on the chromosome or on a compatible plasmid. The copy number of the pUEX vectors is significantly higher than that of the pEX vectors (four compared to one).

ii. pKK233-2

Plasmid pKK233-2 (4.6 kb; 59) which carries a strong promoter, p_{trc}, a hybrid of the *trp* and *lac* operon promoters, can be used for the efficient synthesis of foreign proteins or fragments thereof, in an unfused state. The hybrid promoter is composed of the 5'-end (-35 region) of the *trp* promoter and the 3'-end (-10 region and operator region) of the *lac* operon. The promoter carries one additional nucleotide between the -10 region and -35 region, compared with the *tac* promoter. The increase of the 16 bp spacing of the *tac* promoter results in the consensus 17 bp spacing of the *trc* promoter. The *in vivo* activities of the two promoters are the same. The *trc* promoter is fully repressed in bacterial strains carrying the mutant allele, I^q, of the *lacI* repressor gene. The promoter can be induced by addition to the growth medium of IPTG.

The vector pKK233-2 (*Figure 22*) harbours the S–D sequence of the *lacZ* gene, followed by a translation-initiation codon ATG at a distance of eight nucleotides, one nucleotide shorter than in the genuine *lac* sequence. The ATG codon overlaps a unique *Nco*I site. DNA fragments can, therefore, be fused to the ATG codon if an *Nco*I site coincides with their start codon, by blunt-end ligation of fragments to the filled-in *Nco*I site of the vector, or after addition of an *Nco*I linker to the DNA fragment (see Chapter 7, *Protocols 11* and *12*). The vector also carries unique promoter-distal cloning sites, for *Pst*I and *Hind*III, located immediately downstream from the *Nco*I site. Two tandemly arranged transcrip- tion terminators were inserted behind the promoter-distal cloning sites, to prevent read-through of transcription into the *amp* gene. Plasmid pKK233-2 conveys resistance to ampicillin.

iii. pIN-III vectors

The pIN-III (7.4 kb; 60) series of vectors contains the strong, constitutive promoter of the *lpp* gene. Expression from this promoter was made regulatable by placing the controllable promoter/operator region from the *lac* operon at a site in between the *lpp* promoter and the RBS of the *lpp* gene. All vectors harbour the *lacI* gene. Thus, expression from the p_{lpp}–p_{lac} promoters will only occur in the

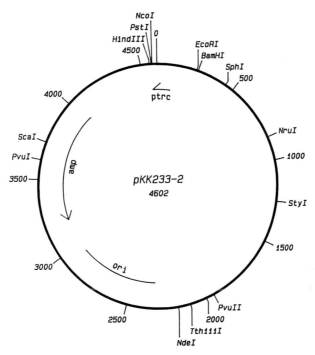

Figure 22. Structure of pKK233-2. Plasmid pKK233-2 comprises a 3.9 kb DNA fragment from pBR322 with the *amp* gene and the *ori* region (nucleotides 375–4246). At the latter position, the pBR322 region is fused to a DNA fragment of 424 bp carrying the *rrnB* terminator region. The *Pst*I site in the *amp* gene has been eliminated by *in vitro* mutagenesis. The nucleotide sequences of the constituents of pKK233-2 are known and can be retrieved from UWGCG Vecbase. The sequence of the translation-initiation region and cloning sites is given below.

pKK233-2   AGGAAACAGACC ATG GCT GCA GCC AAG CTT

S–D   *Nco*I   *Pst*I   *Hind*III

presence of the inducer of the *lac* operon, IPTG. A set of unique cloning sites has been inserted downstream from the translation–initiation codon of the *lpp* gene. In the pIN-III-A series (*Figure 23*) the region with the cloning sites was joined to the third codon of the *lpp* gene, in the pIN-III-B series the cloning sites were placed directly after the region encoding the cleavage site of pre-lipoprotein, and in the pIN-III-C series nine amino acids into the mature lipoprotein. Each series comprises three vectors, with cloning sites in the three possible reading frames.

To further enhance the versatility of the vectors, a DNA fragment was inserted between the promoter of the *lacZ* gene and the cloning sites of pIN-III-A3, encoding the signal sequence of the *ompA* gene, an outer membrane protein of *E. coli*. The presence of this DNA fragment allows the formation of fusions with the signal sequence of OmpA. Such fusion proteins are expected to be transported through the inner membrane, into the periplasmic space, unless the structure of

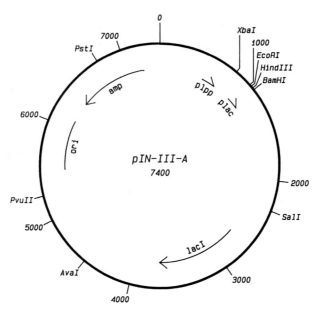

Figure 23. Structure of pIN-III-A. Plasmid pIN-III-AI comprises a *Pst*I-*Xba*I fragment containing the *lpp* promoter and the 5'-end of the *amp* gene, a 113 bp *Bst*NI-*Hae*III fragment carrying the p_lac sequence including the beginning of the *lacZ* gene, an *Xba*I-*Sal*I fragment with the 5'- and 3'-ends of the *lpp* gene and a 22 bp linker sequence, a 2.4 kb *Eco*RI fragment carrying the *lacI* gene and a *Sal*I-*Pst*I fragment of pBR322 with the 3'-end of the *amp* gene. The nucleotide sequence of the vectors is known. The sequence of the translation-initiation region with cloning sites is given for plasmids of the pIN-III-A series.

| pIN-III-A1 | ATG AAA GGA ATT TCC AAG CTT GGA TCC GGC |
| | *Eco*RI · *Hin*dIII · *Bam*HI |

| pIN-III-A2 | ATG AAA GGG AAG GAA TTC CAA GCT TGG ATC CGG |
| | *Eco*RI · *Hin*dIII · *Bam*HI |

| pIN-III-A3 | ATG AAA GGG GGA ATT CCA AGC TTG GAT CCG |
| | *Eco*RI · *Hin*dIII · *Bam*HI |

the protein itself prohibits transport. Secretion of proteins into the periplasmic space may be advantageous for proteins that would interfere with cell metabolism, if present within the cytoplasm. Hybrid proteins expressed in a pIN-III-C vector contain the first nine amino acids of the lipoprotein. Such fusions are expected to be transferred to the outer membrane. When a protein is fused to the B site, the hybrid protein is also expected to pass the inner membrane, but may be still bound to it because the lipoprotein signal sequence has not been cleaved off.

All vectors harbour the *amp* gene as selection marker, have translation stop codons beyond the cloning sites, in three reading frames, and have a transcription-terminator sequence further downstream.

4.1.4 Vectors permitting easy purification of the protein product

After expression of a foreign protein in *E. coli*, further experimentation with the protein may be hampered because the amounts produced are too low and the protein is not pure enough. To overcome these problems, a variety of expression vectors has been designed which permit easy purification of the synthesized protein. The principle of the method is the formation of a fusion between the foreign protein and a protein which can easily be purified on the basis of substrate or immunoaffinity chromatography. Carriers that are commonly used for this purpose are Staphylococcal protein A, which can be purified by immunoaffinity chromatography to IgG, and β-galactosidase, which can be purified by either substrate or immunoaffinity chromatography. Although these methods give over 90% yields and a high degree of purification, an obvious disadvantage is that the protein is obtained as a fusion protein, which might be less desirable. A further improvement of the methodology has been found in the introduction, at the junction of the foreign protein and carrier protein, of an amino acid sequence which can be specifically cleaved by a protease. After purification of the fusion protein, the active foreign protein can be released by treatment with the protease, which preferably should be attached to a solid matrix.

i. pSS20∗

Plasmid pSS20∗ (5380 bp; *Figure 24*; 61) is presented as an example of this type of vector. The vector harbours the structural gene for β-galactosidase with a sequence encoding the recognition site for the protease collagenase, at its 3'-end. Further downstream from the collagenase recognition site lie unique cloning sites for *Hind*III, *Pst*I, and *Sal*I. Expression of a gene cloned into one of these sites is controlled by the *lac* promoter. Expression can be induced by the addition of IPTG. A fusion protein can be purified by substrate chromatography on APTG-Sepharose 4B or by immunoaffinity chromatography with antibodies raised againt the carrier protein. Cleavage of the fusion protein with collagenase yields the desired protein, with a N-terminal extension of 40 amino acids derived from the collagenase recognition sequence. In the examples described in the literature the extension had no effect on the activity of the proteins produced. This may not, however, always be the case.

ii. pRIT vectors

Plasmids of the pRIT series (7.1 kb; *Figure 25*; 62) can be used to express protein sequences fused to Staphylococcal protein A. The protein A sequence includes five IgG-binding domains which strongly bind to IgG molecules immobilized on a solid matrix. Fusion proteins can be highly purified with more than 95% recovery by passing a crude bacterial extract through a column of IgG-Sepharose 4B. Expression of fusion proteins is regulated by the promoter of the protein A gene, which is functional in *E. coli*. The protein A sequence includes a signal

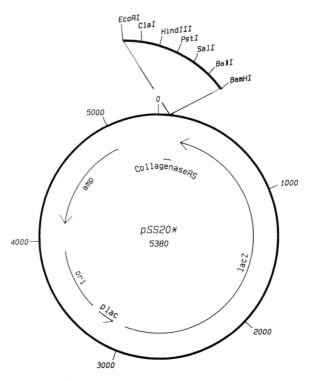

Figure 24. Structure of pSS20∗. Plasmid pSS20∗ comprises pUR290 (81) with the insertion of a 180 bp *Sau*3AI fragment into the *Bam*HI site. Plasmid pUR290 comprises the *Hind*III-*Pvu*II fragment of pBR322 containing the *amp* gene and *ori* region, and an *Hae*II-*Hind*III fragment containing the *lacZ* gene. The C-terminal sequence of the *lacZ* gene was modified to obtain a polylinker sequence as part of the *lacZ* reading frame. The nucleotide sequences of the constituent parts of the vector are known. These sequences can be retrieved from the EMBL database. The sequence of the collagenase recognition site and cloning sites are given below.

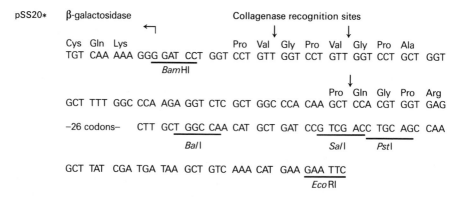

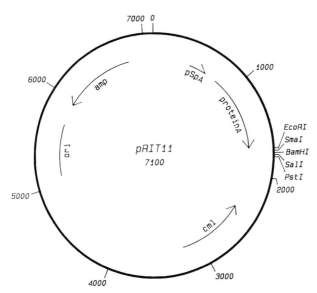

Figure 25. Structure of pRIT11. Plasmid pRIT11 comprises a 1.2 kb *Taq*I-*Eco*RI fragment containing the protein A gene, a 0.2 kb *Eco*RI-*Pvu*II polylinker sequence, a 2.9 kb *Hind*III fragment of plasmid pC194 with the *cml* gene and the *S. aureus* origin of replication, and a 2.8 kb *Pvu*II-*Cla*I fragment of the pEMBL plasmids containing the *amp* gene and *ori* region. The nucleotide sequences of the polylinker regions are shown below.

pRIT11	GAA TTC CCG GGG ATC CGT CGA CCT GCA GC
	*Eco*RI *Sma*I *Bam*HI *Sal*I *Pst*I
pRIT12	GGA ATT CCC GGG GAT CCG TCG ACC TGC AGC
	*Eco*RI *Sma*I *Bam*HI *Sal*I *Pst*I
pRIT13	AGG AAT TCC CGG GGA TCC GTC GAC CTG CAG
	*Eco*RI *Sma*I *Bam*HI *Sal*I *Pst*I

sequence which directs the fusion protein to the periplasmic space, where it is less vulnerable to proteolytic degradation.

The three pRIT plasmids, pRIT11, pRIT12, and pRIT13 have a polylinker region with multiple cloning sites, in the three possible reading frames. The pRIT plasmids also contain an origin of replication and selection marker (*cml*: chloramphenicol resistance) for replication and selection, respectively, in Gram-positive organisms.

4.2 Vectors for *in vitro* transcription

A special class of expression vectors comprises vectors that carry a promoter from bacteriophage SP6, T3, or T7. Such vectors are primarily used for the *in vitro* transcription of cloned genes. The promoters from these bacteriophage are highly specific; for example, RNA polymerase from SP6 does not promote RNA synthesis from any DNA sequence except the SP6 promoter (see *Appendix 6*). The same holds for the two other promoters. RNA synthesis initiated at these promoters is very efficient. From 1 µg of template DNA (in general, plasmid DNA which has been linearized by cutting with a restriction enzyme) one can easily obtain several micrograms of RNA. The vectors are, therefore, very useful for making radioactive probes, to be used in RNA–DNA hybridization experiments or for *in vitro* transcription-translation experiments. Initially plasmid vectors were used for these purposes. More recently also plasmid vectors with a replicon of a single-stranded phage and bacteriophage λ vectors have been constructed that can be used for this type of experiment.

4.2.1 Plasmid vectors for *in vitro* transcription

Vectors of the pGEM series (63) are given as examples of plasmid vectors. They carry two promoters, those of bacteriophage SP6 and T7, which are opposing each other and are separated by a polylinker region with multiple cloning sites. The orientation of the promoters permits the synthesis from a cloned gene of either sense or antisense RNA.

Plasmids of the pGEM (*Figure 26*) series are identical, except for the nature and orientation of the cloning sites in the polylinker region. Plasmids pGEM-1 (2865 bp) and pGEM-2 (2869 bp) have unique promoter-distal cloning sites for *Acc*I, *Ava*I, *Bam*HI, *Eco*RI, *Hin*dIII, *Pst*I, *Sac*I, *Sal*I, *Sma*I, and *Xba*I, but in a different orientation. Plasmids pGEM-3 (2867 bp) and pGEM-4 (2871 bp) have the same set of cloning sites, but, in addition, unique promoter-distal cloning sites for *Kpn*I and *Sph*I. A second *Sph*I site in the vector has been removed. In plasmid pGEM-3 the orientation of the cloning sites is the same as in pGEM-1. In pGEM-4 the order of sites is the same as in pGEM-2.

i. *Experimental approaches*

In this section an application of the use of this type of expression vectors is described. To get a better insight into the determinants that affect the morphogenesis of poliovirus, the processing of the polio protein P1 was studied (64). The present example describes the *in vitro* synthesis of poliovirus proteins P1 and 3CD under the control of the T7 promoter. The cDNA sequence encoding P1 was cloned beyond the T7 promoter in the vector pT7-6, which is very similar to those described above. In an analogous fashion, the poliovirus cDNA sequence 3CD, presumed to encode a protease activity, was cloned into the vector beyond the T7 promoter. The vectors were linearized by digestion with restriction enzymes that cut the vector after the polio sequences. The linear DNA molecules were transcribed with T7 RNA polymerase and the resulting RNA was translated in a rabbit reticulocyte lysate. To analyse the processing, protein P1 was labelled with [^{35}S]methionine, while protein 3CD was unlabelled. The protein products were incubated at 37°C and analysed by polyacrylamide gel electrophoresis. *Figure 27* shows that protein P1 encoded by plasmid pLOP324 is faithfully transcribed and translated *in vitro*. In addition, and more

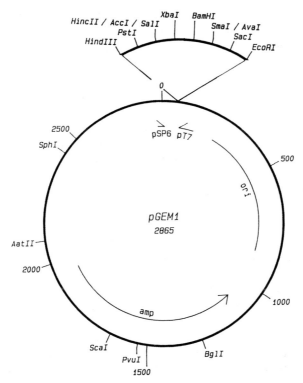

Figure 26. Structure of pGEM1. Plasmids of the pGEM series comprise the 2.2 kb *Sph*l-*Pvu*ll fragment of pUC12 (16) and a 0.6 kb fragment containing the SP6 and T7 promoters and polylinker sequence. The nucleotide sequence of the vectors is given in the manufacturer's catalogue. The nucleotide sequences of the region in the vector including the transcription-start sites and the cloning sites are given for pGEM-1 and pGEM-2.

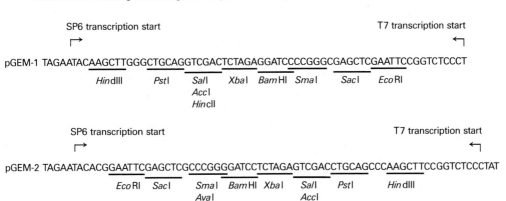

Plasmids of the pSP70 series (2.4 kb; 82) have essentially the same properties and structure as the pGEM plasmids, but contain different sets of cloning sites between the SP6 and T7 promoters.

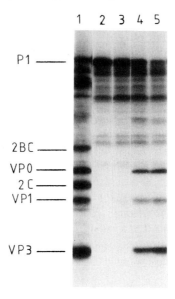

Figure 27. Analysis of processing activity of protein 3CD towards protein P1. An RNA transcript from a plasmid containing the gene encoding protein P1 was translated in a rabbit reticulocyte lysate in the presence of [^{35}S]methionine. The protein products were then incubated with proteins 3C or 3CD which were also synthesized in an *in vitro* transcription–translation system. Samples taken at time points 15 min (lane 4) and 80 min incubation (lanes 2, 3 and 5) were subjected to electrophoresis. The reactions were programmed as follows: lane 1, virion RNA; lanes 2, 3, 4, and 5, RNA transcribed from a plasmid with the P1 gene. Lane 2 received an equal volume of control lysate (no RNA added); lane 3 received protein 3C; lanes 4 and 5 received protein 3CD. The results clearly show that a plasmid containing the P1 gene under the control of the T7 promoter can be faithfully transcribed and translated, to yield protein P1 which can be correctly processed into VP0, VP1, and VP3. The results have been reproduced with permission from *Journal of General Virology*, **69**, 1627 (1988).

importantly, protein 3CD, which was also made *in vitro*, is capable of correctly processing P1, to yield the capsid proteins VP0, VP1, and VP3.

4.2.2 Vectors with a replicon of single-stranded bacteriophage

During the last few years a new generation of vectors has appeared which have incorporated elements from other vectors previously shown to be of great value in studies on control of gene expression. Most of these vectors are not freely available but have to be purchased, because they have been constructed in the laboratories of commercial companies. Since the structure of these vectors is relatively simple, they can be readily imitated, if necessary, from bits and pieces of vectors that have no commercial value.

All vectors described in the following paragraphs have in common that they are small in size, ranging from 2.9 to 3.5 kb, can replicate both as double-stranded and single-stranded DNA, carry a gene to select for transformants (*amp*) and a second gene (*lacZ'*) allowing selection of recombinant plasmids, and an array of cloning sites flanked at one side, or in most cases at both sides, by a strong promoter from bacteriophage SP6, T3, or T7. Most of these vectors exist as pairs which are identical except for the orientation of the polylinker region with cloning sites. The promoter, p_{lac}, which controls expression of the *lacZ'* gene in the vector, can also be used for *in vivo* expression of an inserted DNA fragment, provided that the reading frames of the fragment and initiator codon ATG of the *lacZ'* gene are aligned.

Since the bacteriophage promoters have an opposing polarity, RNA transcripts can be made *in vitro* from the sense and antisense strand, without the need to re-clone the fragment in another orientation. For every type of vector, two configurations exist, differing only in the orientation of the origin of replication of bacteriophage fl. Thus both strands of a DNA fragment can be easily obtained as separate clones. The vectors are particularly useful for determination of the nucleotide sequence of a cloned DNA fragment, either on double-stranded or single-stranded DNA. A second major application of these vectors is site-specific mutagenesis of cloned DNA fragments, which can be performed on single-stranded recombinant vectors. Because the entire nucleotide sequence of the vectors is known, the design of such a mutagenesis experiment is greatly facilitated by computer programs. The third application, the mapping of restriction enzyme sites in very large inserts, makes use of DNA probes which are complementary to the nucleotide sequences of the phage promoters to determine the distance of a particular restriction enzyme site from the phage promoters.

i. pBluescript vectors

Plasmids pBluescript SK + and pBluescript SK −, and pBluescript KS + and pBluescript KS− (2964 bp; *Figure 28*; 65) carry the promoters from bacteriophage T3 and T7, and a polylinker region with 21 unique cloning sites between the promoters. In the SK version of the vectors, the *Sac*I site in the polylinker region lies close to the T3 promoter, while the *Kpn*I site is located near the T7 promoter. In the KS version of the vectors, the orientation of the polylinker region is reversed. The vectors also exist in the + and − configuration.

ii. pGEM vectors

Plasmids pGEM-3Zf+ and pGEM-3Zf− (3199 bp; *Figure 29*; 66) carry SP6 and T7 promoters at both sides of a set of cloning sites, derived from pUC19. The pair of vectors otherwise have the same properties as the vectors described above.

4.2.3 Bacteriophage λ vectors; cosmid vectors

i. λZAP

A vector which potentially is very powerful is the λZAP vector (*Figure 30*; 67). It has been designed for the construction of gene libraries. Inserted DNA fragments can directly be sequenced and used for *in vitro* RNA synthesis without the need to first subclone DNA fragments. The vector combines the advantages of the λgt phage to make gene libraries (Section 2.2.2.*i*), those of plasmid vectors that can

Survey of cloning vectors for Escherichia coli

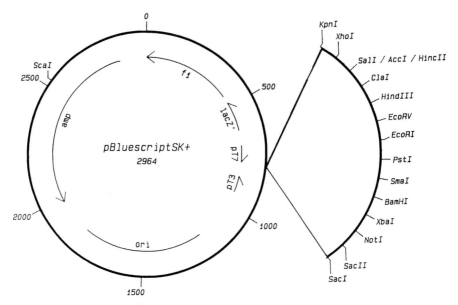

Figure 28. Structure of pBluescriptSK+. Plasmids of the pBluescript series comprise a 0.5 kb fragment from fl (nucleotides 3–458), sequences from pUC19 containg the *lacZ'* gene (nt 460–624), *ori* region and *amp* gene (nt 795–2964), and a synthetic DNA fragment (nt 626–791) with the T3 and T7 promoters and the polylinker region. The nucleotide sequence of the vectors is known and is detailed in the manufacturer's catalogue. The sequence of the polylinker with the cloning sites is given below.

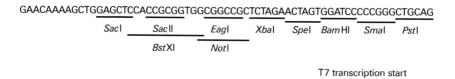

pBluescript SK+   T3 transcription start

⌐→

GGAAACAGCTATGACCATGATTACGCCAAGCTCGGAATTAACCCTCACTAAAGG

GAACAAAAGCTGGAGCTCCACCGCGGTGGCGGCCGCTCTAGAACTAGTGGATCCCCCGGGCTGCAG
            *Sac*I   *Sac*II   *Eag*I   *Xba*I *Spe*I *Bam*HI *Sma*I   *Pst*I
                    *Bst*XI   *Not*I

                                            T7 transcription start
                                                        ←⌐

GAATTCGATATCAAGCTTATCGATACCGTCGACCTCGAGGGGGGGGCCCGGTACCCAATTCGCCCTATA
*Eco*RI *Eco*RV *Hin*dIII   *Cla*I   *Sal*I   *Xho*I

GTGAGTCGTATTACAATTCACTGGCCGTCGTTTTAC

Plasmids pBS+ and pBS− (3204 bp; 83) and plasmids pT7/T3*a*-18 and pT7/T3*a*-19 (2.9 kb; 84) are essentially the same as pBluescript, but contain a different set of cloning sites between the T7 and T3 promoters.

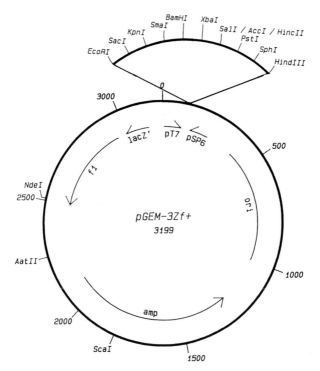

Figure 29. Structure of pGEM-3Zf +. Plasmids pGEM-3Zf + and pGEM-3Zf − comprise a 0.1 kb DNA fragment with the SP6 and T7 promoters and the polylinker region, a 160 bp DNA fragment with the *lacZ'* region, a 0.5 kb DNA fragment with the *ori* region of fl, a 2.2 kb DNA fragment from pBR322 with the *amp* gene, and a 0.2 kb DNA fragment with the *lac* promoter-operator region. The nucleotide sequences of the vectors are given in the manufacturer's catalogue. The sequences of the promoters and cloning sites are given below.

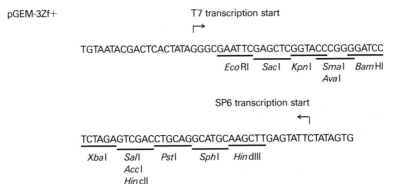

229

λ ZAP

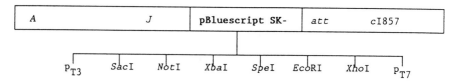

Figure 30. Structure of *λ*ZAP. DNA fragments inserted into one of the unique cloning sites (*Sac*I-*Xho*I) can easily be subcloned as pBluescript plasmids (see *Figure 28*) and transcribed *in vitro* from promoters (p_{T3}, p_{T7}) which flank the region containing the cloning sites. Symbols *A, J, att,* and *c*I857 refer to *λ* genes.

replicate as single-stranded DNA and those of vectors containing strong bacteriophage promoters. Phage *λ*ZAP is an insertion vector with a capacity for DNA fragments up to 10.8 kb in length. It harbours a complete copy of the plasmid pBluescript SK −, which is flanked on one side by the initiator region and on the other side by the terminator region of the origin of replication of fl. The presence of these regions allows for the *in vivo* excision from the phage vector of a complete copy of a recombinant pBluescript SK − vector. Because the excised plasmid contains an intact fl *ori* region, the recombinant vector can be multiplied and encapsidated into single-stranded phage particles. Since the pBluescript SK − part of the vector contains the *lacZ'* region with multiple cloning sites, a cDNA fragment inserted into one of these sites can be identified on the basis of the transition from blue to colourless plaques. After excision of the plasmid vector the insert can easily be sequenced and can be used for *in vitro* transcription–translation experiments. The polylinker of the plasmid part of the vector contains 21 cloning sites, six of which are unique.

ii. Lorist vectors
The possibility to apply cosmid vectors for the isolation of large regions of a eukaryotic genome has recently been enlarged by the introduction into cosmid vectors of the elements to generate RNA transcripts from cloned DNA fragments. The Lorist family of vectors (68, 69) was designed to facilitate the isolation and ordering of overlapping recombinant clones. These vectors are based on the origin of replication of bacteriophage *λ*, contributing to a more uniform copy number and yield of recombinant cosmid vector. The vectors carry two strong promoters, p_{T7} and p_{SP6}, which are facing each other and flank a region with two or more unique cloning sites. *In vitro* transcription with either T7 or SP6 RNA polymerase of DNA cloned into one of these sites yields RNA probes that are specific for the adjacent end of the insert. In most cases these probes correspond to unique genome sequences and can be used to screen for overlapping clones.

All Lorist vectors (*Figure 31*) carry the neomycin-resistance gene as a selection marker. In Lorist2 (5564 bp) and Lorist6 (5156 bp), two transcription terminators have been placed on both sides of the cloning sites, to insulate the DNA replication functions of the vector from the cloned sequences.

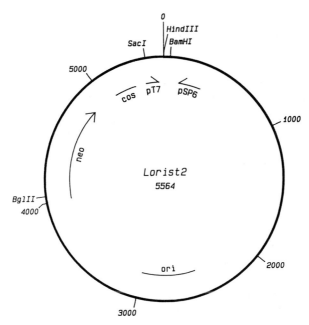

Figure 31. Structure of Lorist2. A synthetic *trp* terminator sequence, t_{trpa} was introduced at the *Eco*RI site at nucleotide 5121 of Lorist B and a synthetic *rrnC* terminator at the *Nhe*I site at nucleotide 252. The synthetic sequences are arranged so that the *trp* terminator, the T7 promoter, the cloning sites, the SP6 promoter and the *rrnC* terminator form a portable *Eco*RI fragment. The nucleotide sequences of the constituents of the vectors are known and can be retrieved from the EMBL database. The arrangement of the *Bam*HI and *Hind*III cloning sites with respect to the transcription start sites used by T7 and SP6 RNA polymerase in Lorist2 are shown below.

4.3 Promoter-probe vectors

4.3.1 General properties

To study regulation of gene expression, vectors have been constructed that permit the isolation and characterization of signals that control gene expression at the level of transcription. In this section, vectors will be described that can be used for the isolation of promoter-bearing DNA fragments, to study the effect of factors on promoter activity, to compare the activity of a promoter with that of other promoters, and to study the effect of mutations in the promoter on its activity.

The basic feature of a promoter-probe vector is the presence of a genetic marker that is deprived of its promoter and thus is inactive, but can be activated by placing a promoter-bearing DNA fragment in front of it. The genetic markers that are used for this type of vector are antibiotic resistance markers, which, if activated, confer resistance on bacteria to that antibiotic. Alternatively, markers are used that will complement a deficiency, if activated. The most commonly employed markers are tetracycline resistance or chloramphenicol resistance (chloramphenicol acetyl transferase: *cml*) markers on the one hand and genes encoding galactokinase (*galK*) and β-galactosidase (*lacZ*) on the other hand. These markers have in common that the gene function can be readily selected. All except the *cml* gene have positive and negative selection which make them more flexible. All markers except the *tet* gene code for readily assayable products. An advantage of the *galK* and *lacZ* markers over the antibiotic resistance markers, is that differences in promoter strength can be determined quantitatively *in vitro* and qualitatively *in vivo*.

Most promoter-probe vectors harbour a second, independent selectable marker besides that used for characterization of the promoter, and one or more unique cloning sites before the first selection marker.

4.3.2 Specific vectors

i. pKO vectors

Plasmids of the pKO (3.9 kb; 70) series harbour the *amp* gene to select for transformants, and the *galK* gene, without its promoter but with its genuine RBS, as marker for cloning promoters. Plasmids of the pKO series cannot complement *E. coli* GalK⁻ host bacteria. When a DNA fragment containing a promoter is inserted into one of the cloning sites before *galK*, the gene is transcribed and the plasmid complements the GalK⁻ phenotype of the host, resulting in growth on minimal medium with galactose, and in the formation of red colonies on McConkey-galactose agar plates.

To minimize the effect of upstream translation initiation, translation-stop codons are present. They are located at 174, 121, and 74 bp before the *galK* coding sequences, beyond the cloning site, in the three reading frames. These stop codons uncouple from the *galK* transcript all translation initiating in the DNA insert.

Plasmids of the pKO series can also be used for isolation of mutated promoter sequences. Since recombinant pKO containing a promoter will kill GalE⁻T⁻K⁻ bacteria in medium containing galactose, promoter-down mutations can easily be isolated by selection of bacteria growing in the presence of galactose.

The plasmids of the pKO series (pKO-1–pKO-11) are identical except for differences in the cloning sites before the *galK* gene. The structure of one of the vectors, pKO-2 (71) is given in *Figure 32*.

ii. pNM vectors

Plasmids pNM480, pNM481, and pNM482 (8.6 kb; *Figure 33*; 72) can be used for cloning DNA fragments containing promoters together with translation-initiation sequences, allowing the analysis of expression from a cloned promoter

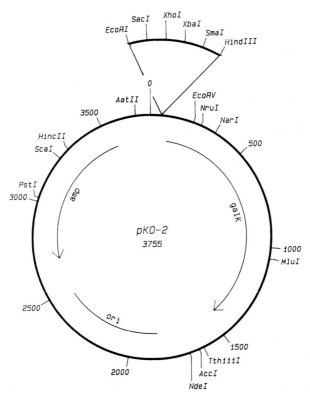

Figure 32. Structure of pKO2. The vector comprises the *Eco*RI-*Pvu*II fragment of pBR322 containing the *amp* gene and *ori* region, a 290 bp *Eco*RI-*Hind*III fragment of bacteriophage λ with *O*-gene sequences and a 1.3 kb fragment containing the *E. coli galK* gene. The nucleotide sequence of the vector is known and can be retrieved from UWGCG Vecbase. The nucleotide sequence of a relevant part of the vector with cloning sites is given.

pKO-2   GAATTCTCATGTTTGACAGCTTATCATCGGAGCTCTCGAGTCTAGAATCGATCCCGGG

           *Eco*RI   *Sac*I *Xho*I *Xba*I *Cla*I *Sma*I

        AAGCTTAACTAACTAACAGCT

        *Hin*dIII

by means of synthesis of an active β-galactosidase fusion protein. The plasmids contain the *E. coli lac* operon, comprising the *lacZ* gene, the *lac Y* gene coding for lactose permease and the *lacA* gene encoding transacetylase. The vectors lack the *lac* promoter, operator, RBS and the first eight codons of *lacZ*. DNA fragments inserted into one of the unique cloning sites before the *lacZ* gene form gene fusions, if they contain promoters and translation-initiation signals, and providing that the translation-initiation codon ATG is in frame with the *lacZ* coding region. To facilitate alignment, the reading frames in the three vectors

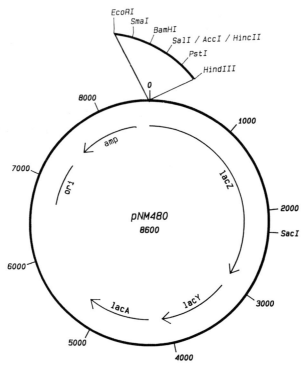

Figure 33. Structure of pNM480. Plasmids pNM480, pNM481, and pNM482 comprise a 6.2 kb *Eco*RI-*Sal*I fragment containing the *lac* genes and a 2.4 kb *Pvu*II fragment from pUC8 (15) with the origin of replication, the *amp* gene and polylinker region. The *Acc*I site in the polylinker region was rendered unique by removal of other *Acc*I sites in the *lac* operon. The nucleotide sequences of the vectors are known (72). The nucleotide sequences with the restriction sites of relevant parts of the vectors are given.

pNM480   GA ATT CCC GGG GAT CCG TCG ACC TGC AGC CAA GCT TGC GAT CCC GTC

               *Eco*RI   *Sma*I   *Bam*HI   *Sal*I   *Pst*I   *Hin*dIII

                                     *Acc*I

                                     *Hin*cII

pNM481   GAA TTC CCG GGG ATC CGT CGA CCT GCA GCC AAG CTT GCT CCC GTC

               *Eco*RI   *Sma*I   *Bam*HI   *Sal*I   *Pst*I   *Hin*dIII

                                       *Acc*I

                                     *Hin*cIII

pNM482   G AAT TCC CGG GGA TCC GTC GAC CTG CAG CCA AGC TTC GAT CCC GTC

               *Eco*RI   *Sma*I   *Bam*HI   *Sal*I   *Pst*I   *Hin*dIII

                                       *Acc*I

                                     *Hin*cII

were shifted by one nucleotide. Consequently, in one of the three vectors the ATG is properly aligned with the *lacZ* coding sequence.

4.4 Terminator-probe vectors

4.4.1 General properties

The rationale in designing terminator-probe vectors is analogous to that used for promoter-probe vectors. Between a selectable marker encoding an easily-assayable product and its promoter, cloning sites are engineered so that DNA fragments which contain terminators can be inserted at that site. The presence of the terminator will result in a reduction or even extinction of transcription of the gene beyond the terminator, leading to a change in the phenotype of the host bacteria. The same type of marker genes are used in terminator-probe vectors as in promoter-probe vectors, antibiotic resistance markers (*tet* and *cml*) and the *galK* and *lacZ* markers. The *galK* gene is particularly useful, because vectors that do or do not express the *galK* gene can be discriminated in either GalE⁻T⁻K⁻ or GalK⁻ hosts. The efficiency of a terminator can also be quantitatively determined by measuring galactokinase activities *in vitro*.

In some transcription-terminator vectors two marker genes are placed in tandem, with the cloning site(s) between them. This type of vector is superior to that containing a single marker, because the expression of the first gene serves as an internal control for the efficiency of transcription from the promoter. If both genes encode assayable products, as is the case for the vector described below, the expression ratio of the second versus the first gene is a measure for the termination function of the cloned DNA fragment.

4.4.2 Specific vectors

Plasmid pOL4 (6.0 kb; *Figure 34*; 73) comprises a hybrid operon in which the *cml* and *galK* genes were placed in tandem under the control of the promoter, p_L of bacteriophage λ. The presence of a terminator in a DNA fragment inserted into the unique *Sma*I site between the promoter-proximal gene (*cml*) and promoter-distal gene (*galK*) can be detected by measuring *galK* expression, both *in vivo* and *in vitro*. Expression of both genes is controlled by the product of the *c*I857 repressor gene of bacteriophage λ, which is present on the plasmid. Expression of the two genes can be modulated by varying the temperature between 28°C (fully repressed) and 42°C (fully induced).

5. Conclusions

During the last fifteen years considerable progress has been made in recombinant DNA research. The successful exploitation in the early stages of vectors like pBR322 and pACYC177 has prompted the design of a great variety of vectors. The presence to date of an enormous variety of vectors stems to a great part from the large body of genetic and biochemical knowledge of *E. coli* which is available.

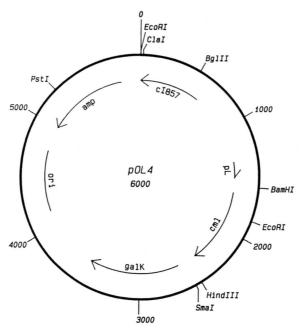

Figure 34. Structure of pOL4. Plasmid pOL4 comprises a *Bg*/II-*Bam*HI fragment containing the promoter p$_L$, a 0.7 kb *Bam*HI-*Hind*III fragment containing the *cml* gene, a 3.6 kb *Eco*RI-*Hind*III fragment from plasmid pKG1800 (70) containing the *galK* gene and the origin of replication, and an *Eco*RI-*Bg*/II fragment containing the *cl857* gene. The nucleotide sequence of the constituent parts of the vector is known and can be retrieved from the EMBL database.

Without this knowledge it would have been impossible to construct such highly sophisticated vectors as we have now at our disposal.

New vectors are continually being constructed. Since for most types of application vectors already exist, most efforts are being spent on further improvement of vectors, so that they can be used for special purposes, such as for the synthesis of proteins that can be easily purified and recovered in an unfused state, for directing the proteins synthesized to a special compartment, or for increasing the ease with which a gene can be cloned and sequenced. It is expected that this type of research will continue for many years as the vectors present today are used to further refine our insight in the mechanisms of control of gene expression in *E. coli*. This knowledge will, no doubt, be exploited in the design of tomorrow's vectors for studying gene expression and for producing economically important proteins. For the latter purpose, research is also needed in order to construct vectors that can be stably maintained under non-selective conditions. If *E. coli* is ever to be used in fermentation processes, antibiotic resistance markers will have to be replaced by other selective traits, and vectors will have to be further stabilized. For plasmid vectors this will require study of the elements that control plasmid replication, copy number and structural and segregational stability. Another route to obtain a genetically-engineered *E. coli* strain that stably expresses a foreign protein at high rate is by incorporation into the chromosome of multiple copies of a vector. At present such strains are unstable, so further research is required, if this pathway is to be pursued.

Acknowledgements

I thank Dr Cees van den Hondel for suggestions concerning the design and for critical reading of the manuscript. Thanks are also due to Drs Pat Crowley and Rob Leer for correcting the manuscript.

References

1. Warren, G.J., Twigg, A.J., and Sheratt, D.J. (1978). *Nature*, **274**, 259.
2. Meacock, P.A. and Cohen, S.N. (1980). *Cell*, **20**, 529.
3. Summers, D.K. and Sherratt, D.J. (1984). *Cell*, **36**, 1097.
4. Cesareni, G., Muesing, M.A., and Poliski, B. (1982). *Proceedings of the National Academy of Sciences of the USA*, **79**, 6313.
5. Ptashne, M. (1986). *A Genetic Switch: Gene Control and Phage λ*. Blackwell Scientific Publications and Cell Press, Oxford.
6. Cohen, S.N., Chang, A.C.Y., Boyer, H.W., and Helling, R.B. (1973). *Proceedings of the National Academy of Sciences of the USA*, **70**, 3240.
7. Pouwels, P.H., Enger-Valk, B.E., and Brammar, W.J. (1985). *Cloning Vectors*. Elsevier Science Publishers, Amsterdam.
8. Uhlin, B.E., Molin, S., Gustafsson, P., and Nordström, K. (1979). *Gene*, **6**, 91.
9. Balbás, P., Soberón, X., Merino, E., Zurita, M., Lomeli, H., Valle, F., Flores, N., and Bolivar, F. (1986). *Gene*, **50**, 3.
10. Bolivar, F., Rodriguez, R.L., Greene, P.J., Betlach, M.C., Heyneker, H.L., and Boyer, H.W. (1977). *Gene*, **2**, 95.
11. Skogman, G., Nilsson, J., and Gustafsson, P. (1983). *Gene*, **23**, 105.
12. *Recombinant DNA Technical Bulletin*, NIH 1 (1978).
13. Soberon, X., Covarrubias, L., and Bolivar, F. (1980). *Gene*, **9**, 287.
14. Zurita, M., Bolivar, F., and Soberon, X. (1984). *Gene*, **28**, 119.
15. Vieira, J. and Messing, J. (1982). *Gene*, **19**, 259.
16. Messing, J. (1983). In *Methods in Enzymology*, Vol. 101 (ed. R. Wu, L. Grossman, and K. Moldave), pp. 20–78. Academic Press, New York.
17. Norrander, J., Kempe, T., and Messing, J. (1983). *Gene*, **26**, 101.
18. Miller, J. (1972). In *Experiments in Molecular Genetics*, pp. 47–55. Cold Spring Harbor Laboratory, Cold Spring Harbor, New York.
19. Sanger, F., Nicklen, S., and Coulson, A.R. (1977) *Proceedings of the National Academy of Sciences of the USA*, **70**, 5463.
20. Maxam, A. and Gilbert, W. (1977). *Proceedings of the National Academy of Sciences of the USA*, **34**, 560.
21. Pridmore, R.D. (1987). *Gene*, **56**, 309.
22. Punt, P.J., Dingemanse, M.A., Jacobs-Meijsing, B.J.M., Pouwels, P.H., and van den Hondel, C.A.M.J.J. (1988). *Gene*, **69**, 49.
23. Chang, A.C.Y. and Cohen, S.N. (1978). *Journal of Bacteriology*, **134**, 1141.

24. Churchward, G., Belin, D., and Nagamine, Y. (1984). *Gene*, **31**, 165.
25. Collins, J. and Hohn, B. (1977). *Proceedings of the National Academy of Sciences of the USA*, **74**, 3259.
26. Borck, K., Beggs, J.D., Brammar, W.J., Hopkins, A.S., and Murray, N.E. (1976). *Molecular and General Genetics*, **146**, 199.
27. Murray, N.E., Brammar, W.J., and Murray, K. (1977). *Molecular and General Genetics*, **150**, 53.
28. Hoyt, M.A., Knight, D.M., Das, A., Miller, H.I., and Echols, H. (1982). *Cell*, **31**, 565.
29. Hendrix, R.W., Roberts, J.W., Stahl, F., and Weisberg, R.A. (ed.) (1983). *Lambda II*. Cold Spring Harbor Laboratory, Cold Spring Harbor, New York.
30. Zissler, J., Signer, E., and Shaefer, F. (1971). In *The Bacteriophage Lambda* (ed. A.D. Hershey, pp. 469–75, Cold Spring Harbor Laboratory, Cold Spring Harbor, New York.
31. Huynh, T.V., Young, R.A., and Davis, R.W. (1985). In *DNA Cloning Techniques: A Practical Approach* (ed. D.M. Glover), pp. 49–78. IRL Press, Oxford.
32. Han, J.H., Stratowa, C., and Rutter, W.J. (1987). *Biochemistry*, **26**, 1617.
33. Han, J.H. and Rutter, W.J. (1987). *Nucleic Acids Research*, **15**, 6304.
34. Frischauff, A.M., Lehrach, H., Poustka, A., and Murray, N. (1983). *Journal of Molecular Biology*, **170**, 827.
35. Ish-Horowicz, D. and Burke, J.F. (1981). *Nucleic Acids Research*, **9**, 2989.
36. Poustka, A., Rackwitz, H-R., Frischauf, A.M., Hohn, B., and Lehrach, H. (1984). *Proceedings of the National Academy of Sciences of the USA*, **81**, 4129.
37. Yannisch-Perron, C., Vieira, J., and Messing, J. (1985). *Gene*, **33**, 103.
38. Geider, K., Hohmeyer, C., Haas, R., and Meyer, T. (1985). *Gene*, **33**, 341.
39. Zoller, M.J. and Smith, M. (1984). *DNA*, **3**, 479.
40. Berk, A.J. and Sharp, P.A. (1977). *Cell*, **12**, 721.
41. Dotto, G.P., Enea, V., and Zinder, N.D. (1981). *Virology*, **114**, 463.
42. Cesareni, G. and Murray, J.A.H. (1987). In *Genetic Engineering* (ed. K. Setlow), Vol. 9, pp. 135–54. Plenum Publishing Corporation, New York.
43. Spratt, B.G., Hedge, P.J., te Heesen, S., Edelman, A., and Broome-Smith, J.K. (1986). *Gene*, **41**, 337.
44. Konings, R.N.H., Verhoeven, E.J.M., and Peeters, B.P.H. (1987). In *Methods in Enzymology* (ed. R. Wu and L. Grossman), Vol. 153, pp. 12–34. Academic Press, New York.
45. Amann, E., Brosius, J., and Ptashne, M. (1983). *Gene*, **25**, 167.
46. De Boer, H.A., Comstock, L.J., and Vasser, M. (1983). *Proceedings of the National Academy of Sciences of the USA*, **80**, 21.
47. Hui, A., Hayflick, J., Dinkelspiel, V., and de Boer, H.A. (1984). *EMBO Journal*, **3**, 623.
48. Schottel, J.L., Sninsky, J.J., and Cohen, S.N. (1984). *Gene*, **28**, 177.
49. Hessing, H.G.M., van Rotterdam, C., and Pouwels, P.H. (1987). *Molecular and General Genetics*, **210**, 256.
50. Schauder, B., Blöcker, H., Frank, R., and McCarthy, J.E.G. (1987). *Gene*, **52**, 279.
51. Stanssens, P., Remaut, E., and Fiers, W. (1985). *Gene*, **36**, 211.
52. Remaut, E., Tsao, H., and Fiers, W. (1983). *Gene*, **22**, 103.
53. Kos, A., Broekhuijsen, M.P., van Gorcom, R.F.M., van Rotterdam, J., Enger-Valk, B.E., van Putten, A.J., Veltkamp, E., van den Hondel, C.A.M.J.J., and Pouwels, P.H. (1984). In *Innovation in Biotechnology* (ed. E.H. Houwink and R.R. van der Meer), pp. 415–24. Elsevier, Amsterdam.

54. Yanofsky, C. (1988). *Journal of Biological Chemistry*, **263**, 609.
55. Ikehara, M., Ohtsuka, E., Tokunaga, T., Taniyama, Y., Iwai, S., Kitano, K., Miyamoto, S., Ohgi, T., Sakuragawa, Y., Fujiyama, K., Ikari, T., Kobayashi, M., Miyake, T., Shibahara, S., Ono, A., Ueda, T., Tanaka, T., Baba, H., Miki, T., Sakurai, A., Oishi, T., Chisaka, O., and Matsubara, K. (1984). *Proceedings of the National Academy of Sciences of the USA,*, **81**, 5956.
56. Tacon, W.C.A., Bonass, W.A., Jenkins, B., and Emtage, J.S. (1983). *Gene*, **23**, 255.
57. Bressan, G. and Stanley, K.K. (1987). *Nucleic Acids Research*, **15**, 10056.
58. Stanley, K.K. and Luzio, J.P. (1984). *EMBO Journal*, **3**, 1429.
59. Amann, E. and Brosius, J. (1985). *Gene*, **40**, 183.
60. Masui, Y., Mizuno, T., and Inouye, M. (1984). *Bio/Technology*, **2**, 81.
61. Scholtissek, S. and Grosse, F. (1988). *Gene*, **62**, 55.
62. Löwenadler, B., Nilsson, B., Abrahmsén, L., Moks, T., Ljungqvist, L., Holmgren, E., Paleus, S., Josephson, S., Philipson, L., and Uhlén, M. (1986). *EMBO Journal*, **5**, 2393.
63. Promega Biotec Catalogue, Transcription Systems (1986/1987), p. 24.
64. Jore, J., de Geus, B., Jackson, R.J., Pouwels, P.H., and Enger-Valk, B.E. (1988). *Journal of General Virology*, **69**, 1627.
65. Stratagene Catalogue (1988), p. 108.
66. Promega Biotec Catalogue, Transcription Systems (1987/1988), p. 34.
67. Short, J.M., Fernandez, J.M., Sorge, J.A., and Huse, W.D. (1988). *Nucleic Acids Research*, **16**, 7583.
68. Cross, S.H. and Little, P.F.R. (1986). *Gene*, **49**, 9.
69. Gibson, T.J., Coulson, A.R., Sulston, J.E., and Little, P.F.R. (1987). *Gene*, **53**, 275.
70. McKenney, K., Shimatake, H., Court, D., Schmeissner, U., Brady, C., and Rosenberg, M. (1982). In *Gene Amplification and Analysis* (ed. J.G. Chirikjian and T.S. Papas), Vol. II, pp. 383–415. Elsevier-North Holland, Amsterdam.
71. De Boer, H.A. (1984). *Gene*, **30**, 251.
72. Minton, N.P. (1984). *Gene*, **31**, 269.
73. Honigman, A., Mahajna, J., Altuvia, S., Koby, S., Teff, D., Locker-Giladi, H., Hyman, H., Kronman, C., and Oppenheim, A. B. (1985). *Gene*, **36**, 131.
74. Appleyard, R.K. (1954). *Genetics*, **39**, 440.
75. Hanahan, D. (1983). *Journal of Molecular Biology*, **166**, 557.
76. Boyer, H.W. and Roulland-Dussoix, D. (1969). *Journal of Molecular Biology*, **41**, 459.
77. Clarke, L. and Carbon, J. (1978). *Journal of Molecular Biology*, **120**, 517.
78. Hohn, B. and Murray, K. (1977). *Proceedings of the National Academy of Sciences of the USA*, **74**, 3259.
79. Twigg, A.J. and Sheratt, D. (1980). *Nature*, **283**, 216.
80. Chen, E.Y. and Seeburg, P.H. (1985). *DNA*, **4**, 165.
81. Rüther, U. and Müller-Hill, B. (1983). *EMBO Journal*, **2**, 1791.
82. Anon (1987). *Promega Notes*, **8**, 6.
83. Anon (1987). *Focus*, **9**, 11.
84. Stratagene Catalogue (1988), p. 106.

A1

Equipment required for molecular biology

T. A. BROWN

Much of the equipment needed for molecular biology will already be present in an active biological research department. The lists that follow are intended for guidance and do not cover every possible experiment that you may wish to do. Details of individual items of equipment are provided elsewhere in the book. A more comprehensive description of the requirements for molecular biology has been published (1).

1. General facilities

- Genetic manipulation containment facilities (see *Appendix 2*)
- Microbiological facilities:
  - autoclaves for sterilizing media
  - clean areas for molecular biology work (laminar flow cabinets are useful but not essential)
  - 37°C room. Incubators can be used as an alternative but it must be possible to incubate Petri dishes, as well as cultures in test-tubes and flasks from 5 ml to 2 litres.
- Glassware cleaning facility
- Darkroom facilities
  - red-light darkroom for autoradiography, including tanks and accessories for developing X-ray films
  - short-wavelength (302 or 366 nm) UV transilluminator
  - Polaroid camera or CCD detector for recording agarose gels
- Cold room
- Sub-zero storage facilities:
  - $-80°C$ freezer
  - liquid nitrogen storage vessels

- Radiochemical handling and disposal facilities, primarily for ^{35}S and ^{32}P
- Wet-ice machine and dry-ice storage

2. Departmental equipment (can be shared by a number of groups)

- Ultracentrifuge capable of 70 000 r.p.m., plus at least one fixed-angle and one swing-out rotor plus accessories. Density gradient centrifugation can take 48 h so one machine is rarely sufficient.
- Refrigerated centrifuge capable of 25 000 r.p.m., plus rotors for 50 to 500 ml buckets
- UV spectrophotometer
- Liquid scintillation counter
- Sonicator with a variety of probes
- Four-point balance
- Platform incubators for 45°C, 55°C, and 65°C
- Vacuum drier (a freeze-drier is not essential)
- Vacuum oven
- 100°C oven
- Oligonucleotide synthesis machine, though one machine can quickly become overloaded if several groups are carrying out PCR experiments

3. Laboratory equipment (required by each group)

- Bench-top centrifuge, capable of 5000 r.p.m. but not refrigerated
- Power supplies:
  - 3000 V for DNA sequencing
  - 200 V for agarose gels, electroelution, etc.
- Gel apparatus:
  - vertical gel apparatus (e.g. 21×40 cm) for DNA sequencing
  - horizontal gel apparatus (e.g. 10×10 cm, 15×10 cm, and 15×15 cm), including equipment dedicated for RNA work
- Thermal cycler for PCR
- Microfuge
- Cassettes and intensifying screens for autoradiography
- Light-box for observing autoradiographs
- Water baths (non-shaking): ideally three, one set at 37°C, one at 65°C, and one varied for requirements

- Gel drier
- Two-point top pan balance
- Vortex mixers
- Magnetic stirrers
- pH meter
- Microwave for melting agar and preparing agarose
- Bag-sealer for setting up hybridization analyses. Commercial ones are very expensive so try your local discount store.
- Automatic pipettes (e.g. Gilson P20, P200, P1000); each researcher will need a set
- Double-distilled water supply
- Fridge
- $-20°C$ freezer
- Fume cupboard
- The following are considered luxury items:
  - UV cross-linking incubator
  - electrotransfer or vacuum blotting unit
  - spin-vac
  - slot blotter or dot blotter

4. Computer facilities

Remarkably, there was a time when molecular biologists did not use computers. Now at least some computing facilities are essential. Ideally, you need a PC that is linked to a mainframe and which allows you to communicate with other systems. You will then have access to all the required facilities for DNA sequence analysis and database searching. As a second-best alternative a free-standing PC can be provided with commercial packages for DNA sequence analysis and database searching.

Reference

1. Blumberg, D.D. (1987). *Methods in Enzymology*, Vol. 152 (ed. S.L. Berger and A.R. Kimmel), pp. 3–20. Academic Press, New York.

Safety

T. A. BROWN

Safety is an individual responsibility and you must ensure that you are aware of safety requirements before embarking on gene cloning experiments. You should consult your departmental and institutional officers about the necessary precautions to take. The following is intended merely as a guide.

1. Genetic manipulation

In most countries genetic manipulation experiments must be carried out in accordance with guide-lines laid down by national authorities (e.g. refs 1 and 2). You must understand and comply with these guide-lines before starting any experiments which involve DNA cloning.

2. Microorganisms

Most molecular biology experiments are carried out with attenuated strains of *E. coli* that are unable to survive outside of the test-tube. Nevertheless, precautions to avoid contamination, as described in Section 1 of Chapter 1 of this volume, should be observed whenever bacteria are being handled.

3. Radiochemicals

Although ^{35}P and ^{32}S are not the most dangerous radionuclides in biological research they are used in large amounts in molecular biology experiments. Your national and local regulations for handling and disposal should be followed at all times. Some molecular biology techniques that involve radiochemicals are messy. Before running a polyacrylamide gel with labelled DNA check that you can dispose of the (highly-radioactive) contents of the lower reservoir without spillage. Before you carry out your first hybridization experiment make sure you can seal and open plastic bags without causing a mess.

4. Organic solvents

The most dangerous solvent used in molecular biology is phenol. If possible purchase phenol that does not need redistilling (see Section 1.1 of Chapter 3 in this volume). Always wear gloves when you are carrying out phenol extractions and make sure the gloves do actually provide a barrier to phenol: some lightweight disposable gloves do not.

5. Mutagens/carcinogens

Most chemicals that bind to DNA in the test-tube are mutagenic and/or carcinogenic. An example is ethidium bromide, which is used in large amounts to detect DNA in agarose gels. Some manuals recommend adding ethidium bromide to the running buffer, but in general terms this should be avoided. It does not provide any major advantage over post-staining (unless you want to follow the electrophoresis) and creates a major health hazard. Check that your local regulations allow ethidium bromide solutions to be disposed of down a sink. If not refer to refs 3–5 for decontamination procedures. Remember that gels stained with ethidium bromide are dangerous and should not be disposed with the general lab waste.

6. Toxic chemicals

Several toxic chemicals are used in molecular biology. The most dangerous is probably acrylamide because of the possibility of creating aerial dust contamination when handling the powder. Avoid weighing out acrylamide if possible and handle the powder in a fume hood. Refer to Section 4.1.2 of Chapter 5 in this volume for further information.

7. Ultraviolet radiation

The most dangerous piece of equipment used by molecular biologists is the UV transilluminator. Standard laboratory models have recommended exposure levels of 4.5 sec or less at 12 inches. Always use a suitable screen when observing agarose gels and wear face and eye protection when the screen is removed for photography. Be particularly careful that your wrists and neck are protected when you are excising bands from a gel.

8. High-voltage electricity

The voltages used in electrophoresis can kill. Commercial gel apparatus is designed so that lids cannot be removed before the power supply is disconnected.

Home-made equipment should incorporate similar safeguards. Broken gel tanks should not be used. Remember that liquids can conduct electricity, and that leaky apparatus is potentially lethal.

References

1. Health and Safety Executive, Advisory Committee on Genetic Manipulation (1988). *Note 7: Guidelines for the Categorisation of Genetic Manipulation Experiments.* HMSO, London.
2. Guidelines for Research Involving Recombinant DNA Molecules (1984). *Federal Register*, **49**, Part VI, No. 227, 46266.
3. Bensaude, O. (1988). *Trends in Genetics*, **4**, 89.
4. Lunn, G. and Sansone, E.B. (1987). *Analytical Biochemistry*, **162**, 453.
5. Quillardet, P. and Hofnung, M. (1988). *Trends in Genetics*, **4**, 89.

A3

Important *Escherichia coli* strains

T. A. BROWN

1. Genotypes of important strains

See also Section 3.1 of Chapter 2, and *Table 1* in Chapter 9, of this volume.

1.1 Strains for general purpose cloning

C600	*supE44 thi1 thr1 leuB6 lacY1 tonA21*
DH1	*gyrA96 recA1 relA1 endA1 thi1 hsdR17 supE44*
HB101	*proA2 recA13 ara14 lacY1 galK2 xyl5 mtl1 rpsL20 supE44 hsdS20*
JA221	*hsdR lacY leuB6 trpE5 recA1*
RR1	*proA2 ara14 lacY1 galK2 xyl5 mtl1 rpsL20 supE44 hsdS20*

1.2 Strains for single-stranded M13 vectors

These strains can also be used for plasmid vectors that utilize Lac selection. MV1184 and XL1-BLUE can be used with M13KO7 helper phage for phagemid cloning. SURE and XL1-BLUE are marketed by Stratagene. For a more comprehensive list see ref. 1.

JM83	*ara* Δ*(pro-lac) rpsL thi* (ϕ80 *lacZ*Δ*M15)*
JM109	*recA1 endA1 gyrA96 thi hsdR17 supE44 relA1* Δ*(lac-proAB)* [F′ *traD36* $proAB^+$ $lacI^q$ *lacZ*Δ*M15*]
JM110	*rpsL thr leu thi lacY galK galT ara tonA tsx dam dcm supE44* Δ*(lac-proAB)* [F′ *traD36* $proAB^+$ $lacI^q$ *lacZ*Δ*M15*]
MV1184	*ara* Δ*(lac-proAB) rpsL thi* (ϕ80 *lacZ*Δ*M15)* Δ(*srl-recA)306*::Tn*10* F′[*traD36* $proAB^+$ $lacI^q$ *lacZ*Δ*M15*]
SURE	*recB recJ sbcC201 uvrC umuC*::Tn*5*(kan^r) *lac* Δ*(hsdRMS) endA1 gyrA96 thi relA1 supE44* F′[$proAB^+$ $lacI^q$ *lacZ*Δ*M15* Tn*10*)
XL1-BLUE	*supE44 hsdR17 recA1 endA1 gyrA46 thi relA1 lac* F′[$proAB^+$ $lacI^q$ *lacZ*Δ*M15* Tn*10*]

1.3 Strains for λ vectors

Y1088   *supE supF metB trpR hsdR tonA21 strA ΔlacU169 proC*::Tn5(pMC9)
Y1090   *supF ΔlacU169 Δlon araD139 strA trpC22*::Tn10(pMC9)

1.4 Strains for λ packaging extracts

BHB2688   N205 *recA* (λ*imm434 c*Its *b2 red3 Eam4 Sam7*)/λ
BHB2690   N205 *recA* (λ*imm434 c*Its *b2 red3 Dam15 Sam7*)/λ
SMR10   *su* (λ*cos2 ΔB xis1 red3 gam*am210 *c*Its857 *nin5 Sam7*)/λ

1.5 Brief explanation of genotypes

1.5.1 Recombination deficiency

The major recombination pathways in *E. coli* are coded by the *rec* genes. Strains that are Rec$^+$ are less suitable for cloning as they can multimerize plasmids and rearrange insert DNA. There are at least nine different *rec* genes but in practice a *recA* strain is sufficiently recombination-deficient to act as a host for cloned DNA. However, some types of higher eukaryotic DNA present special problems and require a more highly modified strain such as SURE.

1.5.2 Host-controlled restriction-modification

Most strains used for cloning are derived from *E. coli* K, which possesses the *EcoK* restriction-modification system. This is coded by the *hsd* genes, with an *hsdM* strain being methylation-deficient, *hsdR* being restriction-deficient, and *hsdS* lacking both functions. Strains used for cloning are usually *hsdR* or *hsdS* so that there is no possibility of *EcoK* restriction sites in the cloned DNA being cleaved.

   More recently other restriction systems (McrA, McrB, and Mrr) have been discovered in *E. coli*. The status of many strains with respect to these systems is uncertain (1).

1.5.3 Suppressor mutations

SupE and *SupF* are amber suppressor mutations that result in insertion of glutamine and tyrosine respectively at UAG codons. These mutations provide a form of biological containment in that phages carrying amber mutations can survive only in a suppressing strain.

1.5.4 Lac selection

Plasmids carrying the *lacZ'* gene require special hosts. Usually these carry a chromosomal deletion called Δ(*lac-proAB*), which spans the lactose operon and surrounding region. The deletion is partially complemented by an engineered F' plasmid, which carries *proAB*$^+$ (rescuing proline auxotrophy), and *lacZ*ΔM15, which is the *lac* operon minus the *lacZ'* segment. The F' plasmid also carries *lacI^q*, a mutation in the *lac* operator that results in over-expression of the operon.

The strains therefore carry all of the sequence information needed for utilization of lactose except for the α-peptide coded by *lacZ'*.

The presence of the *proAB*$^+$ genes on the F' plasmid means that retention of F' can be selected for by maintaining the bacteria on a proline-deficient minimal medium. This is important because F' bacteria are unstable and can revert to F$^-$ if stored for long periods.

Reference

1. Brown, T.A. (1991). *Molecular Biology Labfax*. BIOS, Oxford.

A4

Recipes and general procedures

T. A. BROWN

1. *Escherichia coli* culture media

1.1 Liquid media

Recipes for 1 litre of media:

DYT: 16 g bacto-tryptone, 10 g bacto-yeast extract, 5 g NaCl.

LB: 10 g bacto-tryptone, 5 g bacto-yeast extract, 10 g NaCl.

Nutrient broth: 25 g bacto-nutrient broth.

SOB: 20 g bacto-tryptone, 5 g bacto-yeast extract, 0.5 g NaCl.

YT: 8 g bacto-tryptone, 5 g bacto-yeast extract, 5 g NaCl.

With all media the pH should be checked and if necessary adjusted to 7.0–7.2 with NaOH. Media should be sterilized by autoclaving at 121°C, 15 lb in^{-1}, for 20 min.

1.2 Solid media

To prepare solid media add the required amount of bacto-agar before autoclaving. For agar plates add 15 g agar per litre. For top agar add 6 g agar per litre.

LTS is DYT plus 6 g agar per litre. YTS is YT plus 6 g agar per litre.

1.3 Supplements

1.3.1 Supplements for λ phage

Maltose: Prepare a 20% (w/v) stock solution and filter-sterilize. Store at room temperature and add to autoclaved media (final concentration 0.2%) immediately before use.

MgSO$_4$: Prepare a 1 M stock solution, autoclave and add to autoclaved media (final concentration usually 10 mM) immediately before use.

1.3.2 Antibiotics

All antibiotics must be prepared as filter-sterilized stock solutions and added to autoclaved media immediately before use. Agar media must be cooled to 50°C before antibiotic supplementation as many are heat-sensitive.

Ampicillin
    stock: 50 mg ml^{-1} in water
    working concentration: 40–60 μg ml^{-1} (plates)
                                   25 μg ml^{-1} (broth)

Kanamycin
    stock: 35 mg ml^{-1} in water
    working concentration: 25–50 μg ml^{-1} (plates)
                                   25–70 μg ml^{-1} (broth)

Tetracycline
    stock: 5 mg ml^{-1} in ethanol
    working concentration: 15 μg ml^{-1} (plates)
                                   7.5 μg ml^{-1} (broth)

Note that tetracycline is light-sensitive and all solutions and plates should be wrapped in foil or stored in the dark. Magnesium ions antagonize the activity of tetracycline.

2. Buffers

Phosphate-buffered saline (PbS): 8.0 g NaCl, 0.34 g KH_2PO_4, 1.21 g K_2HPO_4 per 1 litre. pH should be 7.3. Sterilize by autoclaving.

SM (λ storage and dilution): 5.8 g NaCl, 2 g $MgSO_4.7H_2O$, 50 ml 1 M Tris–HCl pH 7.5, 5 ml 2% (w/v) gelatin solution, water to make 1 litre. Sterilize by autoclaving.

TE buffers: 10 mM Tris–HCl, 1 mM EDTA (pH 8.0). The pH of the Tris–HCl determines the pH of the TE buffer.

3. Preparation of RNase A

Commercial supplies of RNase A usually contain DNase activity. To prepare a DNase-free stock solution:

1. Dissolve solid RNase A to 10 mg ml^{-1} in 10 mM Tris–HCl pH 7.5, 15 mM NaCl.

2. Heat to 100°C for 15 min.

3. Cool slowly to room temperature. Store in aliquots at -20°C.

4. Preparation of siliconized glassware

DNA will adsorb to non-siliconized glassware, resulting in substantial losses. To prepare siliconized glassware:

1. Rinse the glassware in dimethyldichlorosilane (2% solution in 1,1,1-trichloroethane). The solution can be re-used several times. Avoid contact with the skin. Use a fume hood.

2. Allow the glassware to dry thoroughly (up to 2 h).

3. Rinse each item twice in double-distilled water.

Glass wool can be treated in exactly the same way. Plasticware does not usually need to be siliconized though some applications require it.

If a dessicator can be dedicated to siliconization then instead of step 1 evaporate 1 ml of siliconizing solution under vacuum inside the dessicator with the glassware to be treated.

WARNING: Siliconizing solutions are toxic and flammable.

Restriction and methylation

T. A. BROWN

1. Restriction endonucleases

Almost 1500 restriction endonucleases are now known and over 150 are commercially-available. Complete lists plus details of restriction sites and reaction conditions have been published (1).

2. Site-specific methylases

At the last count 117 site-specific methylases had been characterized. A full list including recognition sequences and other relevant information is available (1).

3. Restriction fragment patterns

Restriction fragments are frequently used as size markers for agarose and polyacrylamide gel electrophoresis. *Figures 1–6* are restriction endonuclease digests that have been computer-generated assuming that all digests are run in 1.4% agarose. The scale on the left is in base pairs. The data were kindly supplied by C. Keller, P. Myers, and R.J. Roberts, Cold Spring Harbor Laboratories.

Reference

1. Brown, T.A. (1991). *Molecular Biology Labfax*. BIOS, Oxford.

Restriction and methylation

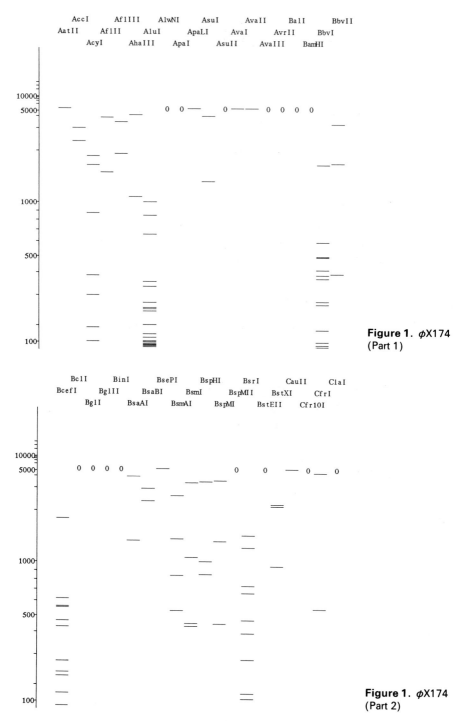

Figure 1. φX174 (Part 1)

Figure 1. φX174 (Part 2)

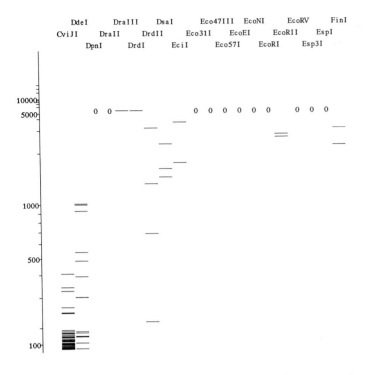

Figure 1. φX174
(Part 3)

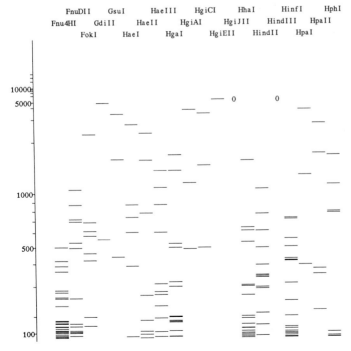

Figure 1. φX174
(Part 4)

259

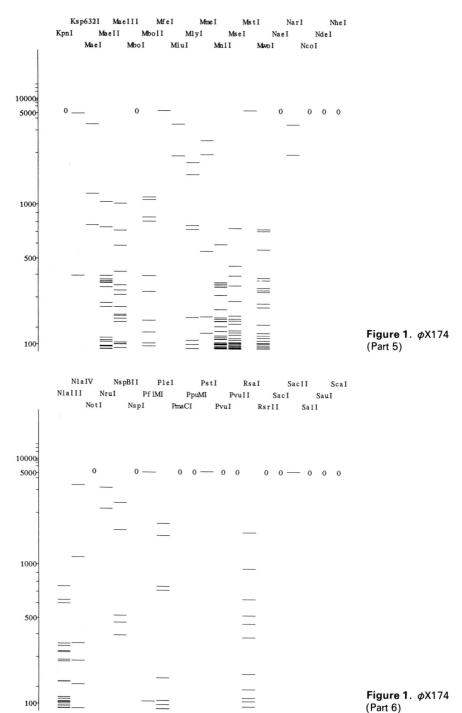

Figure 1. φX174
(Part 5)

Figure 1. φX174
(Part 6)

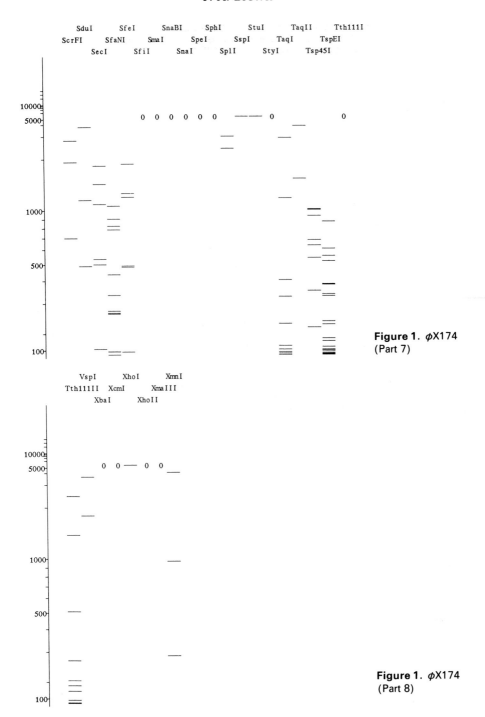

Figure 1. φX174
(Part 7)

Figure 1. φX174
(Part 8)

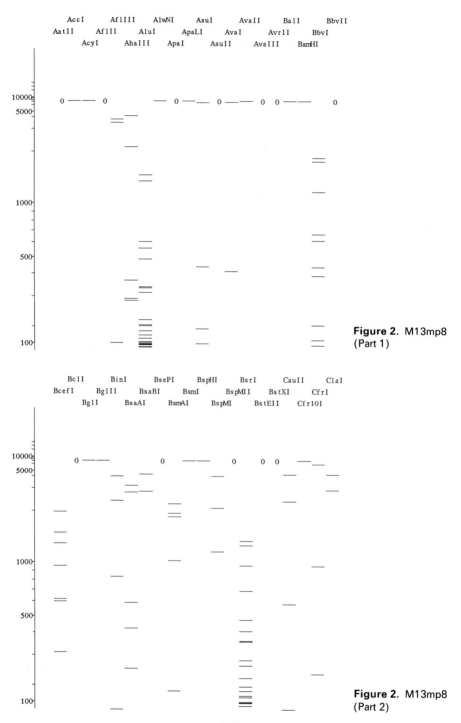

Figure 2. M13mp8
(Part 1)

Figure 2. M13mp8
(Part 2)

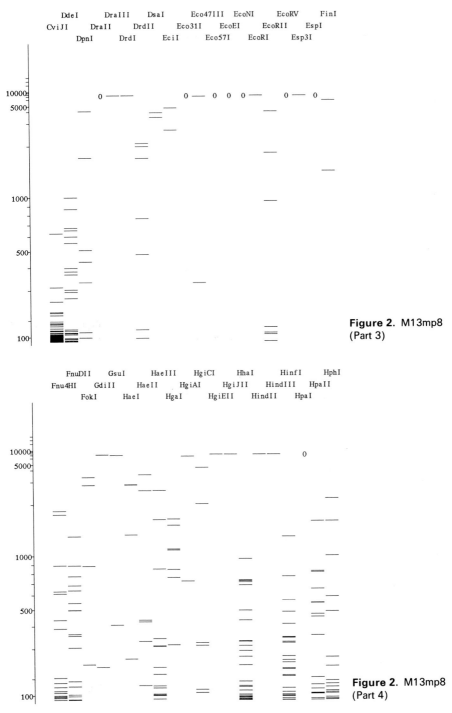

Figure 2. M13mp8
(Part 3)

Figure 2. M13mp8
(Part 4)

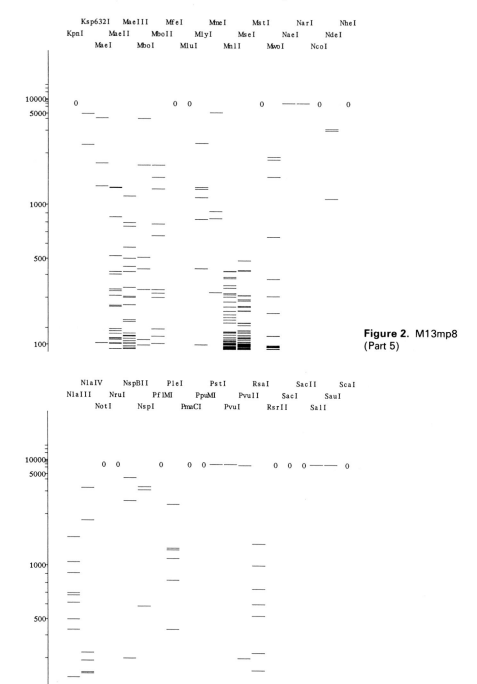

Figure 2. M13mp8 (Part 5)

Figure 2. M13mp8 (Part 6)

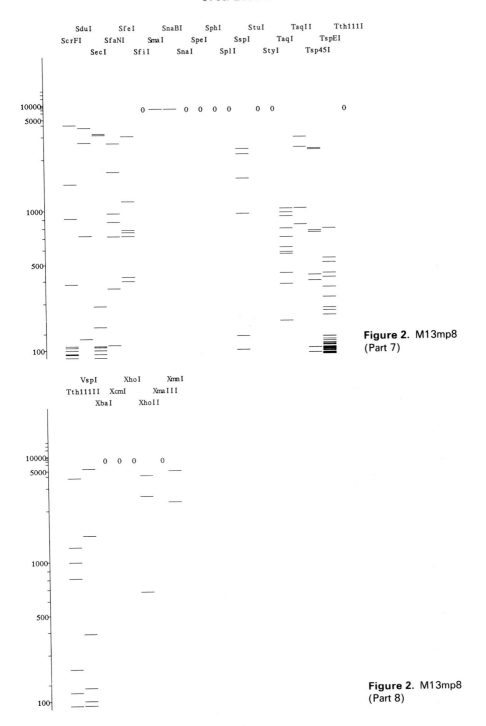

Figure 2. M13mp8
(Part 7)

Figure 2. M13mp8
(Part 8)

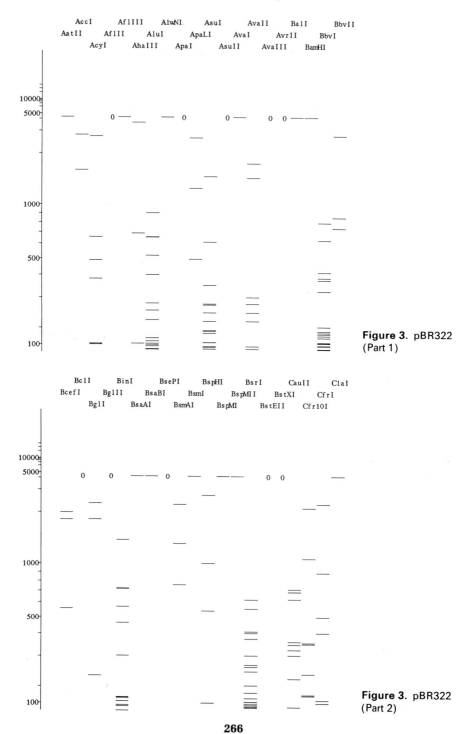

Figure 3. pBR322 (Part 1)

Figure 3. pBR322 (Part 2)

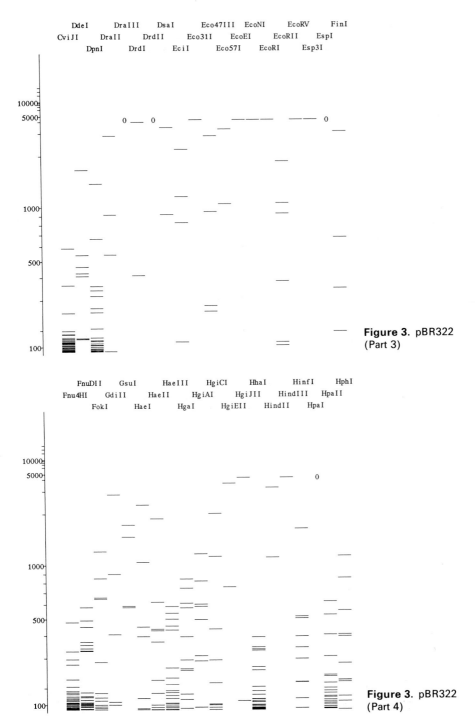

Figure 3. pBR322 (Part 3)

Figure 3. pBR322 (Part 4)

267

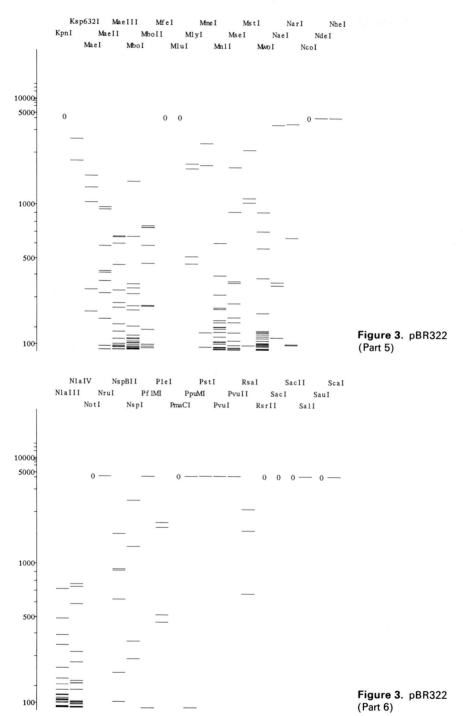

Figure 3. pBR322 (Part 5)

Figure 3. pBR322 (Part 6)

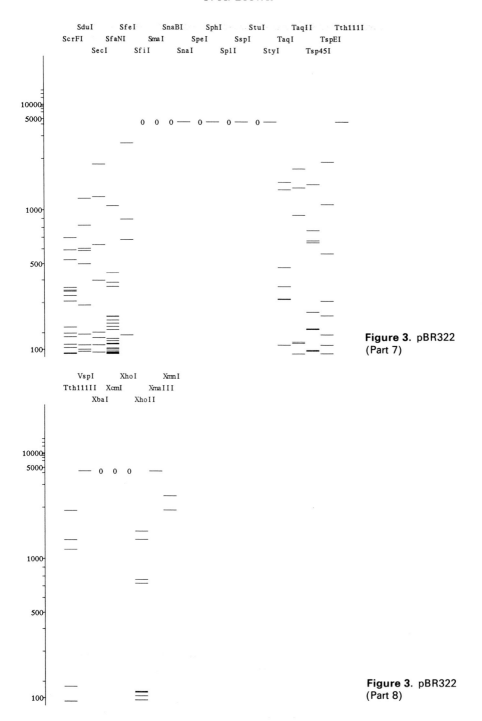

Figure 3. pBR322 (Part 7)

Figure 3. pBR322 (Part 8)

Restriction and methylation

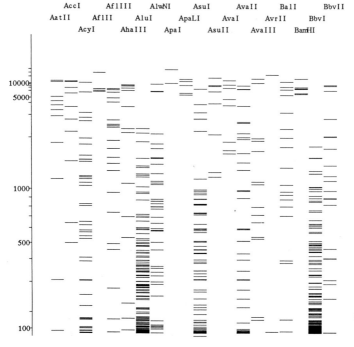

Figure 4. Phage λ (Part 1)

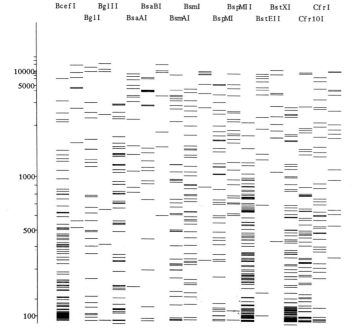

Figure 4. Phage λ (Part 2)

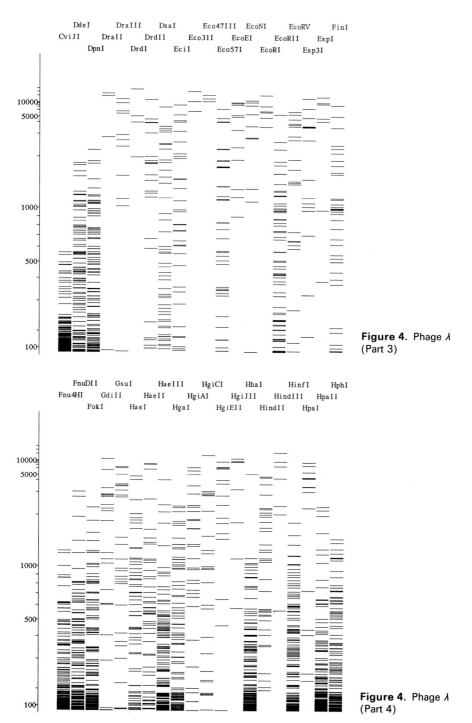

Figure 4. Phage λ
(Part 3)

Figure 4. Phage λ
(Part 4)

271

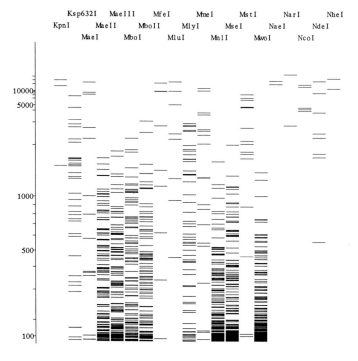

Figure 4. Phage *λ*
(Part 5)

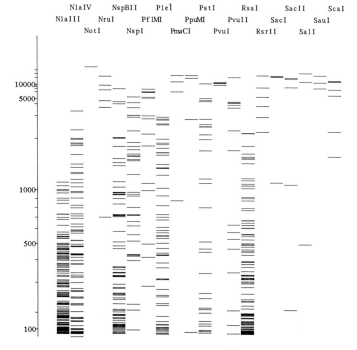

Figure 4. Phage *λ*
(Part 6)

272

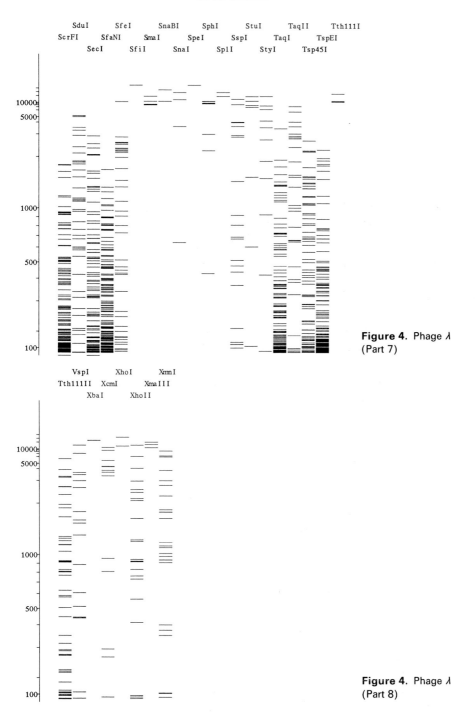

Figure 4. Phage λ
(Part 7)

Figure 4. Phage λ
(Part 8)

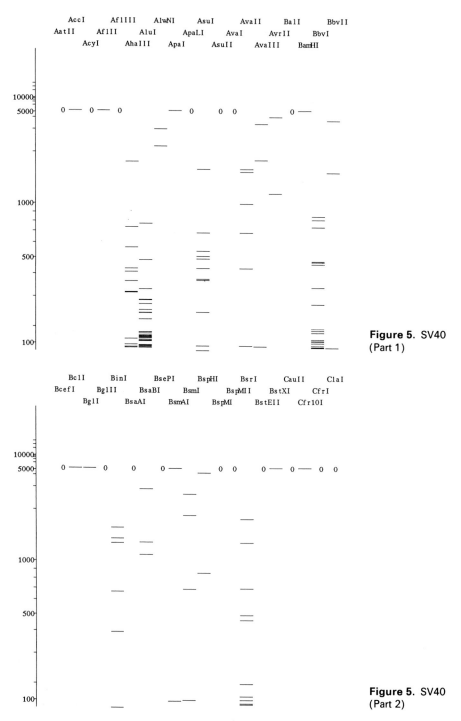

Figure 5. SV40 (Part 1)

Figure 5. SV40 (Part 2)

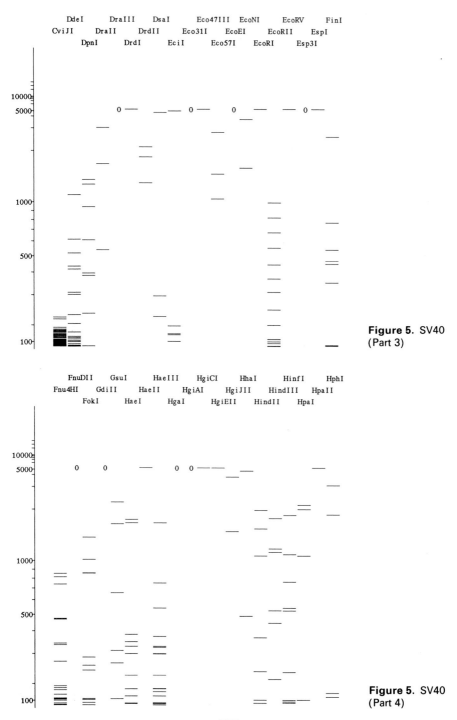

Figure 5. SV40 (Part 3)

Figure 5. SV40 (Part 4)

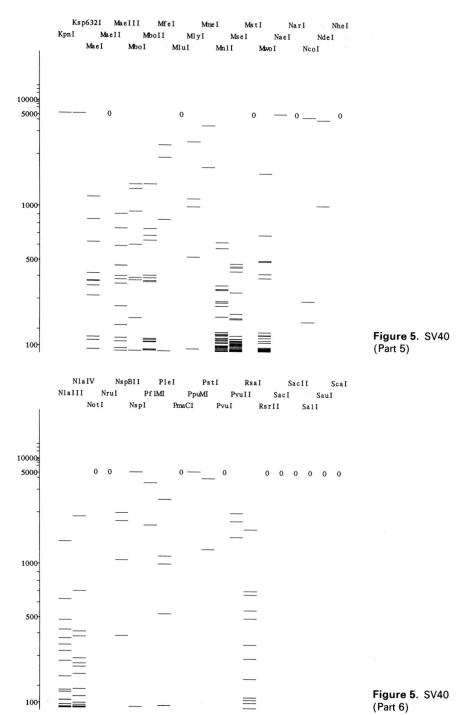

Figure 5. SV40 (Part 5)

Figure 5. SV40 (Part 6)

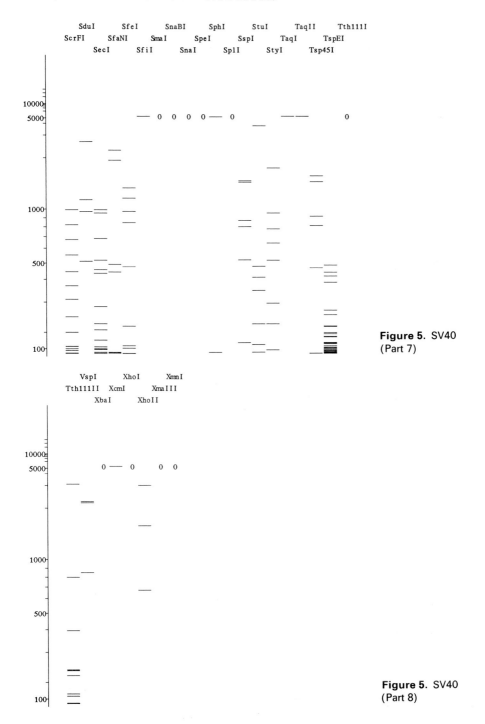

Figure 5. SV40 (Part 7)

Figure 5. SV40 (Part 8)

Restriction and methylation

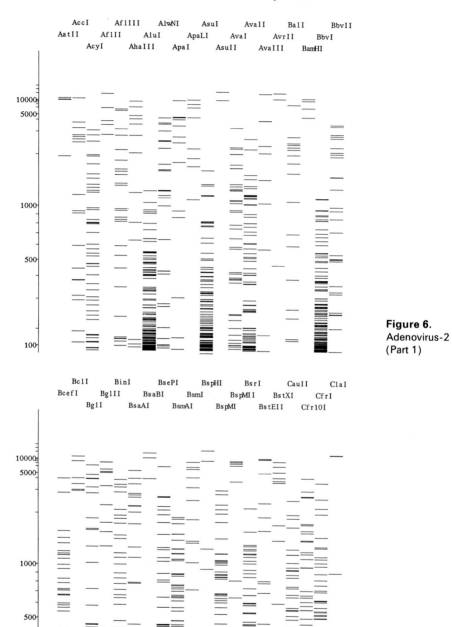

Figure 6.
Adenovirus-2
(Part 1)

Figure 6.
Adenovirus-2
(Part 2)

T. A. Brown

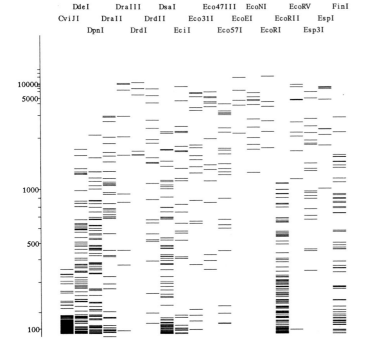

Figure 6.
Adenovirus-2
(Part 3)

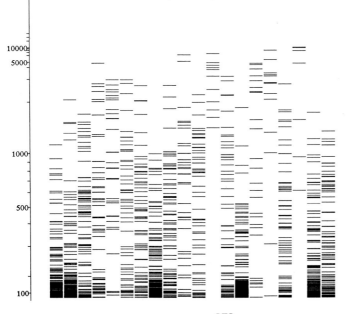

Figure 6.
Adenovirus-2
(Part 4)

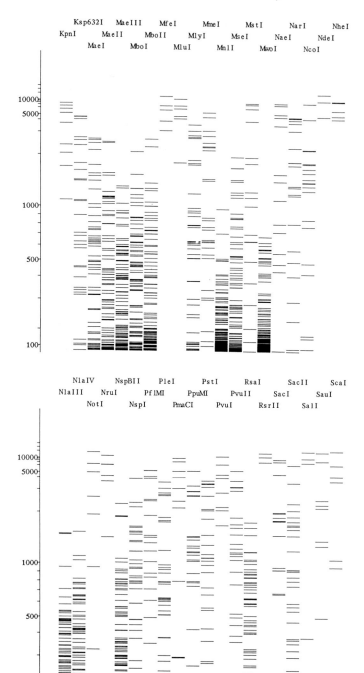

Figure 6.
Adenovirus-2
(Part 5)

Figure 6.
Adenovirus-2
(Part 6)

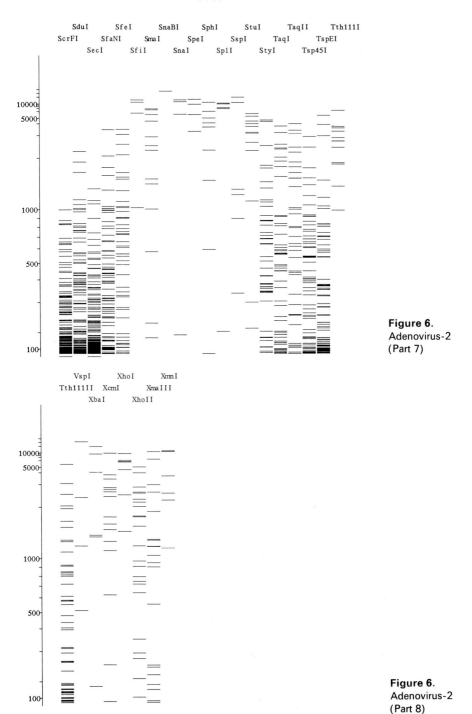

Figure 6.
Adenovirus-2
(Part 7)

Figure 6.
Adenovirus-2
(Part 8)

DNA and RNA modification enzymes

T. A. BROWN

Many different enzymes for DNA and RNA manipulations are now available from commercial suppliers. A comprehensive list is provided in ref. 1 and most are mentioned in various chapters of this book. The following is a summary.

1. DNA polymerases

DNA polymerase I (pol I) possesses a $5' \rightarrow 3'$ DNA-dependent DNA polymerase activity as well as $5' \rightarrow 3'$ and $3' \rightarrow 5'$ exonuclease activities. It is used for DNA labelling by nick translation and for second-strand cDNA synthesis.

Klenow polymerase is derived from pol I by cleavage of the enzyme segment responsible for the $5' \rightarrow 3'$ exonuclease. The enzyme is used for a number of applications, notably DNA sequencing, DNA labelling by random priming or end-filling, and conversion of 5'-overhangs to blunt ends.

Special DNA polymerases for sequencing include modified versions of the Klenow polymerase, and modified and unmodified forms of the T7 DNA polymerase. The ideal is an enzyme that is rapid, error-free, and has high processivity.

T4 DNA polymerase has a highly-active $3' \rightarrow 5'$ exonuclease which enables it to carry out an end-replacement reaction that is useful in DNA labelling. It is also used to convert 3'-overhangs into blunt ends.

***Taq* DNA polymerase** and similar enzymes are heat-stable and so can be used in PCR and certain specialized DNA sequencing applications.

Reverse transcriptases are obtained from avian myeloblastosis virus (AMV) or Moloney murine leukaemia virus (M-MuLV). They are RNA-dependent DNA polymerases, so copy an RNA template into DNA, the basis to cDNA synthesis.

2. RNA polymerases

Promoter-specific RNA polymerases include the SP6, T3, and T7 enzymes. They have a high-specificity for their target promoter sequences and so are used for the *in vitro* synthesis of large amounts of RNA from cloned genes.

3. Nucleases

Bal 31 nuclease possesses a complex set of activities that enable it to progressively shorten double-stranded blunt-ended molecules.

S1 nuclease is single-strand specific and more active on DNA than RNA, so can be used to trim non-hybridized regions from DNA–RNA duplexes. Its use in S1 nuclease mapping allows the termini of transcripts to be located, along with intron boundaries.

RNase H degrades the RNA component of a DNA-RNA hybrid and plays an important role in cDNA synthesis.

DNase I is an endodeoxyribonuclease that has a number of important applications. It can be used to introduce nicks in double-stranded DNA molecules prior to labelling by nick translation, and can detect protein binding sites in nuclease protection experiments.

4. Ligases

The **T4 DNA ligase** is usually used in construction of recombinant DNA molecules, but the **E. coli ligase** (which requires NAD rather than ATP as the co-factor) is also used. There is also a **T4 RNA ligase**.

5. End-modification enzymes

Alkaline phosphatase removes 5′-phosphate groups from single- and double-stranded DNA molecules, which prevents self-ligation.

T4 polynucleotide kinase adds phosphates to 5′-hydroxyl termini, which may be necessary before linkers or adapters can be used in cloning experiments.

Terminal deoxynucleotidyl transferase will add a single-stranded tail on to a (usually) blunt-ended molecule, the basis to homopolymer tailing.

Reference

1. Brown, T.A. (1991). *Molecular Biology Labfax*. BIOS, Oxford.

Commercial suppliers

T. A. BROWN

Amersham International, Lincoln Place, Green End, Aylesbury, Buckingham-shire HP20 2TP, UK.
Applied Biosystems, 850 Lincoln Centre Drive, Foster City, CA 94404, USA.
Appligene, 74 Route du Rhin, BP 72, Illkirch Cedex, France.
Beckman, PO Box 10200, Palo Alto, CA 94304, USA.
BIO 101, Box 2284, La Jolla, CA 92038-2284, USA.
Bio-Rad, 3300 Regatta Boulevard, Richmond, CA 94804, USA.
Biotechnology Research Enterprises, GPO Box 498, Adelaide, South Australia 5001, Australia.
Boehringer-Mannheim, PO Box 310 120, D-6800 Mannheim 31, Germany.
Calbiochem, PO Box 12087, San Diego, CA 92112-4180, USA.
Clontech Laboratories, 4030 Fabian Way, Palo Alto, CA 94303, USA.
Du Pont Nemours, Postfach 401240, 6072 Dreieich, Germany.
FMC, 5 Maple Street, Rockland, ME 04841, USA.
Gelman, 600 South Wagner Road, Ann Arbor, MI 48106, USA.
Gen-Probe, 9620 Chesapeake Drive, San Diego, CA 92123, USA.
Gibco-BRL, PO Box 6009, Gaithersburg, MD 20877, USA.
Gilson, 3000 West Beltline Highway, PO Box 27, Middleton, WI 53562, USA.
Hoefer, PO Box 77387, 654 Minnesota Street, San Francisco, CA 94107, USA.
Hybaid, 111–113 Waldegrave Road, Teddington, Middlesex TW11 8LL, UK.
International Biotechnologies, PO Box 9558, 25 Science Park, New Haven, CT 06535, USA.
Medac, Fehlandtstrasse 3, D2000 Hamburg 36, Germany.
Miles Laboratories, 1127 Myrtle Street, Elkhart, Indiana 46514, USA.
New England Biolabs, 32 Tozer Road, Beverly, MA 01915-5510, USA.
New England Nuclear, 548 Albany Street, Boston, MA 02118, USA.
Perkin Elmer Cetus, 761 Main Avenue, Norwalk, CT 06859, USA.
Pharmacia-LKB, Bjorkgatan 30, 751 82 Uppsala, Sweden.
Promega, 2800 Woods Hollow Road, Madison, WI 53711, USA.
Qiagen, Studio City, CA 91604, USA.
Renner, Riedstrasse 6, D6701 Darmstadt, Germany.
Sartorius, 140 Wilbur Place, Bohemia, NY 11716, USA.
Savant, 110–113 Bi-county Boulevard, Farmingdale, NY 11735, USA.

Schleicher and Schuell, PO Box 4, D3354 Dassel, Germany.
Sigma, PO Box 14508, St Louis, MI 63178, USA.
Stratagene, 11099 North Torrey Pines Road, La Jolla, CA 92037, USA.
Stehelin, Spalentorweg 62, 4003 Basel, Switzerland.
Techne, 3700 Brunswick Pike, Princeton, NJ 08540-6192, USA.
United States Biochemical, PO Box 22400, Cleveland, Ohio 44122, USA.

Index

ORDER OTHER TITLES OF INTEREST TODAY

Price list for: UK, Europe, Rest of World (excluding US and Canada)

138.	Plasmids (2/e) Hardy, K.G. (Ed)		
......	Spiralbound hardback	0-19-963445-9	£30.00
......	Paperback	0-19-963444-0	£19.50
136.	RNA Processing: Vol. II Higgins, S.J. & Hames, B.D. (Eds)		
......	Spiralbound hardback	0-19-963471-8	£30.00
......	Paperback	0-19-963470-X	£19.50
135.	RNA Processing: Vol. I Higgins, S.J. & Hames, B.D. (Eds)		
......	Spiralbound hardback	0-19-963344-4	£30.00
......	Paperback	0-19-963343-6	£19.50
134.	NMR of Macromolecules Roberts, G.C.K. (Ed)		
......	Spiralbound hardback	0-19-963225-1	£32.50
......	Paperback	0-19-963224-3	£22.50
133.	Gas Chromatography Baugh, P. (Ed)		
......	Spiralbound hardback	0-19-963272-3	£40.00
......	Paperback	0-19-963271-5	£27.50
132.	Essential Developmental Biology Stern, C.D. & Holland, P.W.H. (Eds)		
......	Spiralbound hardback	0-19-963423-8	£30.00
......	Paperback	0-19-963422-X	£19.50
131.	Cellular Interactions in Development Hartley, D.A. (Ed)		
......	Spiralbound hardback	0-19-963391-6	£30.00
......	Paperback	0-19-963390-8	£18.50
129	Behavioural Neuroscience: Volume II Sahgal, A. (Ed)		
......	Spiralbound hardback	0-19-963458-0	£32.50
......	Paperback	0-19-963457-2	£22.50
128	Behavioural Neuroscience: Volume I Sahgal, A. (Ed)		
......	Spiralbound hardback	0-19-963368-1	£32.50
......	Paperback	0-19-963367-3	£22.50
127.	Molecular Virology Davison, A.J. & Elliott, R.M. (Eds)		
......	Spiralbound hardback	0-19-963358-4	£35.00
......	Paperback	0-19-963357-6	£25.00
126.	Gene Targeting Joyner, A.L. (Ed)		
......	Spiralbound hardback	0-19-963407-6	£30.00
......	Paperback	0-19-9634036-8	19.50
125.	Glycobiology Fukuda, M. & Kobata, A. (Eds)		
......	Spiralbound hardback	0-19-963372-X	£32.50
......	Paperback	0-19-963371-1	£22.50
124.	Human Genetic Disease Analysis (2/e) Davies, K.E. (Ed)		
......	Spiralbound hardback	0-19-963309-6	£30.00
......	Paperback	0-19-963308-8	£18.50
122.	Immunocytochemistry Beesley, J. (Ed)		
......	Spiralbound hardback	0-19-963270-7	£35.00
......	Paperback	0-19-963269-3	£22.50
123.	Protein Phosphorylation Hardie, D.G. (Ed)		
......	Spiralbound hardback	0-19-963306-1	£32.50
......	Paperback	0-19-963305-3	£22.50
121.	Tumour Immunobiology Gallagher, G., Rees, R.C. & others (Eds)		
......	Spiralbound hardback	0-19-963370-3	£40.00
......	Paperback	0-19-963369-X	£27.50
120.	Transcription Factors Latchman, D.S. (Ed)		
......	Spiralbound hardback	0-19-963342-8	£30.00
......	Paperback	0-19-963341-X	£19.50
119.	Growth Factors McKay, I. & Leigh, I. (Eds)		
......	Spiralbound hardback	0-19-963360-6	£30.00
......	Paperback	0-19-963359-2	£19.50
118.	Histocompatibility Testing Dyer, P. & Middleton, D. (Eds)		
......	Spiralbound hardback	0-19-963364-9	£32.50
......	Paperback	0-19-963363-0	£22.50
117.	Gene Transcription Hames, B.D. & Higgins, S.J. (Eds)		
......	Spiralbound hardback	0-19-963292-8	£35.00
......	Paperback	0-19-963291-X	£25.00
116.	Electrophysiology Wallis, D.I. (Ed)		
......	Spiralbound hardback	0-19-963348-7	£32.50
......	Paperback	0-19-963347-9	£22.50
115.	Biological Data Analysis Fry, J.C. (Ed)		
......	Spiralbound hardback	0-19-963340-1	£50.00
......	Paperback	0-19-963339-8	£27.50
114.	Experimental Neuroanatomy Bolam, J.P. (Ed)		
......	Spiralbound hardback	0-19-963326-6	£32.50
......	Paperback	0-19-963325-8	£22.50
113.	Preparative Centrifugation Rickwood, D. (Ed)		
......	Spiralbound hardback	0-19-963208-1	£45.00
......	Paperback	0-19-963211-1	£25.00
......	Paperback	0-19-963099-2	£25.00
112.	Lipid Analysis Hamilton, R.J. & Hamilton, Shiela (Eds)		
......	Spiralbound hardback	0-19-963098-4	£35.00
......	Paperback	0-19-963099-2	£25.00
111.	Haemopoiesis Testa, N.G. & Molineux, G. (Eds)		
......	Spiralbound hardback	0-19-963366-5	£32.50
......	Paperback	0-19-963365-7	£22.50
110.	Pollination Ecology Dafni, A.		
......	Spiralbound hardback	0-19-963299-5	£32.50
......	Paperback	0-19-963298-7	£22.50
109.	In Situ Hybridization Wilkinson, D.G. (Ed)		
......	Spiralbound hardback	0-19-963328-2	£30.00
......	Paperback	0-19-963327-4	£18.50
108.	Protein Engineering Rees, A.R., Sternberg, M.J.E. & others (Eds)		
......	Spiralbound hardback	0-19-963139-5	£35.00
......	Paperback	0-19-963138-7	£25.00
107.	Cell-Cell Interactions Stevenson, B.R., Gallin, W.J. & others (Eds)		
......	Spiralbound hardback	0-19-963319-3	£32.50
......	Paperback	0-19-963318-5	£22.50
106.	Diagnostic Molecular Pathology: Volume I Herrington, C.S. & McGee, J. O'D. (Eds)		
......	Spiralbound hardback	0-19-963237-5	£30.00
......	Paperback	0-19-963236-7	£19.50
105.	Biomechanics-Materials Vincent, J.F.V. (Ed)		
......	Spiralbound hardback	0-19-963223-5	£35.00
......	Paperback	0-19-963222-7	£25.00
104.	Animal Cell Culture (2/e) Freshney, R.I. (Ed)		
......	Spiralbound hardback	0-19-963212-X	£30.00
......	Paperback	0-19-963213-8	£19.50
103.	Molecular Plant Pathology: Volume II Gurr, S.J., McPherson, M.J. & others (Eds)		
......	Spiralbound hardback	0-19-963352-5	£32.50
......	Paperback	0-19-963351-7	£22.50
102	Signal Transduction Milligan, G. (Ed)		
......	Spiralbound hardback	0-19-963296-0	£30.00
......	Paperback	0-19-963295-2	£18.50
101.	Protein Targeting Magee, A.I. & Wileman, T. (Eds)		
......	Spiralbound hardback	0-19-963206-5	£32.50
......	Paperback	0-19-963210-3	£22.50
100.	Diagnostic Molecular Pathology: Volume II: Cell and Tissue Genotyping Herrington, C.S. & McGee, J.O'D. (Eds)		
......	Spiralbound hardback	0-19-963239-1	£30.00
......	Paperback	0-19-963238-3	£19.50
99.	Neuronal Cell Lines Wood, J.N. (Ed)		
......	Spiralbound hardback	0-19-963346-0	£32.50
......	Paperback	0-19-963345-2	£22.50

98. **Neural Transplantation** Dunnett, S.B. & Björklund, A. (Eds)
...... Spiralbound hardback 0-19-963286-3 **£30.00**
...... Paperback 0-19-963285-5 **£19.50**

97. **Human Cytogenetics: Volume II: Malignancy and Acquired Abnormalities (2/e)** Rooney, D.E. & Czepulkowski, B.H. (Eds)
...... Spiralbound hardback 0-19-963290-1 **£30.00**
...... Paperback 0-19-963289-8 **£22.50**

96. **Human Cytogenetics: Volume I: Constitutional Analysis (2/e)** Rooney, D.E. & Czepulkowski, B.H. (Eds)
...... Spiralbound hardback 0-19-963288-X **£30.00**
...... Paperback 0-19-963287-1 **£22.50**

95. **Lipid Modification of Proteins** Hooper, N.M. & Turner, A.J. (Eds)
...... Spiralbound hardback 0-19-963274-X **£32.50**
...... Paperback 0-19-963273-1 **£22.50**

94. **Biomechanics-Structures and Systems** Biewener, A.A. (Ed)
...... Spiralbound hardback 0-19-963268-5 **£42.50**
...... Paperback 0-19-963267-7 **£25.00**

93. **Lipoprotein Analysis** Converse, C.A. & Skinner, E.R. (Eds)
...... Spiralbound hardback 0-19-963192-1 **£30.00**
...... Paperback 0-19-963231-6 **£19.50**

92. **Receptor-Ligand Interactions** Hulme, E.C. (Ed)
...... Spiralbound hardback 0-19-963090-9 **£35.00**
...... Paperback 0-19-963091-7 **£27.50**

91. **Molecular Genetic Analysis of Populations** Hoelzel, A.R. (Ed)
...... Spiralbound hardback 0-19-963278-2 **£32.50**
...... Paperback 0-19-963277-4 **£22.50**

90. **Enzyme Assays** Eisenthal, R. & Danson, M.J. (Eds)
...... Spiralbound hardback 0-19-963142-5 **£35.00**
...... Paperback 0-19-963143-3 **£25.00**

89. **Microcomputers in Biochemistry** Bryce, C.F.A. (Ed)
...... Spiralbound hardback 0-19-963253-7 **£30.00**
...... Paperback 0-19-963252-9 **£19.50**

88. **The Cytoskeleton** Carraway, K.L. & Carraway, C.A.C. (Eds)
...... Spiralbound hardback 0-19-963257-X **£30.00**
...... Paperback 0-19-963256-1 **£19.50**

87. **Monitoring Neuronal Activity** Stamford, J.A. (Ed)
...... Spiralbound hardback 0-19-963244-8 **£30.00**
...... Paperback 0-19-963243-X **£19.50**

86. **Crystallization of Nucleic Acids and Proteins** Ducruix, A. & Giegé, R. (Eds)
...... Spiralbound hardback 0-19-963245-6 **£35.00**
...... Paperback 0-19-963246-4 **£25.00**

85. **Molecular Plant Pathology: Volume I** Gurr, S.J., McPherson, M.J. & others (Eds)
...... Spiralbound hardback 0-19-963103-4 **£30.00**
...... Paperback 0-19-963102-6 **£19.50**

84. **Anaerobic Microbiology** Levett, P.N. (Ed)
...... Spiralbound hardback 0-19-963204-9 **£32.50**
...... Paperback 0-19-963262-6 **£22.50**

83. **Oligonucleotides and Analogues** Eckstein, F. (Ed)
...... Spiralbound hardback 0-19-963280-4 **£32.50**
...... Paperback 0-19-963279-0 **£22.50**

82. **Electron Microscopy in Biology** Harris, R. (Ed)
...... Spiralbound hardback 0-19-963219-7 **£32.50**
...... Paperback 0-19-963215-4 **£22.50**

81. **Essential Molecular Biology: Volume II** Brown, T.A. (Ed)
...... Spiralbound hardback 0-19-963112-3 **£32.50**
...... Paperback 0-19-963113-1 **£22.50**

80. **Cellular Calcium** McCormack, J.G. & Cobbold, P.H. (Eds)
...... Spiralbound hardback 0-19-963131-X **£35.00**
...... Paperback 0-19-963130-1 **£25.00**

79. **Protein Architecture** Lesk, A.M.
...... Spiralbound hardback 0-19-963054-2 **£32.50**
...... Paperback 0-19-963055-0 **£22.50**

78. **Cellular Neurobiology** Chad, J. & Wheal, H. (Eds)
...... Spiralbound hardback 0-19-963106-9 **£32.50**
...... Paperback 0-19-963107-7 **£22.50**

77. **PCR** McPherson, M.J., Quirke, P. & others (Eds)
...... Spiralbound hardback 0-19-963226-X **£30.00**
...... Paperback 0-19-963196-4 **£19.50**

76. **Mammalian Cell Biotechnology** Butler, M. (Ed)
...... Spiralbound hardback 0-19-963207-3 **£30.00**
...... Paperback 0-19-963209-X **£19.50**

75. **Cytokines** Balkwill, F.R. (Ed)
...... Spiralbound hardback 0-19-963218-9 **£35.00**
...... Paperback 0-19-963214-6 **£25.00**

74. **Molecular Neurobiology** Chad, J. & Wheal, H. (Eds)
...... Spiralbound hardback 0-19-963108-5 **£30.00**
...... Paperback 0-19-963109-3 **£19.50**

73. **Directed Mutagenesis** McPherson, M.J. (Ed)
...... Spiralbound hardback 0-19-963141-7 **£30.00**
...... Paperback 0-19-963140-9 **£19.50**

72. **Essential Molecular Biology: Volume I** Brown, T.A. (Ed)
...... Spiralbound hardback 0-19-963110-7 **£32.50**
...... Paperback 0-19-963111-5 **£22.50**

71. **Peptide Hormone Action** Siddle, K. & Hutton, J.C.
...... Spiralbound hardback 0-19-963070-4 **£32.50**
...... Paperback 0-19-963071-2 **£22.50**

70. **Peptide Hormone Secretion** Hutton, J.C. & Siddle, K. (Eds)
...... Spiralbound hardback 0-19-963068-2 **£35.00**
...... Paperback 0-19-963069-0 **£25.00**

69. **Postimplantation Mammalian Embryos** Copp, A.J. & Cockroft, D.L. (Eds)
...... Spiralbound hardback 0-19-963088-7 **£15.00**
...... Paperback 0-19-963089-5 **£12.50**

68. **Receptor-Effector Coupling** Hulme, E.C. (Ed)
...... Spiralbound hardback 0-19-963094-1 **£30.00**
...... Paperback 0-19-963095-X **£19.50**

67. **Gel Electrophoresis of Proteins (2/e)** Hames, B.D. & Rickwood, D. (Eds)
...... Spiralbound hardback 0-19-963074-7 **£35.00**
...... Paperback 0-19-963075-5 **£25.00**

66. **Clinical Immunology** Gooi, H.C. & Chapel, H. (Eds)
...... Spiralbound hardback 0-19-963086-0 **£32.50**
...... Paperback 0-19-963087-9 **£22.50**

65. **Receptor Biochemistry** Hulme, E.C. (Ed)
...... Paperback 0-19-963093-3 **£25.00**

64. **Gel Electrophoresis of Nucleic Acids (2/e)** Rickwood, D. & Hames, B.D. (Eds)
...... Spiralbound hardback 0-19-963082-8 **£32.50**
...... Paperback 0-19-963083-6 **£22.50**

63. **Animal Virus Pathogenesis** Oldstone, M.B.A. (Ed)
...... Spiralbound hardback 0-19-963100-X **£15.00**
...... Paperback 0-19-963101-8 **£12.50**

62. **Flow Cytometry** Ormerod, M.G. (Ed)
...... Paperback 0-19-963053-4 **£22.50**

61. **Radioisotopes in Biology** Slater, R.J. (Ed)
...... Spiralbound hardback 0-19-963080-1 **£32.50**
...... Paperback 0-19-963081-X **£22.50**

60. **Biosensors** Cass, A.E.G. (Ed)
...... Spiralbound hardback 0-19-963046-1 **£30.00**
...... Paperback 0-19-963047-X **£19.50**

59. **Ribosomes and Protein Synthesis** Spedding, G. (Ed)
...... Spiralbound hardback 0-19-963104-2 **£15.00**
...... Paperback 0-19-963105-0 **£12.50**

58. **Liposomes** New, R.R.C. (Ed)
...... Spiralbound hardback 0-19-963076-3 **£35.00**
...... Paperback 0-19-963077-1 **£22.50**

57. **Fermentation** McNeil, B. & Harvey, L.M. (Eds)
...... Spiralbound hardback 0-19-963044-5 **£30.00**
...... Paperback 0-19-963045-3 **£19.50**

56. **Protein Purification Applications** Harris, E.L.V. & Angal, S. (Eds)
...... Spiralbound hardback 0-19-963022-4 **£30.00**
...... Paperback 0-19-963023-2 **£18.50**

55. **Nucleic Acids Sequencing** Howe, C.J. & Ward, E.S. (Eds)
...... Spiralbound hardback 0-19-963056-9 **£30.00**
...... Paperback 0-19-963057-7 **£19.50**

54. **Protein Purification Methods** Harris, E.L.V. & Angal, S. (Eds)
...... Spiralbound hardback 0-19-963002-X **£30.00**
...... Paperback 0-19-963003-8 **£22.50**

53. **Solid Phase Peptide Synthesis** Atherton, E. & Sheppard, R.C.
...... Spiralbound hardback 0-19-963066-6 **£15.00**
...... Paperback 0-19-963067-4 **£12.50**

52. **Medical Bacteriology** Hawkey, P.M. & Lewis, D.A. (Eds)
...... Paperback 0-19-963009-7 **£25.00**

51. **Proteolytic Enzymes** Beynon, R.J. & Bond, J.S. (Eds)
...... Spiralbound hardback 0-19-963058-5 **£30.00**
...... Paperback 0-19-963059-3 **£19.50**

50. **Medical Mycology** Evans, E.G.V. & Richardson, M.D. (Eds)
...... Spiralbound hardback 0-19-963010-0 **£37.50**
...... Paperback 0-19-963011-9 **£25.00**

49. **Computers in Microbiology** Bryant, T.N. & Wimpenny, J.W.T. (Eds)
...... Paperback 0-19-963015-1 **£12.50**

48.	**Protein Sequencing** Findlay, J.B.C. & Geisow, M.J. (Eds)		
......	Spiralbound hardback	0-19-963012-7	**£15.00**
......	Paperback	0-19-963013-5	**£12.50**
47.	**Cell Growth and Division** Baserga, R. (Ed)		
......	Spiralbound hardback	0-19-963026-7	**£15.00**
......	Paperback	0-19-963027-5	**£12.50**
46.	**Protein Function** Creighton, T.E. (Ed)		
......	Spiralbound hardback	0-19-963006-2	**£32.50**
......	Paperback	0-19-963007-0	**£22.50**
45.	**Protein Structure** Creighton, T.E. (Ed)		
......	Spiralbound hardback	0-19-963000-3	**£32.50**
......	Paperback	0-19-963001-1	**£22.50**
44.	**Antibodies: Volume II** Catty, D. (Ed)		
......	Spiralbound hardback	0-19-963018-6	**£30.00**
......	Paperback	0-19-963019-4	**£19.50**
43.	**HPLC of Macromolecules** Oliver, R.W.A. (Ed)		
......	Spiralbound hardback	0-19-963020-8	**£30.00**
......	Paperback	0-19-963021-6	**£19.50**
42.	**Light Microscopy in Biology** Lacey, A.J. (Ed)		
......	Spiralbound hardback	0-19-963036-4	**£30.00**
......	Paperback	0-19-963037-2	**£19.50**
41.	**Plant Molecular Biology** Shaw, C.H. (Ed)		
......	Paperback	1-85221-056-7	**£12.50**
40.	**Microcomputers in Physiology** Fraser, P.J. (Ed)		
......	Spiralbound hardback	1-85221-129-6	**£15.00**
......	Paperback	1-85221-130-X	**£12.50**
39.	**Genome Analysis** Davies, K.E. (Ed)		
......	Spiralbound hardback	1-85221-109-1	**£30.00**
......	Paperback	1-85221-110-5	**£18.50**
38.	**Antibodies: Volume I** Catty, D. (Ed)		
......	Paperback	0-947946-85-3	**£19.50**
37.	**Yeast** Campbell, I. & Duffus, J.H. (Eds)		
......	Paperback	0-947946-79-9	**£12.50**
36.	**Mammalian Development** Monk, M. (Ed)		
......	Hardback	1-85221-030-3	**£15.00**
......	Paperback	1-85221-029-X	**£12.50**
35.	**Lymphocytes** Klaus, G.G.B. (Ed)		
......	Hardback	1-85221-018-4	**£30.00**
34.	**Lymphokines and Interferons** Clemens, M.J., Morris, A.G. & others (Eds)		
......	Paperback	1-85221-035-4	**£12.50**
33.	**Mitochondria** Darley-Usmar, V.M., Rickwood, D. & others (Eds)		
......	Hardback	1-85221-034-6	**£32.50**
......	Paperback	1-85221-033-8	**£22.50**
32.	**Prostaglandins and Related Substances** Benedetto, C., McDonald-Gibson, R.G. & others (Eds)		
......	Hardback	1-85221-032-X	**£15.00**
......	Paperback	1-85221-031-1	**£12.50**
31.	**DNA Cloning: Volume III** Glover, D.M. (Ed)		
......	Hardback	1-85221-049-4	**£15.00**
......	Paperback	1-85221-048-6	**£12.50**
30.	**Steroid Hormones** Green, B. & Leake, R.E. (Eds)		
......	Paperback	0-947946-53-5	**£19.50**
29.	**Neurochemistry** Turner, A.J. & Bachelard, H.S. (Eds)		
......	Hardback	1-85221-028-1	**£15.00**
......	Paperback	1-85221-027-3	**£12.50**
28.	**Biological Membranes** Findlay, J.B.C. & Evans, W.H. (Eds)		
......	Hardback	0-947946-84-5	**£15.00**
......	Paperback	0-947946-83-7	**£12.50**
27.	**Nucleic Acid and Protein Sequence Analysis** Bishop, M.J. & Rawlings, C.J. (Eds)		
......	Hardback	1-85221-007-9	**£35.00**
......	Paperback	1-85221-006-0	**£25.00**
26.	**Electron Microscopy in Molecular Biology** Sommerville, J. & Scheer, U. (Eds)		
......	Hardback	0-947946-64-0	**£15.00**
......	Paperback	0-947946-54-3	**£12.50**
25.	**Teratocarcinomas and Embryonic Stem Cells** Robertson, E.J. (Ed)		
......	Paperback	1-85221-004-4	**£19.50**
24.	**Spectrophotometry and Spectrofluorimetry** Harris, D.A. & Bashford, C.L. (Eds)		
......	Hardback	0-947946-69-1	**£15.00**
......	Paperback	0-947946-46-2	**£12.50**
23.	**Plasmids** Hardy, K.G. (Ed)		
......	Paperback	0-947946-81-0	**£12.50**
22.	**Biochemical Toxicology** Snell, K. & Mullock, B. (Eds)		
......	Paperback	0-947946-52-7	**£12.50**
19.	**Drosophila** Roberts, D.B. (Ed)		
......	Hardback	0-947946-66-7	**£32.50**
......	Paperback	0-947946-45-4	**£22.50**
17.	**Photosynthesis: Energy Transduction** Hipkins, M.F. & Baker, N.R. (Eds)		
......	Hardback	0-947946-63-2	**£15.00**
......	Paperback	0-947946-51-9	**£12.50**
16.	**Human Genetic Diseases** Davies, K.E. (Ed)		
......	Hardback	0-947946-76-4	**£15.00**
......	Paperback	0-947946-75-6	**£12.50**
14.	**Nucleic Acid Hybridisation** Hames, B.D. & Higgins, S.J. (Eds)		
......	Hardback	0-947946-61-6	**£15.00**
......	Paperback	0-947946-23-3	**£12.50**
13.	**Immobilised Cells and Enzymes** Woodward, J. (Ed)		
......	Hardback	0-947946-60-8	**£15.00**
12.	**Plant Cell Culture** Dixon, R.A. (Ed)		
......	Paperback	0-947946-22-5	**£19.50**
11a.	**DNA Cloning: Volume I** Glover, D.M. (Ed)		
......	Paperback	0-947946-18-7	**£12.50**
11b.	**DNA Cloning: Volume II** Glover, D.M. (Ed)		
......	Paperback	0-947946-19-5	**£12.50**
10.	**Virology** Mahy, B.W.J. (Ed)		
......	Paperback	0-904147-78-9	**£19.50**
9.	**Affinity Chromatography** Dean, P.D.G., Johnson, W.S. & others (Eds)		
......	Paperback	0-904147-71-1	**£19.50**
7.	**Microcomputers in Biology** Ireland, C.R. & Long, S.P. (Eds)		
......	Paperback	0-904147-57-6	**£18.00**
6.	**Oligonucleotide Synthesis** Gait, M.J. (Ed)		
......	Paperback	0-904147-74-6	**£18.50**
5.	**Transcription and Translation** Hames, B.D. & Higgins, S.J. (Eds)		
......	Paperback	0-904147-52-5	**£12.50**
3.	**Iodinated Density Gradient Media** Rickwood, D. (Ed)		
......	Paperback	0-904147-51-7	**£12.50**

Sets

	Essential Molecular Biology: 2 vol set Brown, T.A. (Ed)		
......	Spiralbound hardback	0-19-963114-X	**£58.00**
......	Paperback	0-19-963115-8	**£40.00**
	Antibodies: 2 vol set Catty, D. (Ed)		
......	Paperback	0-19-963063-1	**£33.00**
	Cellular and Molecular Neurobiology: 2 vol set Chad, J., & Wheal, H. (Eds)		
......	Spiralbound hardback	0-19-963255-3	**£56.00**
......	Paperback	0-19-963254-5	**£38.00**
	Protein Structure and Protein Function: 2 vol set Creighton, T.E. (Ed)		
......	Spiralbound hardback	0-19-963064-X	**£55.00**
......	Paperback	0-19-963065-8	**£38.00**
	DNA Cloning: 2 vol set Glover, D.M. (Ed)		
......	Paperback	1-85221-069-9	**£30.00**
	Molecular Plant Pathology: 2 vol set Gurr, S.J., McPherson, M.J. & others (Eds)		
......	Spiralbound hardback	0-19-963354-1	**£56.00**
......	Paperback	0-19-963353-3	**£37.00**
	Protein Purification Methods, and Protein Purification Applications: 2 vol set Harris, E.L.V. & Angal, S. (Eds)		
......	Spiralbound hardback	0-19-963048-8	**£48.00**
......	Paperback	0-19-963049-6	**£32.00**
	Diagnostic Molecular Pathology: 2 vol set Herrington, C.S. & McGee, J. O'D. (Eds)		
......	Spiralbound hardback	0-19-963241-3	**£54.00**
......	Paperback	0-19-963240-5	**£35.00**
	RNA Processing: 2 vol set Higgins, S.J. & Hames, B.D. (Eds)		
......	Spiralbound hardback	0-19-963473-4	**£54.00**
......	Paperback	0-19-963472-6	**£35.00**
	Receptor Biochemistry; Receptor-Effector Coupling; Receptor-Ligand Interactions: 3 vol set Hulme, E.C. (Ed)		
......	Paperback	0-19-963097-6	**£62.50**
	Human Cytogenetics: 2 vol set (2/e) Rooney, D.E. & Czepulkowski, B.H. (Eds)		
......	Hardback	0-19-963314-2	**£58.50**
......	Paperback	0-19-963313-4	**£40.50**
	Behavioural Neuroscience: 2 vol set Sahgal, A. (Ed)		
......	Spiralbound hardback	0-19-963460-2	**£58.00**
......	Paperback	0-19-963459-9	**£40.00**
	Peptide Hormone Secretion/Peptide Hormone Action: 2 vol set Siddle, K. & Hutton, J.C. (Eds)		
......	Spiralbound hardback	0-19-963072-0	**£55.00**
......	Paperback	0-19-963073-9	**£38.00**

ORDER FORM for UK, Europe and Rest of World

(Excluding USA and Canada)

Qty	ISBN	Author	Title	Amount
			P&P	
			*VAT	
			TOTAL	

Please add postage and packing: £1.75 for UK orders under £20; £2.75 for UK orders over £20; overseas orders add 10% of total.
* EC customers please note that VAT must be added (excludes UK customers)

Name ....

Address ....

....

.... Post code ....

[ ] Please charge £ .... to my credit card
Access/VISA/Eurocard/AMEX/Diners Club (circle appropriate card)

Card No ....   Expiry date ....

Signature ....

Credit card account address if different from above:

....

.... Postcode ....

[ ] I enclose a cheque for £.....................

Please return this form to: OUP Distribution Services, Saxon Way West, Corby, Northants NN18 9ES, UK

OR ORDER BY CREDIT CARD HOTLINE: Tel +44-(0)536-741519 or
Fax +44-(0)536-746337

ORDER OTHER TITLES OF INTEREST TODAY

Price list for: USA and Canada

128.	**Behavioural Neuroscience: Volume I** Sahgal, A. (Ed)		
......	Spiralbound hardback	0-19-963368-1	**$57.00**
......	Paperback	0-19-963367-3	**$37.00**
127.	**Molecular Virology** Davison, A.J. & Elliott, R.M. (Eds)		
......	Spiralbound hardback	0-19-963358-4	**$49.00**
......	Paperback	0-19-963357-6	**$32.00**
126.	**Gene Targeting** Joyner, A.L. (Ed)		
......	Spiralbound hardback	0-19-963407-6	**$49.00**
......	Paperback	0-19-9634036-8	**$34.00**
124.	**Human Genetic Disease Analysis (2/e)** Davies, K.E. (Ed)		
......	Spiralbound hardback	0-19-963309-6	**$54.00**
......	Paperback	0-19-963308-8	**$33.00**
123.	**Protein Phosphorylation** Hardie, D.G. (Ed)		
......	Spiralbound hardback	0-19-963306-1	**$65.00**
......	Paperback	0-19-963305-3	**$45.00**
122.	**Immunocytochemistry** Beesley, J. (Ed)		
......	Spiralbound hardback	0-19-963270-7	**$62.00**
......	Paperback	0-19-963269-3	**$42.00**
121.	**Tumour Immunobiology** Gallagher, G., Rees, R.C. & others (Eds)		
......	Spiralbound hardback	0-19-963370-3	**$72.00**
......	Paperback	0-19-963369-X	**$50.00**
120.	**Transcription Factors** Latchman, D.S. (Ed)		
......	Spiralbound hardback	0-19-963342-8	**$48.00**
......	Paperback	0-19-963341-X	**$31.00**
119.	**Growth Factors** McKay, I. & Leigh, I. (Eds)		
......	Spiralbound hardback	0-19-963360-6	**$48.00**
......	Paperback	0-19-963359-2	**$31.00**
118.	**Histocompatibility Testing** Dyer, P. & Middleton, D. (Eds)		
......	Spiralbound hardback	0-19-963364-9	**$60.00**
......	Paperback	0-19-963363-0	**$41.00**
117.	**Gene Transcription** Hames, B.D. & Higgins, S.J. (Eds)		
......	Spiralbound hardback	0-19-963292-8	**$72.00**
......	Paperback	0-19-963291-X	**$50.00**
116.	**Electrophysiology** Wallis, D.I. (Ed)		
......	Spiralbound hardback	0-19-963348-7	**$56.00**
......	Paperback	0-19-963347-9	**$39.00**
115.	**Biological Data Analysis** Fry, J.C. (Ed)		
......	Spiralbound hardback	0-19-963340-1	**$80.00**
......	Paperback	0-19-963339-8	**$60.00**
114.	**Experimental Neuroanatomy** Bolam, J.P. (Ed)		
......	Spiralbound hardback	0-19-963326-6	**$59.00**
......	Paperback	0-19-963325-8	**$39.00**
113.	**Preparative Centrifugation** Rickwood, D. (Ed)		
......	Spiralbound hardback	0-19-963208-1	**$78.00**
......	Paperback	0-19-963211-1	**$44.00**
111.	**Haemopoiesis** Testa, N.G. & Molineux, G. (Eds)		
......	Spiralbound hardback	0-19-963366-5	**$59.00**
......	Paperback	0-19-963365-7	**$39.00**
110.	**Pollination Ecology** Dafni, A.		
......	Spiralbound hardback	0-19-963299-5	**$56.95**
......	Paperback	0-19-963298-7	**$39.95**
109.	**In Situ Hybridization** Wilkinson, D.G. (Ed)		
......	Spiralbound hardback	0-19-963328-2	**$58.00**
......	Paperback	0-19-963327-4	**$36.00**
108.	**Protein Engineering** Rees, A.R., Sternberg, M.J.E. & others (Eds)		
......	Spiralbound hardback	0-19-963139-5	**$64.00**
......	Paperback	0-19-963138-7	**$44.00**
107.	**Cell-Cell Interactions** Stevenson, B.R., Gallin, W.J. & others (Eds)		
......	Spiralbound hardback	0-19-963319-3	**$55.00**
......	Paperback	0-19-963318-5	**$38.00**
106.	**Diagnostic Molecular Pathology: Volume I** Herrington, C.S. & McGee, J. O'D. (Eds)		
......	Spiralbound hardback	0-19-963237-5	**$50.00**
......	Paperback	0-19-963236-7	**$33.00**
105.	**Biomechanics-Materials** Vincent, J.F.V. (Ed)		
......	Spiralbound hardback	0-19-963223-5	**$70.00**
......	Paperback	0-19-963222-7	**$50.00**
104.	**Animal Cell Culture (2/e)** Freshney, R.I. (Ed)		
......	Spiralbound hardback	0-19-963212-X	**$55.00**
......	Paperback	0-19-963213-8	**$35.00**
103.	**Molecular Plant Pathology: Volume II** Gurr, S.J., McPherson, M.J. & others (Eds)		
......	Spiralbound hardback	0-19-963352-5	**$65.00**
......	Paperback	0-19-963351-7	**$45.00**
102.	**Signal Transduction** Milligan, G. (Ed)		
......	Spiralbound hardback	0-19-963296-0	**$60.00**
......	Paperback	0-19-963295-2	**$38.00**
101.	**Protein Targeting** Magee, A.I. & Wileman, T. (Eds)		
......	Spiralbound hardback	0-19-963206-5	**$75.00**
......	Paperback	0-19-963210-3	**$50.00**
100.	**Diagnostic Molecular Pathology: Volume II: Cell and Tissue Genotyping** Herrington, C.S. & McGee, J.O'D. (Eds)		
......	Spiralbound hardback	0-19-963239-1	**$60.00**
......	Paperback	0-19-963238-3	**$39.00**
99.	**Neuronal Cell Lines** Wood, J.N. (Ed)		
......	Spiralbound hardback	0-19-963346-0	**$68.00**
......	Paperback	0-19-963345-2	**$48.00**
98.	**Neural Transplantation** Dunnett, S.B. & Bj‹148›rklund, A. (Eds)		
......	Spiralbound hardback	0-19-963286-3	**$69.00**
......	Paperback	0-19-963285-5	**$42.00**
97.	**Human Cytogenetics: Volume II: Malignancy and Acquired Abnormalities (2/e)** Rooney, D.E. & Czepulkowski, B.H. (Eds)		
......	Spiralbound hardback	0-19-963290-1	**$75.00**
......	Paperback	0-19-963289-8	**$50.00**
96.	**Human Cytogenetics: Volume I: Constitutional Analysis (2/e)** Rooney, D.E. & Czepulkowski, B.H. (Eds)		
......	Spiralbound hardback	0-19-963288-X	**$75.00**
......	Paperback	0-19-963287-1	**$50.00**
95.	**Lipid Modification of Proteins** Hooper, N.M. & Turner, A.J. (Eds)		
......	Spiralbound hardback	0-19-963274-X	**$75.00**
......	Paperback	0-19-963273-1	**$50.00**
94.	**Biomechanics-Structures and Systems** Biewener, A.A. (Ed)		
......	Spiralbound hardback	0-19-963268-5	**$85.00**
......	Paperback	0-19-963267-7	**$50.00**
93.	**Lipoprotein Analysis** Converse, C.A. & Skinner, E.R. (Eds)		
......	Spiralbound hardback	0-19-963192-1	**$65.00**
......	Paperback	0-19-963231-6	**$42.00**
92.	**Receptor-Ligand Interactions** Hulme, E.C. (Ed)		
......	Spiralbound hardback	0-19-963090-9	**$75.00**
......	Paperback	0-19-963091-7	**$50.00**
91.	**Molecular Genetic Analysis of Populations** Hoelzel, A.R. (Ed)		
......	Spiralbound hardback	0-19-963278-2	**$65.00**
......	Paperback	0-19-963277-4	**$45.00**

90. **Enzyme Assays** Eisenthal, R. & Danson, M.J. (Eds)
...... Spiralbound hardback 0-19-963142-5 **$68.00**
...... Paperback 0-19-963143-3 **$48.00**
89. **Microcomputers in Biochemistry** Bryce, C.F.A. (Ed)
...... Spiralbound hardback 0-19-963253-7 **$60.00**
...... Paperback 0-19-963252-9 **$40.00**
88. **The Cytoskeleton** Carraway, K.L. & Carraway, C.A.C. (Eds)
...... Spiralbound hardback 0-19-963257-X **$60.00**
...... Paperback 0-19-963256-1 **$40.00**
87. **Monitoring Neuronal Activity** Stamford, J.A. (Ed)
...... Spiralbound hardback 0-19-963244-8 **$60.00**
...... Paperback 0-19-963243-X **$40.00**
86. **Crystallization of Nucleic Acids and Proteins** Ducruix, A. & Giegé, R. (Eds)
...... Spiralbound hardback 0-19-963245-6 **$60.00**
...... Paperback 0-19-963246-4 **$50.00**
85. **Molecular Plant Pathology: Volume I** Gurr, S.J., McPherson, M.J. & others (Eds)
...... Spiralbound hardback 0-19-963103-4 **$60.00**
...... Paperback 0-19-963102-6 **$40.00**
84. **Anaerobic Microbiology** Levett, P.N. (Ed)
...... Spiralbound hardback 0-19-963204-9 **$75.00**
...... Paperback 0-19-963262-6 **$45.00**
83. **Oligonucleotides and Analogues** Eckstein, F. (Ed)
...... Spiralbound hardback 0-19-963280-4 **$65.00**
...... Paperback 0-19-963279-0 **$45.00**
82. **Electron Microscopy in Biology** Harris, R. (Ed)
...... Spiralbound hardback 0-19-963219-7 **$65.00**
...... Paperback 0-19-963215-4 **$45.00**
81. **Essential Molecular Biology: Volume II** Brown, T.A. (Ed)
...... Spiralbound hardback 0-19-963112-3 **$65.00**
...... Paperback 0-19-963113-1 **$45.00**
80. **Cellular Calcium** McCormack, J.G. & Cobbold, P.H. (Eds)
...... Spiralbound hardback 0-19-963131-X **$75.00**
...... Paperback 0-19-963130-1 **$50.00**
79. **Protein Architecture** Lesk, A.M.
...... Spiralbound hardback 0-19-963054-2 **$65.00**
...... Paperback 0-19-963055-0 **$45.00**
78. **Cellular Neurobiology** Chad, J. & Wheal, H. (Eds)
...... Spiralbound hardback 0-19-963106-9 **$73.00**
...... Paperback 0-19-963107-7 **$43.00**
77. **PCR** McPherson, M.J., Quirke, P. & others (Eds)
...... Spiralbound hardback 0-19-963226-X **$55.00**
...... Paperback 0-19-963196-4 **$40.00**
76. **Mammalian Cell Biotechnology** Butler, M. (Ed)
...... Spiralbound hardback 0-19-963207-3 **$60.00**
...... Paperback 0-19-963209-X **$40.00**
75. **Cytokines** Balkwill, F.R. (Ed)
...... Spiralbound hardback 0-19-963218-9 **$64.00**
...... Paperback 0-19-963214-6 **$44.00**
74. **Molecular Neurobiology** Chad, J. & Wheal, H. (Eds)
...... Spiralbound hardback 0-19-963108-5 **$56.00**
...... Paperback 0-19-963109-3 **$36.00**
73. **Directed Mutagenesis** McPherson, M.J. (Ed)
...... Spiralbound hardback 0-19-963141-7 **$55.00**
...... Paperback 0-19-963140-9 **$45.00**
72. **Essential Molecular Biology: Volume I** Brown, T.A. (Ed)
...... Spiralbound hardback 0-19-963110-7 **$65.00**
...... Paperback 0-19-963111-5 **$45.00**
71. **Peptide Hormone Action** Siddle, K. & Hutton, J.C.
...... Spiralbound hardback 0-19-963070-4 **$70.00**
...... Paperback 0-19-963071-2 **$50.00**
70. **Peptide Hormone Secretion** Hutton, J.C. & Siddle, K. (Eds)
...... Spiralbound hardback 0-19-963068-2 **$70.00**
...... Paperback 0-19-963069-0 **$50.00**
69. **Postimplantation Mammalian Embryos** Copp, A.J. & Cockroft, D.L. (Eds)
...... Spiralbound hardback 0-19-963088-7 **$70.00**
...... Paperback 0-19-963089-5 **$50.00**
68. **Receptor-Effector Coupling** Hulme, E.C. (Ed)
...... Spiralbound hardback 0-19-963094-1 **$70.00**
...... Paperback 0-19-963095-X **$45.00**
67. **Gel Electrophoresis of Proteins (2/e)** Hames, B.D. & Rickwood, D. (Eds)
...... Spiralbound hardback 0-19-963074-7 **$75.00**
...... Paperback 0-19-963075-5 **$50.00**
66. **Clinical Immunology** Gooi, H.C. & Chapel, H. (Eds)
...... Spiralbound hardback 0-19-963086-0 **$69.95**
...... Paperback 0-19-963087-9 **$50.00**

65. **Receptor Biochemistry** Hulme, E.C. (Ed)
...... Paperback 0-19-963093-3 **$50.00**
64. **Gel Electrophoresis of Nucleic Acids (2/e)** Rickwood, D. & Hames, B.D. (Eds)
...... Spiralbound hardback 0-19-963082-8 **$75.00**
...... Paperback 0-19-963083-6 **$50.00**
63. **Animal Virus Pathogenesis** Oldstone, M.B.A. (Ed)
...... Spiralbound hardback 0-19-963100-X **$68.00**
...... Paperback 0-19-963101-8 **$40.00**
62. **Flow Cytometry** Ormerod, M.G. (Ed)
...... Paperback 0-19-963053-4 **$50.00**
61. **Radioisotopes in Biology** Slater, R.J. (Ed)
...... Spiralbound hardback 0-19-963080-1 **$75.00**
...... Paperback 0-19-963081-X **$45.00**
60. **Biosensors** Cass, A.E.G. (Ed)
...... Spiralbound hardback 0-19-963046-1 **$65.00**
...... Paperback 0-19-963047-X **$43.00**
59. **Ribosomes and Protein Synthesis** Spedding, G. (Ed)
...... Spiralbound hardback 0-19-963104-2 **$75.00**
...... Paperback 0-19-963105-0 **$45.00**
58. **Liposomes** New, R.R.C. (Ed)
...... Spiralbound hardback 0-19-963076-3 **$70.00**
...... Paperback 0-19-963077-1 **$45.00**
57. **Fermentation** McNeil, B. & Harvey, L.M. (Eds)
...... Spiralbound hardback 0-19-963044-5 **$65.00**
...... Paperback 0-19-963045-3 **$39.00**
56. **Protein Purification Applications** Harris, E.L.V. & Angal, S. (Eds)
...... Spiralbound hardback 0-19-963022-4 **$54.00**
...... Paperback 0-19-963023-2 **$36.00**
55. **Nucleic Acids Sequencing** Howe, C.J. & Ward, E.S. (Eds)
...... Spiralbound hardback 0-19-963056-9 **$59.00**
...... Paperback 0-19-963057-7 **$38.00**
54. **Protein Purification Methods** Harris, E.L.V. & Angal, S. (Eds)
...... Spiralbound hardback 0-19-963002-X **$60.00**
...... Paperback 0-19-963003-8 **$40.00**
53. **Solid Phase Peptide Synthesis** Atherton, E. & Sheppard, R.C.
...... Spiralbound hardback 0-19-963066-6 **$58.00**
...... Paperback 0-19-963067-4 **$39.95**
52. **Medical Bacteriology** Hawkey, P.M. & Lewis, D.A. (Eds)
...... Paperback 0-19-963009-7 **$50.00**
51. **Proteolytic Enzymes** Beynon, R.J. & Bond, J.S. (Eds)
...... Spiralbound hardback 0-19-963058-5 **$60.00**
...... Paperback 0-19-963059-3 **$39.00**
50. **Medical Mycology** Evans, E.G.V. & Richardson, M.D. (Eds)
...... Spiralbound hardback 0-19-963010-0 **$69.95**
...... Paperback 0-19-963011-9 **$50.00**
49. **Computers in Microbiology** Bryant, T.N. & Wimpenny, J.W.T. (Eds)
...... Paperback 0-19-963015-1 **$40.00**
48. **Protein Sequencing** Findlay, J.B.C. & Geisow, M.J. (Eds)
...... Spiralbound hardback 0-19-963012-7 **$56.00**
...... Paperback 0-19-963013-5 **$38.00**
47. **Cell Growth and Division** Baserga, R. (Ed)
...... Spiralbound hardback 0-19-963026-7 **$62.00**
...... Paperback 0-19-963027-5 **$38.00**
46. **Protein Function** Creighton, T.E. (Ed)
...... Spiralbound hardback 0-19-963006-2 **$65.00**
...... Paperback 0-19-963007-0 **$45.00**
45. **Protein Structure** Creighton, T.E. (Ed)
...... Spiralbound hardback 0-19-963000-3 **$65.00**
...... Paperback 0-19-963001-1 **$45.00**
44. **Antibodies: Volume II** Catty, D. (Ed)
...... Spiralbound hardback 0-19-963018-6 **$58.00**
...... Paperback 0-19-963019-4 **$39.00**
43. **HPLC of Macromolecules** Oliver, R.W.A. (Ed)
...... Spiralbound hardback 0-19-963020-8 **$54.00**
...... Paperback 0-19-963021-6 **$45.00**
42. **Light Microscopy in Biology** Lacey, A.J. (Ed)
...... Spiralbound hardback 0-19-963036-4 **$62.00**
...... Paperback 0-19-963037-2 **$38.00**
41. **Plant Molecular Biology** Shaw, C.H. (Ed)
...... Paperback 1-85221-056-7 **$38.00**
40. **Microcomputers in Physiology** Fraser, P.J. (Ed)
...... Spiralbound hardback 1-85221-129-6 **$54.00**
...... Paperback 1-85221-130-X **$36.00**
39. **Genome Analysis** Davies, K.E. (Ed)
...... Spiralbound hardback 1-85221-109-1 **$54.00**
...... Paperback 1-85221-110-5 **$36.00**
38. **Antibodies: Volume I** Catty, D. (Ed)
...... Paperback 0-947946-85-3 **$38.00**
37. **Yeast** Campbell, I. & Duffus, J.H. (Eds)
...... Paperback 0-947946-79-9 **$36.00**

36.	**Mammalian Development** Monk, M. (Ed)		
......	Hardback	1-85221-030-3	**$60.00**
......	Paperback	1-85221-029-X	**$45.00**
35.	**Lymphocytes** Klaus, G.G.B. (Ed)		
......	Hardback	1-85221-018-4	**$54.00**
34.	**Lymphokines and Interferons** Clemens, M.J., Morris, A.G. & others (Eds)		
......	Paperback	1-85221-035-4	**$44.00**
33.	**Mitochondria** Darley-Usmar, V.M., Rickwood, D. & others (Eds)		
......	Hardback	1-85221-034-6	**$65.00**
......	Paperback	1-85221-033-8	**$45.00**
32.	**Prostaglandins and Related Substances** Benedetto, C., McDonald-Gibson, R.G. & others (Eds)		
......	Hardback	1-85221-032-X	**$58.00**
......	Paperback	1-85221-031-1	**$38.00**
31.	**DNA Cloning: Volume III** Glover, D.M. (Ed)		
......	Hardback	1-85221-049-4	**$56.00**
......	Paperback	1-85221-048-6	**$36.00**
30.	**Steroid Hormones** Green, B. & Leake, R.E. (Eds)		
......	Paperback	0-947946-53-5	**$40.00**
29.	**Neurochemistry** Turner, A.J. & Bachelard, H.S. (Eds)		
......	Hardback	1-85221-028-1	**$56.00**
......	Paperback	1-85221-027-3	**$36.00**
28.	**Biological Membranes** Findlay, J.B.C. & Evans, W.H. (Eds)		
......	Hardback	0-947946-84-5	**$54.00**
......	Paperback	0-947946-83-7	**$36.00**
27.	**Nucleic Acid and Protein Sequence Analysis** Bishop, M.J. & Rawlings, C.J. (Eds)		
......	Hardback	1-85221-007-9	**$66.00**
......	Paperback	1-85221-006-0	**$44.00**
26.	**Electron Microscopy in Molecular Biology** Sommerville, J. & Scheer, U. (Eds)		
......	Hardback	0-947946-64-0	**$54.00**
......	Paperback	0-947946-54-3	**$40.00**
24.	**Spectrophotometry and Spectrofluorimetry** Harris, D.A. & Bashford, C.L. (Eds)		
......	Hardback	0-947946-69-1	**$56.00**
......	Paperback	0-947946-46-2	**$39.95**
23.	**Plasmids** Hardy, K.G. (Ed)		
......	Paperback	0-947946-81-0	**$36.00**
22.	**Biochemical Toxicology** Snell, K. & Mullock, B. (Eds)		
......	Paperback	0-947946-52-7	**$40.00**
19.	**Drosophila** Roberts, D.B. (Ed)		
......	Hardback	0-947946-66-7	**$67.50**
......	Paperback	0-947946-45-4	**$46.00**
17.	**Photosynthesis: Energy Transduction** Hipkins, M.F. & Baker, N.R. (Eds)		
......	Hardback	0-947946-63-2	**$54.00**
......	Paperback	0-947946-51-9	**$36.00**
16.	**Human Genetic Diseases** Davies, K.E. (Ed)		
......	Hardback	0-947946-76-4	**$60.00**
......	Paperback	0-947946-75-6	**$34.00**
14.	**Nucleic Acid Hybridisation** Hames, B.D. & Higgins, S.J. (Eds)		
......	Hardback	0-947946-61-6	**$60.00**
......	Paperback	0-947946-23-3	**$36.00**
12.	**Plant Cell Culture** Dixon, R.A. (Ed)		
......	Paperback	0-947946-22-5	**$36.00**

11a.	**DNA Cloning: Volume I** Glover, D.M. (Ed)		
......	Paperback	0-947946-18-7	**$36.00**
11b.	**DNA Cloning: Volume II** Glover, D.M. (Ed)		
......	Paperback	0-947946-19-5	**$36.00**
10.	**Virology** Mahy, B.W.J. (Ed)		
......	Paperback	0-904147-78-9	**$40.00**
9.	**Affinity Chromatography** Dean, P.D.G., Johnson, W.S. & others (Eds)		
......	Paperback	0-904147-71-1	**$36.00**
7.	**Microcomputers in Biology** Ireland, C.R. & Long, S.P. (Eds)		
......	Paperback	0-904147-57-6	**$36.00**
6.	**Oligonucleotide Synthesis** Gait, M.J. (Ed)		
......	Paperback	0-904147-74-6	**$38.00**
5.	**Transcription and Translation** Hames, B.D. & Higgins, S.J. (Eds)		
......	Paperback	0-904147-52-5	**$38.00**
3.	**Iodinated Density Gradient Media** Rickwood, D. (Ed)		
......	Paperback	0-904147-51-7	**$36.00**

Sets

	Essential Molecular Biology: 2 vol set Brown, T.A. (Ed)		
......	Spiralbound hardback	0-19-963114-X	**$118.00**
......	Paperback	0-19-963115-8	**$78.00**
	Antibodies: 2 vol set Catty, D. (Ed)		
......	Paperback	0-19-963063-1	**$70.00**
	Cellular and Molecular Neurobiology: 2 vol set Chad, J. & Wheal, H. (Eds)		
......	Spiralbound hardback	0-19-963255-3	**$133.00**
......	Paperback	0-19-963254-5	**$79.00**
	Protein Structure and Protein Function: 2 vol set Creighton, T.E. (Ed)		
......	Spiralbound hardback	0-19-963064-X	**$114.00**
......	Paperback	0-19-963065-8	**$80.00**
	DNA Cloning: 2 vol set Glover, D.M. (Ed)		
......	Paperback	1-85221-069-9	**$92.00**
	Molecular Plant Pathology: 2 vol set Gurr, S.J., McPherson, M.J. & others (Eds)		
......	Spiralbound hardback	0-19-963354-1	**$110.00**
......	Paperback	0-19-963353-3	**$75.00**
	Protein Purification Methods, and Protein Purification Applications: 2 vol set Harris, E.L.V. & Angal, S. (Eds)		
......	Spiralbound hardback	0-19-963048-8	**$98.00**
......	Paperback	0-19-963049-6	**$68.00**
	Diagnostic Molecular Pathology: 2 vol set Herrington, C.S. & McGee, J. O'D. (Eds)		
......	Spiralbound hardback	0-19-963241-3	**$105.00**
......	Paperback	0-19-963240-5	**$69.00**
	Receptor Biochemistry; Receptor-Effector Coupling; Receptor-Ligand Interactions: 3 vol set Hulme, E.C. (Ed)		
......	Paperback	0-19-963097-6	**$130.00**
	Human Cytogenetics: (2/e): 2 vol set Rooney, D.E. & Czepulkowski, B.H. (Eds)		
......	Hardback	0-19-963314-2	**$130.00**
......	Paperback	0-19-963313-4	**$90.00**
	Peptide Hormone Secretion/Peptide Hormone Action: 2 vol set Siddle, K. & Hutton, J.C. (Eds)		
......	Spiralbound hardback	0-19-963072-0	**$135.00**
......	Paperback	0-19-963073-9	**$90.00**

ORDER FORM for USA and Canada

Qty	ISBN	Author	Title	Amount
			S&H	
	CA and NC residents add appropriate sales tax			
			TOTAL	

Please add shipping and handling: US $2.50 for first book, (US $1.00 each book thereafter)

Name ....

Address ....

....

.... Zip ....

[ ] Please charge $ .... to my credit card
Mastercard/VISA/American Express (circle appropriate card)

Acct. ....   Expiry date ....

Signature ....

Credit card account address if different from above:

....

.... Zip ....

[ ] I enclose a cheque for US $............

Mail orders to: Order Dept. Oxford University Press, 2001 Evans Road, Cary, NC 27513